Wired Together

Wired Together

The Montreal Neurological Institute and the Origins of Neuroscience

YVAN PRKACHIN

The University of Chicago Press
Chicago and London

The University of Chicago Press, Chicago 60637
The University of Chicago Press, Ltd., London
© 2026 by The University of Chicago
All rights reserved. No part of this book may be used or reproduced in any
manner whatsoever without written permission, except in the case of brief
quotations in critical articles and reviews. For more information, contact the
University of Chicago Press, 1427 E. 60th St., Chicago, IL 60637.
Published 2026
Printed in the United States of America

35 34 33 32 31 30 29 28 27 26   1 2 3 4 5

ISBN-13: 978-0-226-84543-2 (cloth)
ISBN-13: 978-0-226-84546-3 (paper)
ISBN-13: 978-0-226-84544-9 (ebook)
DOI: https://doi.org/10.7208/chicago/9780226845449.001.0001

Library of Congress Cataloging-in-Publication Data

Names: Prkachin, Yvan author
Title: Wired together : the Montreal Neurological Institute and the origins
    of neuroscience / Yvan Prkachin.
Other titles: Montreal Neurological Institute and the origins of neuroscience
Description: Chicago ; London : The University of Chicago Press, 2025. |
    Includes bibliographical references and index.
Identifiers: LCCN 2025030752 | ISBN 9780226845432 cloth | ISBN
    9780226845463 paperback | ISBN 9780226845449 ebook
Subjects: LCSH: Penfield, Wilder, 1891–1976. | Institut neurologique de
    Montréal—History. | Institut neurologique de Montréal—Influence. |
    Neurosciences—Research—Québec (Province)—Montréal—History. |
    Neurosciences—History.
Classification: LCC RC339.C2 P75
LC record available at https://lccn.loc.gov/2025030752

♾ This paper meets the requirements of ANSI/NISO Z39.48-1992
(Permanence of Paper).

Authorized Representative for EU General Product Safety Regulation
(GPSR) queries: **Easy Access System Europe**—Mustamäe tee 50, 10621
Tallinn, Estonia, gpsr.requests@easproject.com
Any other queries: https://press.uchicago.edu/press/contact.html

Contents

Archival Abbreviations

D E C F   Donald Ewen Cameron Fonds (MG1098), McGill University Archives

D H F   Donald Olding Hebb Fonds (MG1045), McGill University Archives

D H P   David Hubel Papers (HMS C253), Francis A. Countway Library of Medicine

F O S P   Francis Otto Schmitt Papers (MC-0154), MIT Archives and Special Collections

H J F   Herbert H. Jasper Fonds (MG4253), McGill University Archives

M H P   Molly Harrower Papers (M842), Cummins Center for the History of Psychology, University of Akron

S O H C   Schmitt Oral History Collection (MC-0226), MIT Archives and Special Collections

W F F   William Feindel Fonds (MG4264), McGill University Archives

W P P   Wilder Penfield Papers, Osler Library

Introduction

Prologue: Burned Toast

An operating room in Montreal, 1934. A French Canadian woman, Mrs. Gold, lies on her side on the operating table. She suffers from epilepsy, and her seizures are usually preceded by an "aura"—a unique sensory hallucination that, in this case, manifests as the smell of burned toast. The man operating on her is the neurosurgeon Wilder Penfield, and he has a plan. "Every time she has a seizure, she smells something burning," Penfield intones in a voice-over. "Now, if we can provoke that smell by probing the surface of the brain, we'll find the source of the seizures." Penfield's electric probe moves over the surface of Mrs. Gold's exposed brain, and its gentle electric current brings forth different responses from the conscious Mrs. Gold. "Mrs. Gold, do you feel anything?" Penfield asks. "I can see the most wonderful lights," Mrs. Gold responds. Penfield moves his probe to a different region of the brain. "Did you pour cold water on my hand, Dr. Penfield?" Moving the probe again, Penfield asks: "Now what?" Music swells, and Mrs. Gold exclaims excitedly: "Burned toast . . . Dr. Penfield, I can smell burned toast!"

The scene described above is familiar to most Canadians of a particular generation. It comes from the short film *Wilder Penfield* in the *Heritage Minutes* series produced by the Canadian Broadcasting Corporation. *Heritage Minutes* aired regularly on Canadian television in the 1990s as part of an effort to inculcate a sense of national pride. "Dr. Wilder Penfield," intones the French Canadian patient in voice-over. "He cured my seizures and hundreds more. They say he drew the road map of the human brain. We just called him the greatest Canadian alive."[1] The Canadians who grew up with this image of Penfield would likely agree with that sentiment. At the very least, they would all know it. In 2018, this Canadian scientific icon was even the subject of that emblem of American popular culture the Google Doodle, which for January 27, 2018, celebrated Penfield's 127th birthday with a cartoon of the man himself, a brain, and a nose inhaling the odor of burned toast.[2]

For all its drama and charm, the "burned toast" scene and its image of Penfield's operating room conceals as much as it reveals. For instance, while the depiction of the operation is relatively accurate, merely localizing the source of the patient's seizures was only the first step. Penfield would have next likely removed portions of her brain to control her seizures, a procedure euphemistically referred to as a *cortical resection*. In 1934, Penfield's surgeries worked only some of the time. Patients often experienced relief from their seizures, but sometimes they did not. Sometimes they relapsed. And sometimes they suffered long and difficult recoveries (perhaps even Mrs. Gold), plagued by memory disturbances and other cognitive problems. In 1934, it would have been difficult to know what the effects of Penfield's "radical treatment" for epilepsy would be.

Five years later and another scene from Penfield's operating theater. This one is real. At the Montreal Neurological Institute (MNI), the neurosurgical clinic he opened in 1934 at the foot of the city's major geographic landmark, Mount Royal, Penfield is performing another operation. In Operating Room 1, he is perched over the exposed brain of a male patient, operating on the left temporal lobe. The patient, a thirty-two-year-old American schoolteacher, had been in a car accident in Philadelphia three years earlier. Fourteen months after the accident, he began to suffer from epileptic seizures. A series of referrals has brought him to the operating room, where Penfield is now searching for the cause of the seizures—a scar on the patient's brain that he is attempting to remove.[3]

Perhaps the most notable contrast between this scene and the one from *Wilder Penfield* is this. In the film, the assistants surrounding Penfield are nameless and largely inconsequential. In the case of the real operation, those who labor along with him are nearly as important as he himself is. Standing beside Penfield is another man, Donald Olding Hebb, a psychologist who speaks continuously with the patient. In the coming months, Hebb will study the patient during his recovery. For the moment, however, because the scar is close to one of the brain's language centers, he is discussing various subjects with the patient to ensure that his ability to speak will not be affected. And standing just off the main operating room floor, peering at his electroencephalograph (EEG), is Herbert Jasper, a young psychologist and physiologist who is the acknowledged master of this new technology. Jasper had initially used the EEG to localize the offending lesion, but subsequent intracranial recording of electric signals has refined the localization

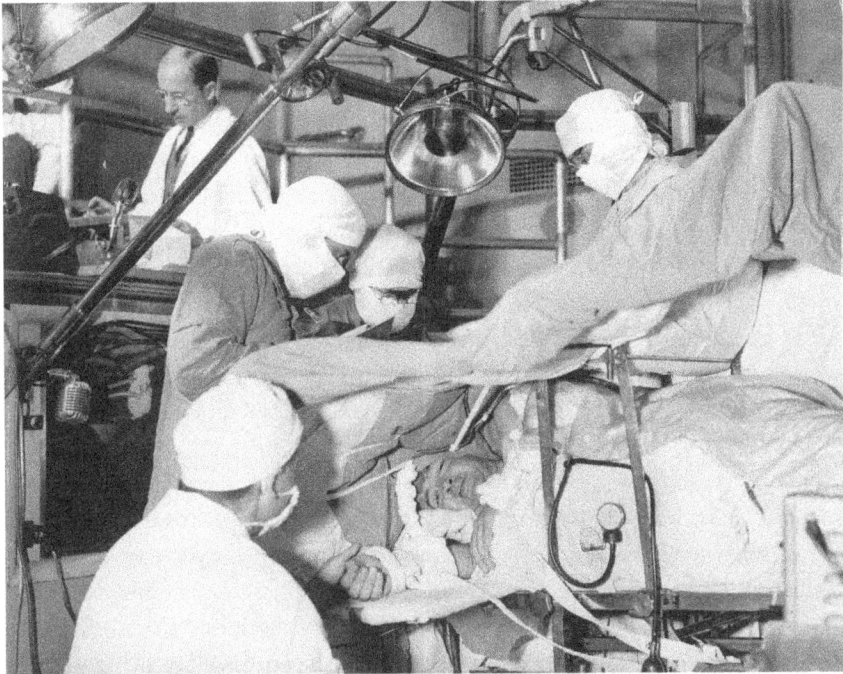

FIGURE I.1. Operating room at the Montreal Neurological Institute, ca. 1950s. Reproduced by permission of the Osler Library of the History of Medicine, McGill University.

of the epileptogenic tissue, and Jasper is monitoring the operation in real time so that he can further perfect his technique for recording brain waves (fig. I.1).

Others may very well have been present in the operating room that day. Since it opened in 1934, dozens of research fellows had come to Montreal to witness the remarkable activities of Penfield's operating room, including the sight of conscious patients reporting the experience of having their brains stimulated with an electric probe. In the coming decades, scientists and others from around the world would flock to Montreal to observe these operations. Many later reported feeling a deep sense of transformation as Penfield's probe seemed to come close to touching the patient's very soul. One observer would even describe it as "Proust on the operating table."[4] But, beyond the hyperbolic pronouncements of interested onlookers, the real men and women who stood over the exposed brain of Penfield's surgical patient that day were accomplishing a more historically significant achievement. Those who labored in the MNI in 1939 were working together and, in doing so, creating something new. In the cosmopolitan and bilingual city of Montreal, a new, interdisciplinary, collaborative scientific field—neuroscience—was being born.

This book is about those other faces in the operating room and the institute, the men and women who made neuroscience possible.

Neuroscience and Its Histories

In the twenty-first century, the word *neuroscience* is ubiquitous, and neuroscience constitutes a significant portion of the modern scientific landscape. Scientific agencies in Europe and the United States regularly commit enormous sums of money for its funding, and the field's flagship professional organization—the Society for Neuroscience (SFN)—is today one of the world's largest scientific organizations, with over thirty-six thousand members from over ninety countries. Yet, before the 1960s, none of them would have referred to themselves as a *neuroscientist*, and the word *neuroscience* was virtually unknown. Historians have documented the emergence and development of many areas of science that study the brain, the nervous system, and the mind. The word *neurology*, for instance, was coined by the seventeenth-century physician Thomas Willis, and the nineteenth century saw the emergence of neurophysiology as a distinct laboratory discipline that studied the workings of the nervous system in animals. By the end of the nineteenth century, the new clinical field of neurology was experiencing its golden age as physicians such as John Hughlings Jackson in England, Jean-Martin Charcot in France, and Silas Weir Mitchell in the United States began to codify the key neurological illnesses and unravel many of their underlying causes. By the early twentieth century, neurophysiology, neuroanatomy, neuropsychiatry, and neuropsychology were all established fields (some larger, some smaller) but mostly confined to separate university departments and clinics. While historians have had much to say about the emergence of these different areas of science, few have examined how they fused in the middle of the twentieth century to create a new, interdisciplinary field.[5] Where, then, did neuroscience come from?

While the word *neuroscience* was likely coined by the physiologist Ralph W. Gerard in the late 1950s, most histories of neuroscience usually attribute its origins to the American biophysicist Francis O. Schmitt. The story goes like this. In 1941, Vannevar Bush, the doyen of American wartime scientific research, and Karl T. Compton, the president of the Massachusetts Institute of Technology (MIT), recruited a young Schmitt to revitalize the university's Biology Department. Schmitt had spent the previous decade at Washington University, St. Louis, working on X-ray diffraction analysis (a technique that would eventually reveal the structure of DNA) and came to MIT with a clear vision: retool the school's biological research to combine biology with physics, chemistry, and mathematics. His success in this field made him one of the primary figures in the consolidation of molecular biology and biophysics in the 1950s.[6]

Inspired by the elegant molecular model of DNA announced by James D. Watson and Francis Crick in 1953 and by his understanding of the unity-of-science movement then popular among philosophers, Schmitt began to organize what he referred to as the *mind-brain* group, a cluster of scientists that he felt could develop a unified approach to understanding the workings of the brain. As one participant and historian of Schmitt's efforts later put it: "[I]f any one man can be credited with the concept and genesis of neuroscience, a field uniting historically disparate disciplines concerned with brain structure and function and behavior, it is Francis O. Schmitt."[7] By 1958, Schmitt was resolved to "investigate the . . . biophysics of central nervous system function . . . by effective utilization of the biophysical and biochemical sciences." He initially called his new field of research *mental biophysics* but later changed the name to *neuroscience* in proposals for funding to NASA and the National Institutes of Health. By 1962, Schmitt had established the Neurosciences Research Program (NRP) at Brandegee Estate in Brookline, Massachusetts, and began amassing members and holding tutorials and work sessions, publishing bulletins, and arranging conferences. In less than a decade, the SFN was established (a development often linked to Schmitt in later hagiographies). Neuroscience, it would then seem, had emerged from the "big science" of Cold War America, right at the heart of the military-industrial-academic complex.

However, the story of neuroscience's origins at MIT turns out to be wrong. Neuroscience was created not by the person who coined the word but by the people who invented its practices and institutions, made its significant discoveries and theories, and built its social networks. The central argument of *Wired Together* is that the MNI was, in fact, the single most significant site for the creation of modern neuroscience in the twentieth century and played a crucial role in "wiring together" the men and women, instruments and ideas that made an interdisciplinary approach to the brain and mind possible. Gerard may have coined the word, Schmitt and MIT may have generated much of its cultural cachet, but the complex, messy, and transformative work of building a new interdisciplinary field of science first took root in Montreal and then spread to the rest of the scientific world.

Wired Together—Genealogies, Weak Ties, and Neuroscience

This book offers a new genealogy of neuroscience. I hope to unearth the vibrant, exciting, and risky scientific and medical world of the 1930s and 1940s in which Wilder Penfield built an institute that allowed for new assemblies of scientists, surgeons, and patients. These assemblies cohered in a new interdisciplinary environment, and I want to explore how that environment became the source for a new interdisciplinary science of the brain—neuroscience.[8] I also want to explore how the legacies of the MNI formed the modern world of neuroscience, often in subtle ways that have been obscured.

Part 1 of this book—"A Genealogy of Neuroscience"—traces how the MNI took shape and enabled new forms of scientific work and collaboration. Penfield's vision for the MNI grew during his training as part of the second generation of professional neurosurgeons, a cohort of men who sought to expand the purview of this young field of medicine. Penfield wanted to create a space that combined the surgical theater and the laboratory under one roof, something that would, he thought, help unravel the causes of several neurological disorders, most notably epilepsy, which he hoped to treat surgically. The emergence of the MNI was also the product of a distinctive set of historical circumstances. Most important was the changed funding priorities of the Rockefeller Foundation (RF), which encouraged investment in psychiatric medicine, laboratory science, and the kind of discipline-crossing initiatives that the MNI embodied. At the same time, the historical circumstances of Montreal enabled Penfield's new approach by providing a large pool of patients and a cosmopolitan hub to draw an international body of scientists.

Penfield's original intention was to make laboratory work the center of the MNI's scientific efforts, but in the end epilepsy surgery made the operating room itself the center of the action. The most exciting science that initially emerged from the MNI came from the operations themselves, as Penfield stimulated the exposed brains of patients who could report their experiences. In effect, he transformed the operating theater into a kind of laboratory for investigating the brain's functions, and this laboratory attracted a host of new scientific actors. These new scientific actors—psychologists, physiologists, chemists, and others—brought their own ideas and instruments and applied them to solving concrete problems in and out of the operating room. The knowledge generated by these new actors functioned as a proof of concept that a new interdisciplinary approach to the brain was possible and as a guiding principle that would eventually motivate one MNI participant (Herbert Jasper) to organize the world's brain researchers under the banner of interdisciplinary brain science. The genealogical history related in part 1 reaches its climax here, with the establishment of the International Brain Research Organization (IBRO) in 1960.

Simultaneously, part 1 of this book will explore what it meant to do interdisciplinary science in the middle of the twentieth century. To do so, here I must introduce a piece of theory. This book draws on the work of other historians who have examined how new fields of science fuse and form new bodies of knowledge and practice. In his studies of physics in the twentieth century, Peter Galison argued that different groups of actors within physics—theorists, experimenters, instrument makers, and other disparate groups—can be considered unique subcultures with their own disciplinary language and epistemologies. How do such disparate groups communicate with each other? To answer this question, Galison drew on studies from linguistic anthropology. Anthropologists have shown that, when two homogeneous linguistic cultures meet and attempt to trade, they frequently develop a trading language that allows for basic translation between them. The participants in these linguistic "trading zones" can agree on the value of particular objects even if they still have broader disagreements on their meaning or use; for example, the term *mass* was used differently by different participants in studies of the electron in the early twentieth century. Yet, despite these differences in understanding, these distinct groups can still hammer out locally understood exchanges, producing a local "trading language." Forced to collaborate on a particular scientific or technological problem, two groups can use this trading language to work out a discrete area of hard-won knowledge. With continued contact, these trading languages can grow into more sophisticated pidgin languages, allowing for complex coordination and exchange among groups and a greater sense of agreed-on meaning. Given enough time and continued contact, these pidgin languages can evolve into fully fledged creole languages possessing a sophisticated vocabulary and grammar.[9] Part 1 of this book shows how the participants at the MNI developed this kind of interdisciplinary trading.

It also examines the limits of this kind of trading and alternative forms of disciplinary interactions in twentieth-century brain science. Alongside the MNI's interdisciplinary trading zone, an alternative approach to the brain was brewing at MIT under the aegis of Francis O. Schmitt. A fervent believer in the transcendent power of molecular biology to unite different areas of science, Schmitt expounded his own approach to the brain in the 1950s, one that was *transdisciplinary* rather than *interdisciplinary*.[10] He aimed to unite the brain sciences through a single transcendent discovery—a memory molecule that would unite biology and psychology. This was his vision for neuroscience, and it informed his transdisciplinary efforts, which focused on scientific communication rather than practice. At the same time, part 1 also examines the limits of these different approaches to scientific unification. Both IBRO and the NRP discovered their limits in the 1960s. For the NRP, Schmitt's transdisciplinary neuroscience was dead on arrival owing to the failure to find a memory molecule. Reductionism proved an insufficient basis for a new brain science, and the NRP had to transition to the role of propagandizer for a transcendent neuroscience. Simultaneously, scaling up the MNI's close interdisciplinary trading to a global level proved illusory, and IBRO had to instead function as a social unifier of neuroscientists rather than neuroscience. The amalgamation of IBRO's and the NRP's different disciplinary strategies in forming the SFN in 1970 remains one of the most important legacies of this period.

The formation of the SFN in 1970 was not the only legacy of the MNI and its new form of brain science. Part 2 of this book—"Weak Ties and Afterlives"—examines the many legacies of the MNI, which ranged from the foundation of cognitive science to the "mind control" experiments of the CIA to the 1981 Nobel Prize in Physiology and Medicine. To understand how the work at the MNI spread to such diverse areas, I need to introduce a second theoretical tool: the insights of the sociologist Mark Granovetter and his concept of the strength of "weak ties." In the 1970s, Granovetter noticed a paradox in human social relations. Communities that consisted of groups made up of "strong ties"—familial relations, close friendships, and insular workplaces— were surprisingly poor at organizing for purposeful social action, adapting to changing circumstances, or influencing other groups. Granovetter explained that sociologists failed to understand this phenomenon because they had spent much of their time examining the strong ties between individuals at the expense of the weak ties that linked individuals and groups. Strong ties bound families and small groups like workplaces together but could never act as bridges to other communities; ideas and innovations seldom diffused from insular, tightly associated groups to the broader social world through strong connections. Instead, through weak ties—casual friendships, workplace acquaintanceships, and temporary employment and apprenticeships—ideas and innovations made their way from one strongly tied group to another.[11]

Granovetter also observed that, through weak ties, organizing efforts in one strongly tied group could link to another, mobilizing entire communities for large-scale social action. By emphasizing the importance of weak ties between groups, he hoped to link the small-scale study of groups (as conducted in social psychology and anthropology) to the study of large-scale social phenomena (such as the economy and the emergence of social movements). It was through weak ties that "small groups aggregate to form large-scale patterns."[12] In Granovetter's estimation, the weak ties between groups of actors made them vibrant and capable of affecting large-scale social change.

Weak ties were part of how the work of the MNI could spread to the world, and part 2 of this book examines the MNI's weak ties in greater detail, showing how the events in Montreal could shape the neuroscientific landscape of the postwar period in ways that have not always been obvious or easy to see. This portion of the book also shows how traversing weak ties can transform the products of an interdisciplinary trading zone, sometimes productively, sometimes disastrously. Each of the three afterlife chapters examines how a participant in Montreal took an element of the MNI trading zone—an instrument, a piece of theory, or an experimental result—and transformed it into something new. In the case of Donald Hebb and David Hubel, these transformations ended productively, with the emergence of a robust neuroscientific theory or a novel tool for experimental practice. And, in the case of D. Ewen Cameron, this transformation ended disastrously as his broken relationship with the MNI left him free to pursue the discredited and dangerous "psychic driving" form of psychiatric therapy.

Taken together, the trading zone of the MNI and the weak ties through which its legacies diffused constituted a complex assembly of historical actors that I describe with a metaphor drawn from an MNI participant during the institute's heyday, one that gives this book its title. One of the scientists I examine in this book is the psychologist Donald Hebb, who spent several years at the MNI and later developed a theory about neural mechanisms of intelligence. The core of his theory, expressed in his 1949 magnum opus *The Organization of Behavior*, was a postulate about how learning might happen as cells in the brain (neurons) become associated with each other as the result of simultaneous activity, forming what he called *cell assemblies*. A foundational concept for the modern study of neural networks, Hebb's theory appears today in the shorthand expression "neurons that fire together, wire together." In much the same way that the neurons of Hebb's cell assemblies formed through repeated simultaneous firing, the men and women of the MNI formed strong assemblies through their interactions. Yet it was also through the weak ties of the MNI to other assemblies that it could affect the macrohistorical evolution of neuroscience. It is in these interactions between strongly bound teamwork and loosely connected actors that neuroscience became "wired together." In this respect, then, this book is about the changing history of scientific teamwork and how that teamwork shifts the historical direction of science.[13]

Chapter Outline

While the story I hope to tell about Montreal and the origin of neuroscience is about more than individual actors, the chapters in this book follow a biographical approach, focusing on specific scientists and physicians. In her study of scientific holism in the German-speaking countries before and during World War II, Anne Harrington also used a biographical approach, "not to reproduce some hagiographic understanding of history of science as a parade of 'Great Men,'" but rather to show how different scientists, engaged in vastly different research programs, still drew on a common set of cultural resources that were unique to Germany in the Weimar years.[14] *Wired Together* inverts this perspective, using biographical studies of its leading actors to show how Montreal brought together different scientists from different cultures and backgrounds who then employed those resources in pursuit of a common project. A biographical approach can then bring some of those background characters in Penfield's operating room into focus, showing how they forged new assemblies. Biographical chapters also allow us to show how the new science that formed at the MNI traveled beyond Montreal and shaped the brain and mind sciences in the decades to come.

Part 1 of this book—"A Genealogy of Neuroscience"—explores the growth of the MNI, the development of its scientific trading zone, and the role of the MNI in establishing a global brain research community.

Chapter 1—"At the Crossroads of Scientific Medicine: Wilder Penfield, the Rockefeller Foundation, and the Genesis of the Montreal Neurological Institute"—traces the origins of Wilder Penfield's vision for a neurological institute that would combine the power of the laboratory with the work of the clinic. The MNI was the product of several key developments in early twentieth-century medicine and science: the professionalization and expansion of neurosurgery, the emergence of laboratory-based medical research and biomedicine, and a new conception of epilepsy as a treatable disorder. Most important, however, was how Penfield sold his idea to a key player in medical philanthropy. In the 1930s, the RF spurred a major change in American psychiatry. The philanthropic foundation, which had played a decisive role in directing the development of American medicine, had pivoted to make psychiatry a significant plank of its funding initiatives. In doing so, it embraced psychobiology, the psychiatric philosophy of the Swiss émigré psychiatrist Adolf Meyer. Psychobiology encouraged psychiatrists to adopt a pragmatic and eclectic approach to mental illness, one in which psychological *and* somatic therapies were part of the same enterprise. It also encouraged the field of neuropsychiatry to direct its attention toward so-called borderland conditions between psychiatry and neurology, such as epilepsy. Penfield proved particularly adept at selling his vision of an interdisciplinary neurosurgical clinic that would attack neuropsychiatric conditions like epilepsy (along with the usual fare of neurosurgeons, such as tumors) to the RF.

This chapter explores these issues by following Penfield's scientific biography from his training in the laboratories of Charles Sherrington in England

and Santiago Ramón y Cajal in Spain through his adoption of the German neurologist Otfrid Foerster's surgical techniques for epilepsy and his early struggles to cultivate his new approach in the competitive surgical world of 1920s New York. It culminates in the founding of the MNI in 1934. This was possible only because Penfield's plan was embraced by Alan Gregg, the head of the RF's Medical Sciences Division and a strong devotee of Meyer's psychobiology. The chapter ends by pointing to the factors that made Montreal an ideal site for Penfield's new approach to neurosurgery and why Penfield thought it would be the perfect place for sparking collaboration among scientists of different nations.

Penfield's surgical approach was radical. Removing portions of brain tissue to relieve epilepsy carried serious risks—the patient could end up paralyzed or mute or lose other critical mental functions. Chapter 2—"The Montreal Method: Molly Harrower, Brenda Milner, and the Neuroscience of Memory"—tells the story of how Penfield expanded the circle of disciplines at the MNI to meet this challenge and the remarkable female psychologists who transformed medical practice and scientific knowledge at the MNI. Through her use of psychological tests, the now-forgotten Molly Harrower established the first psychology department in a neurosurgical clinic and simultaneously developed a trading language with the MNI's surgeons. Harrower laid the institutional foundations for the later work of Brenda Milner, the English psychologist whose work on Penfield's temporal lobe patients and later the amnesiac patient H.M. revolutionized the study of human memory. This chapter serves as a crucial linchpin for the book, demonstrating the dynamics of the Montreal trading zone at its most productive and how it produced one of the most iconic discoveries of neuroscience.

By the 1950s, the MNI trading zone was at its most robust and had inspired many to believe that a new interdisciplinary approach to the brain was possible. At the same time, what form this new scientific field would take remained unclear, with several alternative visions emerging. Chapter 3—"A Tale of Two Sciences: Herbert Jasper, Francis Schmitt, and the Origins of Neuroscience"—tells the story of Herbert Jasper, Penfield's most important scientific collaborator and the man who brought the electroencephalograph into the MNI's operating room. In an era before fMRI and other imaging technologies, when the interior of the skull was still a foreboding place for surgeons to tread, Jasper's device was crucial for confirming diagnoses and locating epileptic lesions. But Jasper was more than a mere diagnostic technician. Through his leadership of the emerging community of electroencephalographers, he spread the MNI's interdisciplinary ethos to the global community of neurological scientists after World War II. This leadership culminated in the creation of IBRO in 1960, a remarkable act of scientific internationalism at the height of the Cold War. However, IBRO found itself unable to "scale up" the MNI's interdisciplinary trading zone to a global level and instead had to transform its ambitions to take advantage of the weak ties of the international EEG community, first by surveying the world's brains science laboratories and clinics. In doing so, it laid the foundations for a global community of neuroscientists in the 1960s.

This chapter also compares the development and ethos of IBRO with its American competitor, the NRP, run by Francis O. Schmitt at MIT. As we will see, Schmitt and his cohort practiced a very different strategy for unifying the brain and mind sciences. A molecular biologist by training and disposition, Schmitt pursued a transdisciplinary quest to discover a memory molecule that he hoped would unify the brain and mind sciences in the same way that the discovery of DNA had unified genetics and molecular biology. The failure of this reductionist quest also forced the NRP to pivot to a new goal, one of public relations for the new field of neuroscience. This chapter reveals simultaneously the true origins of neuroscience and why Schmitt's efforts at MIT became the predominant origin story.

By 1970, the organizing work of IBRO and the transcendent scientific aspirations of the NRP fused in the founding of the SFN, which marked the arrival of neuroscience as a mature scientific field. However, the legacies of the MNI for neuroscience were more complex and pervasive than the founding of a professional organization. Part 2 of this book—"Weak Ties and Afterlives"—tracks three different actors who participated (albeit briefly and not always successfully) in the MNI trading zone and how their different activities shaped the postwar scientific landscape. In each case, the products of the MNI trading zone were transformed as they traversed the weak ties that connected them from Montreal to the rest of the scientific world.

In 1939, Penfield removed most of the frontal lobes of a young man suffering from intractable epilepsy. After the operation, the young man's IQ went up twelve points. Chapter 4—"From Natural Intelligence to Artificial Intelligence: D. O. Hebb, K.M., and the Cognitive Revolution"—tells the story of Donald Hebb, the psychologist who tried to make sense of this bizarre occurrence and, in doing so, transformed how scientists thought about how the brain creates the mind. Hebb's conclusion, expressed in his hugely influential 1949 *The Organization of Behavior*, was that the human brain did not function according to the classic laws of reflex activity. Instead, mind and intelligence were the multifaceted products of the brain's constantly humming neural networks. Hebb's encounter with Penfield's surgical patients evolved into a general theory of brain function that escaped the clinic and exercised a pervasive influence on theories of cognition, behavior, pleasure, pain, and intelligence (both natural and artificial). His story, in turn, is also the story of an extensive network of students and collaborators who extended and modified his work, frequently intersecting with Cold War histories of espionage and brainwashing, debates about race and intelligence, and new engineering projects to build thinking machines.

The MNI's success was built on its interdisciplinary trading zone. But what happens when trading fails? Chapter 5—"Two Solitudes: Psychosurgery and the Troubled Relationship between Wilder Penfield and Ewen Cameron"— follows the dramatic story of the failed collaboration between Penfield and the infamous psychiatrist D. Ewen Cameron as they attempted to develop an alternative to lobotomy in the 1940s. The stakes of this collaboration were high. Would Penfield add a new surgical procedure to his armamentarium and successfully integrate psychiatry into his interdisciplinary enterprise? The fallout from the disastrous collaboration between Penfield and Cameron allows us to analyze why some interdisciplinary endeavors succeed and others fail. What factors prevent the formation of a scientific trading language? This chapter also examines how interdisciplinarity can serve as a cover for reckless and ethically questionable experimentation. After his collaboration with Penfield ended, Cameron went his own way, appropriating the rhetoric of interdisciplinarity to promote his own treatment for schizophrenia, a dangerous and scientifically dubious practice called *psychic driving*. Cobbled together from work done by other scientists (including those of the MNI alumnus Hebb), behaviorist psychology, and heavy doses of psychoactive drugs, this new treatment crossed the boundary from therapy to torture and placed Montreal at the center of the CIA's MK-ULTRA mind control program.

How does the brain see the world? This question animated neurologists, psychologists, philosophers, and artists for decades. In the 1950s and 1960s, the foundational work of David Hubel finally began to unlock the mechanisms of vision within the human brain. Chapter 6—"Eye, Brain, Vision: David Hubel, Microelectrodes, and the Science of Seeing"—tells the story of Hubel's work and its origins at the MNI in the 1950s. The history of Hubel and the scientific instrument he invented—the tungsten microelectrode— also shows how the work of the MNI's assemblies traveled along weak ties, transforming as it interacted with different actors and laboratory cultures. As Hubel and his electrode traveled from Montreal to Washington, DC, Baltimore, and Boston, the device became embedded in different research schools and experimental assemblages, culminating in a Nobel Prize in 1981. While other Nobels had been awarded for neurological discoveries, Hubel's prize could be fairly considered the first Nobel for neuroscience, growing as it did from the interdisciplinary work that had begun in Montreal nearly four decades earlier.

The conclusion of *Wired Together* recounts the fate of the MNI and its many characters. By the mid-1960s, politics and practicalities led to the slow dissolution of the Montreal group. Changes in medical funding in Quebec and Canada altered the flow of patients through the MNI, reducing the amount of scientific work that could happen within its walls. At the same time, the growing French separatist movement made Quebec less cosmopolitan and more inward looking, altering its connections to the rest of the scientific community. The crucial trading zone forged in Montreal began to dissipate, and the MNI's pivotal role in creating modern neuroscience was obscured. I also reflect on how the story of the MNI might reframe our understanding of neuroscience in the modern world. How might we think differently about neuroscience if we recognized it not as a unified field of science but as a hybrid field, a historical product of complex assemblies of ideas, technologies, practices, patients, and historical context? Moreover, while the world of modern neuroscience is remarkably *multidisciplinary*, the story of the MNI highlights the ways in which it may now struggle to be *interdisciplinary* in the remarkable way that it was in midcentury Montreal. What does the history of the MNI suggest about the challenges facing modern neuroscience and neuroscientists as they pursue new knowledge across a bewilderingly complex scientific landscape?

A Genealogy of Neuroscience

At the Crossroads of Scientific Medicine: Wilder Penfield, the Rockefeller Foundation, and the Genesis of the Montreal Neurological Institute

On April 18, 1930, Wilder Penfield wrote the mother of a young man named William Ottmann on whom he had recently performed a complicated brain operation. The patient—at 15 years of age, really only a boy—suffered from debilitating epilepsy, and Penfield aimed to relieve it with a complex and risky procedure. He informed William's mother, Madeline, of the surgery's successful outcome, but he also told her about an idea. Mrs. Ottmann was a wealthy New York society woman who had previously donated money to Penfield for research. Penfield used that money to bring several neurological specialists to Montreal to consult on William's surgery and provide expert and contrasting opinions during the operation—a surgical team of rivals. Now, he hoped to gather "a group of men with me in an Institute for Neurological Investigation [in Montreal] where we can concentrate patients, laboratories and young investigators. I know it is impossible . . . as the whole undertaking would cost a million dollars. But if I ever do get it . . . the problem William has presented us will be the central problem of investigation."[1] Madeline Ottmann had been consistently impressed by the care Penfield offered her son and, within a year, would write him into her will, ensuring $50,000 to continue his efforts.[2]

Only four years later—with the help of patrons with considerably deeper pockets than Madeline Ottmann's—Penfield opened precisely the institute he had proposed. The Montreal Neurological Institute (MNI), which opened its doors on September 27, 1934, contained many public beds, advanced laboratory facilities, and all the high technology associated with cutting-edge surgery in the 1930s—X-ray machines, robust anesthetics, and closed air filtration to minimize surgical infection. Most importantly, two state-of-the-art operating theaters with plate glass windows allowed the institute's research fellows to observe a remarkable sight: as he operated, Penfield's patients remained awake and conscious while he stimulated the surface of their brains with a gentle electric probe. Patients reported flashing lights, sounds, and sensations in their bodies as Penfield passed the probe over their exposed brain, searching for the source of their epilepsy. In time, they would report reliving old memories, lose and regain the ability to speak, and even experience strange doublings of consciousness. "It is essential," Penfield noted, "that each research fellow working in a laboratory may be able to get a clear view of every pathological condition exposed by the operator if he so desires and to return to his proper work without loss of time."[3] To ensure this, a bell was installed to alert laboratory workers that another brain had been exposed. Across the institute, scientists were to filter into the operating theater to observe the remarkable event. While Madeline Ottman's money had briefly enabled Penfield to summon a team of consultants and scientific collaborators, an unprecedented grant from the Rockefeller Foundation (RF) now allowed him to summon those collaborators on command. Unique in the world, the institute that Penfield had described to Madeline Ottmann in 1930 was now a reality and, in the coming decades, would transform the epileptic surgical patient into an object of scientific investigation and the surgical theater into a laboratory and research space. Perhaps just as important, Penfield's MNI would serve as the incubator for a new, interdisciplinary field of science.

To understand the role of Penfield's institute in creating modern neuroscience, we must first understand something of the man and the scientific and medical worlds he occupied. This chapter tells the story of Penfield, his institute, and the origins of its interdisciplinary environment. Penfield's story is that of a young surgeon who experienced and participated in several crucial transformations in science and medicine in the first third of the twentieth century. These included the emergence of a physiological approach to surgery along with new laboratory techniques and therapeutic innovations.

However, as the story of Madeline Ottmann and her son shows, just as important to Penfield's success was his ability to promote his new approach to precisely the wealthy patrons who could make it a reality. His proposal for a new kind of neurosurgical clinic, laid out in 1930 to Ottmann, would eventually find its way to patrons whose plans mirrored his own—most notably, Alan Gregg, the director of the RF's Medical Sciences Division. The RF's campaign to bring psychiatry in line with modern scientific medicine created a funding environment that valued Penfield's plans for an interdisciplinary neurological institute focused on research, and the RF embraced a philosophy for psychiatry that placed neuropsychiatric conditions like epilepsy at the center of its funding priorities. The institute that Penfield created was largely a product of his ability to sell his interdisciplinary vision to an enthusiastic patron, and Gregg's stewardship had made the RF just such a patron.

By following Penfield's scientific biography, we will observe how he combined different scientific styles, technologies, and perspectives. However, we will also see that the institute that he created was as much a product of changes in the funding environment of science and medicine in the early twentieth century as it was of any particular medical or scientific development, and his achievement was as much about *promoting* his vision of interdisciplinarity as it was about practicing it.

A Reluctant Doctor

Born in Spokane, Washington, in 1891, Wilder Penfield did not want to be a doctor. His father, Charles, a country physician, abandoned his mother, Jean, while he was still a boy. This early ordeal left the young Wilder with contempt for the medical profession and an unusually close attachment to his bright and pious mother. Perhaps frustrated by the failure of her marriage, Jean Jefferson Penfield poured much of her intelligence and energy into her son. Given his mother's strong Victorian sense of morality and mission, young Wilder seemed destined for the pulpit rather than the clinic. Jean's commitment to her son was ferocious; by the time he was eight, she had moved the family to Hudson, Wisconsin, and established a private boy's school called the Galahad Academy the purpose of which was to provide Wilder with an exceptional education. Her long-term plan was for her son to win a Rhodes Scholarship (fig. 1.1).[4] Penfield's remarkably close bond with his mother constituted one of the central relationships of his life. It also generated an impressively rich historical source; mother and son exchanged regular letters from 1908 until her death in 1935, and the level of detail and sophistication Penfield included in his letters indicates the high esteem in which he held his mother's intelligence.

Penfield's attraction to science deflected his trajectory away from the pulpit and toward medicine despite the hostility he felt for his absent father's profession. Penfield came of age in an America that was entranced by the achievements of laboratory science. Beginning with the bacteriological revolutions of the late nineteenth century, medicine had clearly benefited from discoveries by expert, discipline-bound, and technologically advanced laboratories. The development of diphtheria antitoxin and a successful vaccine for rabies in the late nineteenth century brought a measure of control to infectious diseases, and the so-called laboratory revolution in medicine was transforming hospital practice and public health. The laboratory also became a cultural symbol of modernity and progress. Sinclair Lewis's 1925 novel *Arrowsmith* cemented the image of the medical scientist in the American

FIGURE 1.1. Wilder Penfield as a young man at Princeton University, ca. 1913. Not a natural ath-
lete, Penfield eventually mastered football as part of his efforts to secure a Rhodes Scholarship. George
Grantham Bain Collection, Library of Congress.

mind, and laboratory education became an integral part of medical training
following the release of the Flexner Report in 1911. By the 1920s and 1930s,
private organizations like the RF and the Carnegie Institution had latched
onto the laboratory revolution as an object of their philanthropic largesse.[5]

Penfield, too, became enamored with science as a young man at Princeton
University, where the biologist E. G. Conklin introduced him to Darwinian
evolution in a way that could be reconciled with the Presbyterian faith of
his mother. Conklin also introduced him to the laboratory. Embryology,
Conklin's specialty, was emerging as a leading area of laboratory investigation, and his lectures on "tiny cells within the living, growing body" that
could be seen through the microscope captivated Penfield. By the time of
his graduation, Penfield's ambitions had turned toward his father's profession
despite his initial resistance. Writing down a list of possible careers in the
Princeton library, Penfield scratched one after another off the list until he
was left with *minister* or *doctor*. He chose the latter, with a special interest in
medical research.[6]

Surgery, too, appealed to Penfield's interest in a type of scientific medicine that would be different from his father's. During his sophomore year at
Princeton in 1911, Penfield and a friend snuck into New York's Presbyterian

Hospital and spent nearly four hours observing operations. Penfield later said to his mother: "Don't think that because I sat through some butchering that I believe I am fitted for a doctor. . . . I was not particularly fascinated with the operations except that the wonderful skill and knowledge of the surgeon appealed to me. I believe that a practitioner (who can have only a little of regular operating work) should not attempt to be a surgeon but he should let the surgeons in the hospital do it all."[7]

Penfield's reverence for the skill and knowledge of the surgeon was in keeping with the new climate in American medicine.[8] By the 1910s, the slow and tentative adoption of aseptic surgical procedures that began in the late nineteenth century was largely complete. The introduction of rubber gloves, autoclaves, and sterilized dressings in the 1890s replaced the carbolic acid sprays and reused sponges and bandages of the mid-nineteenth century, and the operating room assumed a recognizably modern character. Combined with anesthetic procedures to dull operative pain and a growing knowledge of microorganisms, these developments made surgery the primary reason for admission to American hospitals. The increasing safety and efficacy of surgery also emboldened American doctors to explore, probe, and intrude on cavities of the body that had previously seemed forbidden. With the abdomen finally unlocked as a site of surgical intervention, the skull was not far behind.[9] More than any other branch of medicine, surgery appeared to have brought the power of science to the clinic. Surgery combined with laboratory research was an easy choice for a precocious young man with a scientific bent like Penfield. Penfield's growing interest in medical research and surgery was reinforced by a summer course in anatomy at Harvard in 1914, during which he spent hours in the evening working alone in the dissection rooms.[10]

Penfield finally received his long-sought-for Rhodes Scholarship on the eve of World War I and arrived at Oxford nine months later to find it eerily empty; many of its students had gone to fight on the battlefields of Europe. Here, he made two acquaintances that fundamentally altered his life's course. The first was Sir William Osler, a cofounder of Johns Hopkins Hospital and one of the prime movers in the modernization of American medicine. Osler's reputation preceded him. Arriving in Oxford in January 1915, Penfield confided to his mother: "When I look up at the seven volumes of Osler's [*A System of Medicine*] it makes me mentally worship him. It does not seem possible that he is the middle-aged man I saw last Sunday, who, with a room full of tea-drinkers, spent most of his time pretending to bandage up the leg of a young officer to the glee of two little children who had come to see him."[11]

Osler's influence on Penfield was twofold. First, Penfield absorbed Osler's practice of careful and exacting diagnosis, frequently accompanying him on consultations to observe the older man's gentle and probing bedside manner. Second, he shared Osler's abiding interest in neurology and neuropathology. Over the course of his medical career, Osler had been instrumental in spreading the "gospel of clinicopathology" (the careful postmortem correlation of autopsy findings with clinical histories) to North America from its home in the German-speaking institutes of Rudolf Virchow and Karl Rokitansky. He had also closely followed the work of European master neurologists like John Hughlings Jackson and Jean-Martin Charcot and introduced methods for the preservation and dissection of brains into his pathological work. From Osler, Penfield learned how to extract brains and their diagnostic secrets from the recently deceased and to view autopsies as a central (and humane) aspect of medical practice.[12]

A Physiological Surgeon

The other crucial influence at Oxford was Charles Scott Sherrington. Sherrington had emerged from Michael Foster's Cambridge physiology laboratory at the end of the nineteenth century, having absorbed his mentor's evolutionarily inflected approach to experimental animal research.[13] From this foundation, Sherrington would by the turn of the century transform neurophysiology into a theoretically sophisticated and mature scientific field. Between 1884 and 1898, his painstaking experiments with rats, dogs, cats, and monkeys brought a new intellectual order to the study of the nervous system and allowed him to explain the adaptive, integrative nature of animal movement as the product of successively more complex and interacting reflexes. In the nineteenth century, the study of the nervous system had been a piecemeal effort; brain anatomy was separate from reflex motor activity, which in turn was separate from the electrophysiology of the nervous system. Chemistry was nowhere in the picture, and the study of the nervous system's structure was often divorced from the analysis of its function. Sherrington's careful study of the reflex synthesized nearly a century of work, bringing order where there had once been discord. Codified in his 1906 *The Integrative Action of the Nervous System*, his efforts, in the words of one contemporary, "almost single handedly crystallized the special field of neurophysiology."[14]

Sherrington's synthesis was possible because of a vastly improved understanding of the microanatomy of the nervous system itself. The science of histology—the microscopic study of tissues—had grown into an autonomous scientific discipline by the late nineteenth century, with new cellular staining techniques revealing the structure of tissues in greater detail. In the final decades of the nineteenth century, an obscure Spanish microscopist, Santiago

Ramón y Cajal, used the new silver nitrate staining techniques of an equally obscure Italian pathologist, Camillo Golgi, to demonstrate that nerve cells were, in fact, separate rather than part of a continual reticular network as commonly supposed. This experimental demonstration of what the German anatomist H. W. G. Waldeyer christened *neurone theory* won Cajal the Nobel Prize in 1906 and allowed Sherrington to use the nerve cell as the unit for understanding the functions of the nervous system. Sherrington combined this new understanding of the nervous system's microarchitecture with exacting macrolevel animal experimentation to turn the "anatomical facts of the nervous system into physiological language," as one observer put it.[15]

Penfield was first exposed to Sherrington's approach in 1915 during a course at Oxford in mammalian physiology that included exacting instruction in laboratory technique (fig. 1.2). Indeed, he noted that the physiological approach to medicine was becoming particularly dominant in England.[16] "In America," he wrote his mother, "the whole of next year [of medical education at Oxford] would be devoted to Physiology and part of this year. In America they devote a very small part of that time to Physiology. [The English] are extremely thorough and should turn out researchers. But it adds a year."[17]

FIGURE 1.2. Wilder Penfield in Charles Sherrington's laboratory, ca. 1915. After World War I, Penfield would return to Oxford for graduate work in physiology. Reproduced by permission of the Osler Library of the History of Medicine, McGill University.

Elsewhere, he reported: "They [the English] are very short of material and apparati in Anatomy. Just the opposite in Physiology where Osler says they excel."[18] Penfield proved himself an exacting and capable laboratory worker, so much so that Sherrington included his work in his *Mammalian Physiology: A Course of Practical Exercises* (1919).[19] Moreover, it was in Sherrington's laboratory that Penfield developed his approach to surgery; gentle handling of tissue, precision instrumentation, and a talent for keeping experimental animals alive all combined to shape an approach to surgery that blurred the line between medical intervention and scientific experimentation.[20]

Despite his growing fascination with the laboratory, Penfield also desired practical, hands-on medical experience. This impulse nearly cost him his life. Ignoring Osler's warning not to go to the front,[21] in 1916 he traveled to France to gain experience working as a paramedic in a hospital treating frontline soldiers. On his return to England, his ship, the *SS Sussex*, was hit by a German torpedo. He survived and spent several weeks recuperating in Osler's home. After his recovery, he returned to the United States in 1917 to complete his medical education at Johns Hopkins and to marry Helen Kermot, his girlfriend since childhood.

After completing his medical degree in 1918, Penfield spent a brief period observing the surgical practices of Harvey Cushing at the Peter Bent Brigham Hospital.[22] For more than a decade, Cushing had stood as the leader of the new field of neurosurgery. In 1905, he announced the agenda of the new specialty in a pivotal paper in the *Johns Hopkins Hospital Bulletin*. Charting the field's slow progress from its promising start in the nineteenth century, he argued that brain and spinal cord surgery had not advanced as quickly as other fields had because neurosurgeons lacked "certainties of diagnosis and assurances of therapeutic results." Simply put, surgeons were reluctant to operate on the nervous system because their chances of success were poor and their knowledge inadequate. Surgery involving the nervous system—and particularly the brain—remained a grim business, an area of long odds and last-resort operations. Neurologists fared no better. The treatment of neurological illnesses had not advanced significantly in the preceding decades and continued to rest "on the therapeutic tripod of iodine, bromine and electricity," treatments that brought to Cushing's mind the aphorism of Benjamin Franklin: "He is the best physician who knows the worthlessness of most medicines." In Cushing's view, what was needed was for the neurosurgeon to assume much of the diagnostic role then managed by the neurologist. Rather than acting as a technician "called in [with] little knowledge of maladies of this nature and less interest in them," the neurosurgeon should be fully trained in neurological anatomy, pathology, and diagnostics.[23]

In the intervening years, Cushing's leadership had transformed the situation, finally bringing the operative death rate below 10 percent, making the surgical treatment of brain tumors and cysts possible, and placing the specialty at the forefront of American medicine. Neurosurgical societies sprang up in major cities, and American and European surgeons traveled to Cushing's home base in Baltimore to train with him.[24]

At the same time, the list of disorders that neurosurgeons treated remained stagnant—tumors, cysts, and physical trauma were still their bread and butter. The situation would not improve if "the neurologist and surgeon maintain the same distant relations that have heretofore existed between them." To unlock the true therapeutic potential of surgery, surgeons needed to become more scientifically sophisticated: "To successfully cope with the many operative problems offered by the various disorders of the nervous system, a man, after a thorough training in pathology and medicine (in its broadest sense) must study, not only in the neurological clinic but also in the laboratory, the pathology of these afflictions in their histological and—what is still more important—in their experimental aspects."[25] Despite this clarion call to embrace the laboratory, by the time Penfield arrived at the Peter Bent Brigham Hospital in 1918, few neurosurgeons had followed Cushing's advice.

Given his interest in the laboratory and neurophysiology, Penfield might seem a perfect candidate to take Cushing up on his challenge. It is surprising, then, to learn just how cool the relationship between Cushing and the up-and-coming Penfield was—a young man who might seem an ideal candidate for the older man's stewardship. Cushing certainly recognized Penfield's talent, noting to a colleague in 1919 that more interns like him would be desirable.[26] And Penfield acknowledged the contributions of Cushing, whose masterful technique had finally made neurosurgery respectable. At the same time, and crucially, Penfield always credited Sherrington as being his chief surgical inspiration: "It was not the example of Horsley or Cushing that led me into the surgery of the nervous system. It was the inspiration of Sherrington. He was, so it seemed to me from the first, a surgical physiologist, and I hoped then to become a physiological surgeon. As years passed his influence did not grow less but stronger."[27] For Penfield, Sherrington's physiology laboratories provided a more compelling example of what might be possible in neurosurgery even than Cushing's humming practice in the United States did.

Cushing's call to the laboratory was in keeping with a broader trend in late nineteenth- and early twentieth-century surgery as more and more surgeons began to think of their profession as a form of applied physiology rather than the applied anatomy of an earlier generation. Experimental physiology in animal laboratories supplied the basic research and rationale for more aggressive interventions into the body. A new generation of surgeons saw parallels between the surgical theater and the laboratory; both were spaces where controlled intervention in a living organism yielded new knowledge. Surgeons and physiologists both operated on animals and humans, intervening in the body as much to adjust function as to remove diseased tissue or reset bones. The future of medicine lay in bringing the tools and techniques of experimental physiology into the operating theater. By the turn of the century, this laboratory-based physiological approach—intervention not only to remove lesions but also to adjust bodily functions—had birthed surgical initiatives such as organ transplantation.[28] However, while Cushing embraced the concept of applied physiological knowledge, his surgical approach remained conservative. Despite his meticulous technique and rhetoric surrounding laboratory science, Cushing's surgical approach was still largely that of the nineteenth century, that is, anatomical in orientation and conservative in practice: find the tumor, remove it, disturb as little healthy brain tissue as possible, and hope for the best.[29]

Penfield, by contrast, aspired to a more interventionist approach. First, however, he would need greater laboratory experience, so he returned to Sherrington's laboratory in 1919 to expand his neurophysiological knowledge. In Sherrington's lab, he used Cajal's cellular staining techniques to elucidate the Golgi apparatus's structure within nerve cells, earning himself a BSc degree and his first scientific publication in *Brain*.[30]

Penfield also required instruction in the fine diagnostic art of clinical neurology to use his newfound physiological knowledge in the operating room. He began training with Gordon Holmes, an athletic Irishman and talented diagnostician at the National Hospital for the Paralyzed and Epileptic at Queen Square. The leading site for the professionalization of neurology in nineteenth-century England, Queen's Square was now home to Britain's second generation of neurologists, including Holmes and Henry Head. Here, Penfield encountered one of the first obstacles to his growing vision of an interdisciplinary neurosurgical enterprise; the clinic's neurologists shared an abiding dislike of surgery. Queen's Square had, in fact, hired an in-house surgeon, Victor Horsley, in 1886. A pioneer in organ transplantation research, Horsley moved smoothly between laboratory and operating theater, the embodiment of physiological surgery to which Penfield aspired. Yet Horsley's time at Queen's Square had been disappointing. The clinic's neurologists jealously guarded their diagnostic turf, and, without enough referrals, Horsley's attempt to build up a surgical practice floundered.[31] By 1920, neurologists at Queen's Square were actively withholding cases from their new house surgeon, Percey Seargent, for the not-unreasonable belief that operations produced generally poor results and patients frequently died of infections.[32]

At the same time, the possibility of uniting clinical neurology with a physiological approach to neurosurgery excited Penfield: "The surgical cases are tremendously interesting. The best part of it is that I have the chance of studying them carefully and getting the opinion of the best neurologists and then, instead of wondering vaguely whether or not the diagnosis was correct, I have the opportunity of seeing their brain, then seeing them recover, or die, and conclude for myself whether or not the pre-operative conclusions were justified." His first attempt at interdisciplinarity was to embody it in himself, to fuse the disciplines of clinical neurology and surgery in one man by way of physiological knowledge. His interdisciplinarity was also a product of his international experience; he would wed American surgical technique with English physiological and clinical sophistication. "In the next 30 years," he informed his mother, "I believe there will be great strides in our knowledge of the nervous system and our treatment of it and I want to take part in both changes."[33] His optimism would barely survive his return to the United States.

A Terrible Profession

If Penfield left England in 1921 buoyant about the potential of neurosurgery, his next several years in the United States brought him headlong into the grim reality of his chosen profession. Initially intent on joining the staff of the newly formed Henry Ford Hospital in Detroit, he turned down the offer of a thriving neurosurgical practice because it included no possibility for research. An in-house pathologist handled pathology at the Detroit hospital, and animal research was out of the question.[34] Penfield ultimately took a position at New York's Presbyterian Hospital, which had recently become the teaching hospital for Columbia University's medical school.[35]

Penfield began to find his feet as a surgeon at the Presbyterian Hospital, but he also became increasingly demoralized. The high hopes he had for his physiological approach to the brain bore little fruit, and neurosurgery remained a morbid affair. "I was supposed to take over the brain and nerve operations," he lamented to his mother, "which is the most hopeless and difficult field."[36] At the same time, he was suspicious of the motivations of his colleagues: "This is a great mine of clinical material. . . . [I]f the group of which I am a part can only work with the scientific enthusiasm that Osler, and Welch, and Halstead etc. instilled at the beginning of Johns Hopkins thirty years ago, I might live to see a real clinic develop. The trouble has been that the profession of New York has been too commercial. Perhaps the temptation is great for there certainly is a pile of money here."[37]

Professional jealousies and the commercial aspects of medicine in New York also constrained Penfield's access to patients. Frederick Tilney, the leading neurologist in the city, insisted that all neurological cases, including surgical ones, flow through his own New York Neurological Institute (NYNI). Founded in 1909, the NYNI was the first institution in the United States to specialize in neurological illnesses and was initially meant to develop neurological surgery following Cushing's example. However, by the 1920s, it had become the professional base for New York's neurologists, almost all of whom were hostile to surgery. The institute also had no interest in research; there was no dedicated neuropathologist, and the laboratory resources were trivial.[38] Penfield commented on this situation to his mother: "I have been thinking it over and I now see that the thing for me to do is not to go over there [the NYNI] where I may work under neurologists with a certain amount of friction and never get any farther than they are already, which isn't far."[39] When he finally got the chance to operate on a brain tumor patient, the results were disappointing. In the 1920s, neurosurgery was still primarily tumor surgery, and the operations were forbidding, dangerous, and frequently ineffective. In one episode, Penfield was unable to remove a malignant tumor from a female patient who died shortly thereafter. Describing the incident to his mother, he lamented: "[B]rain surgery is a terrible profession. If I did not feel that it will become very different in my lifetime, I should hate it. It seems a bit depressing."[40]

The Sacred Disease

Epilepsy rescued Penfield from his despair, and his efforts to develop a surgical treatment for the condition would alter his own life's course; they would also alter the course of the brain sciences in the twentieth century. More than any other disorder, epilepsy transformed the surgical theater into a laboratory for exploring the functions of the brain. Simultaneously, it occupied a crucial social and cultural position that brought it to the attention of philanthropic organizations that could put Penfield's vision into action. Tumor surgery was a terrible profession—the grim anatomical surgery of the past. Epilepsy surgery was the future—optimistic, forward-looking, and ripe for physiological investigation.

Why epilepsy? The "sacred disease" of antiquity—associated for millennia with demonic possession and witchcraft—had by the late nineteenth century been partially tamed by secular and scientific medicine. In the 1860s and

1870s, the English neurologist John Hughlings Jackson laid out the modern scientific understanding of the disease. *Epilepsy* was "the name for occasional, sudden, excessive, rapid, and local discharge of grey matter," probably electric in nature.[41] His study of epilepsy also contributed to an understanding of brain function. Carefully examining the "march" of his patients' convulsions—for instance, beginning in the thumb, then moving up the hand, arm, and shoulder, across the body to the other arm, and culminating in a full-body convulsion—Jackson concluded that the body must be represented topographically in the brain. The contemporaneous 1870 experiments of the German neurologists Gustav Fritsch and Eduard Hitzig, who stimulated the exposed brain of a dog with an electric probe and discovered the existence of the motor cortex (a topographic representation of motor functions), seemed to confirm Jackson's suspicions. This focal epilepsy joined the aphasias as another neurological condition whose symptoms could be localized to certain areas of the brain. This localizability made epilepsy a condition that could be studied using the tools of modern clinical medicine and might also be amenable to surgical intervention. If epilepsy resulted from localizable, diseased brain tissue, then that tissue could possibly be removed.[42]

At the same time, so many aspects of epilepsy remained mysterious, and this mystery often resulted in stigma and condemnation. Epilepsy was classified into two broad categories: the organic, typically caused by injury, stroke or brain lesions, alcoholism, autointoxication, syphilis, or an amorphous neurological condition known as *reflex irritation*, and the idiopathic, including all epileptic fits with no apparent cause. The idiopathic epilepsies could not be localized by following convulsions of the body and presented a much less coherent symptom pattern, often including unusual behavior. Frustratingly for physicians, most epilepsies fell into the latter category, and patients spent much of their lives as social pariahs, always fearing the onset of a fit. Family care often gave way to confinement in custodial institutions.[43]

The medicalization of epilepsy did little to combat that stigma. While physicians assumed that neuropsychiatric conditions like epilepsy had an organic origin, in keeping with emerging eugenic ideas, they also believed epileptics to be genetically inferior or degenerate. They typically looked on these patients with disdain. As the physician A. E. Osborne put it in 1893: "[There was] an undeniable and alarming increase in the number of epileptics, who are injecting, in a great measure through the propagation of their offspring, an exceedingly dangerous taint to the morals and health of the masses. . . . The Epileptic is an individual of strange characteristics, and of a duality of personality which may quite outdo in viciousness and meekness, criminality and cunning, immorality and simulated innocence, the Jeckyl [sic] and Hyde creation of the novelist."[44] Indeed, the provocatively named *masked epilepsy* was thought to paralyze higher cortical centers that governed morality, leading to violence. The highly visible murder of an elderly Ohio farm couple by the epileptic man Richard Barber in 1888 demonstrated to many that epilepsy was dangerous, both personally and socially.[45]

Despite its medicalization and high social profile, medical science had made little progress in treating epilepsy by the early twentieth century. Treatments ranged from the ineffective to the dangerous. Powerful drugs like bromides and phenobarbital, which had severe side effects, left epileptics disoriented, dull, and drooling. Slumped faces in turn led to the view that epileptics had a distinct physiognomy—bloated, heavy, and lost. Surgical options were not much better. Some surgeons trephined the skull to relieve pressure. Others believed that attacks were triggered by malfunctioning organs, which might be excised. Ovariectomies to control seizures were not uncommon.[46] Epilepsy, then, occupied a key position for Penfield: poorly understood, but potentially medically tractable, ripe for physiological investigation, socially visible, and offering a sizable patient pool.

Penfield's turn to epilepsy also had an important local impetus. He had recently begun to work with the surgeon and pathologist William Clarke at the Columbia College of Physicians and Surgeons. Clarke had joined the college in 1905 to develop a new department of surgical pathology; surgeons increasingly wanted pathological tissue removed during operations to be quickly analyzed to confirm diagnoses, and Clarke was meant to train others in this practice.[47] He was also interested in the nature of wound healing and suggested that Penfield make an experimental study of brain wounds. Penetrating focal wounds to the brain had become both more common and more survivable. Since World War I, many otherwise healthy young men had recovered from such wounds but also frequently developed epilepsy shortly thereafter. Penfield reasoned that understanding the process of brain wound healing might help explain the development of certain focal epilepsies and that a study of experimental brain wounds in animals might lead to a new surgical treatment.

There was a problem, however. When placed under the microscope, Penfield's experimental brain wounds revealed little. The staining methods of Golgi and Cajal (which clearly stained neurons) showed few of the most interesting cell types in the healing scars—notably glial cells (thought to support and nourish the neurons).[48] Penfield tried newer staining methods that had emerged from the Cajal school, a task complicated by the fact that the *Cajal Transactions*, which reported on the activities of the Spanish histologists, appeared only in Spanish. With an impressive tenacity, Penfield translated the work of Cajal's most productive student, Pío del Río-Hortega, and tried his staining methods to make glial cells visible. The results were remarkable but inconsistent; some cells "stood out, clear and complete, as [he] had never seen them before," while others remained murky. Determined to master the new staining method, Penfield convinced his chief of surgery to permit him a six-month sabbatical to travel to the Instituto Cajal in Madrid to learn the latest techniques from Río-Hortega himself. He undertook intensive Spanish-language lessons and, in the spring of 1924, boarded a ship bound for Spain.[49] Aboard the ship, Penfield dispatched a letter to his mother: "I can get no further until I learn something about neuroglia cells in Madrid. . . . I've done nothing but prepare and am still preparing. I do not see the way toward hydrocephalus or epilepsy or any worthwhile problems and so I go on trying to learn, hoping the method will become apparent. But there is not the slightest guarantee that any clue lies in the direction I am taking. I am at the height of my power and still reaching out for new weapons—using none."[50]

Spanish Methods

When Penfield and his wife, Helen, arrived in Spain in April 1924, the country was far from a scientific powerhouse. Despite the economic boom of the immediate postwar period, science there lagged far behind science in the rest of Europe.[51] Following his 1906 Nobel Prize, Santiago Ramón y Cajal convinced the Spanish government to endow him with a small laboratory in Madrid. Here, he trained others in his histological techniques and built up a small but prolific school of histologists over the next two decades. Despite its productivity, the Instituto Cajal had a limited impact on the international scientific community. In his memoirs, Cajal laid the blame for this situation squarely on his decision to publish primarily in Spanish:

> I was always anxious, especially after the State placed in my hands a suitable and well-equipped laboratory, to found a genuinely Spanish school of histologists and biologists. . . . Its discoveries . . . have spread beyond the frontiers and its methods and inventions are applied in foreign laboratories. Moreover, they would be applied more widely if, recognizing the almost total ignorance of the Spanish language among scientists, we were to publish all our works in foreign periodicals. For it must be stated . . . that *hardly a third part of the Spanish histological publications [are] known abroad.*[52]

Cajal's concerns were not unfounded. Since the middle of the nineteenth century, the primary languages of science in the Euro-American world had been German, French, and English. After World War I, growing nationalism among Europe's smaller nations encouraged many to publish in their native tongue, much as Cajal had done. However, most scientific communication was still conducted in the original linguistic triumvirate.[53] The work of Cajal and his school remained locked away in Spanish. In a period of growing scientific nationalism, Penfield's willingness to cross borders and languages displayed an internationalist scientific ethos that would eventually characterize his institute.[54]

The man Penfield came to see, Pío del Río-Hortega, was Cajal's most talented pupil. Indeed, Río-Hortega's talent for histology ultimately caused a rift with his mentor. Cajal had previously succeeded in delineating two major cell types in the nervous system—neurons and glial cells (astrocytes and microglia). Río-Hortega believed that a third type of cell existed, Cajal disagreed strongly, and the debate had eventually soured their relationship.[55]

Penfield set to work acquiring Río-Hortega's tacit knowledge of staining nonneuronal cells. Blocks of rabbit brain were fixed in formalin and then sectioned with a microtome. The resulting paper-thin slices were washed in successive baths of alcohol and carbol-xylol-creosote and then stained with silver carbonate. Penfield's difficulties in New York had mostly been related to the precise timing and heating procedures involved; as he learned in Spain, different timing and heating would lead to cells being stained at different depths. However: "[W]ith care and patience, selective staining might be possible for almost any and all of the cell structures within the brain. This was truly selective photography."[56] In his subsequent months in Spain, Penfield proved himself a capable benchworker and developed a reliable method for staining a cell type that he christened *oligodendroglia*. This provided definitive evidence that Río-Hortega's methods worked and that a third type of glial cell existed.[57] Penfield also continued his study of experimental wounds in the brains of rabbits, this time equipped with Río-Hortega's more precise staining methods (fig. 1.3). The results were impressive; Penfield's staining methods revealed the different types of healing that occurred with different types of wounds to the brain. Importantly, when damaged tissue was removed aseptically during the wound-making process, there was minimal scarring in the injured brain tissue. However, when the damaged tissue was not removed, a thorny nest of scar tissue remained, infested with collagen and other cell types. Notably, this scar tissue would contract over time, and to Penfield this suggested a mechanism by which a brain wound might eventually produce epileptic discharge. Although this remained an educated guess on Penfield's part, what was now indisputable for him was the value of the "Spanish" staining methods, which had finally shed light on the process of brain wound formation.[58]

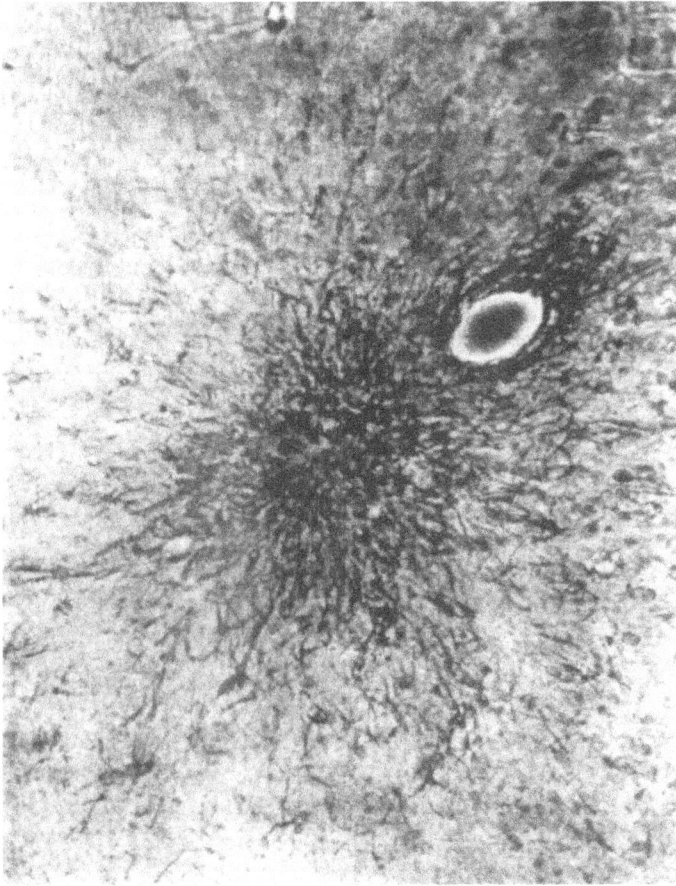

FIGURE 1.3. Using Spanish methods to stain injured brain tissue. From Pío del Río-Hortega and Wilder Penfield, "Cerebral Cicatrix: The Reaction of Neuroglia and Microglia to Brain Wounds," *Bulletin of the Johns Hopkins Hospital* 41, no. 5 (November 1927): 297.

Penfield found that the effect of his time in Spain was twofold. First, it served as a proof of concept that, by mastering the skills of laboratory science and adjacent disciplines, he could expand the practice of neurosurgery. The Spanish methods of cellular staining would form a central plank in his effort to unite the laboratory and the clinic. The experience also reinforced his belief in the value of scientific internationalism and linguistic pluralism. The enemies of scientific and medical progress were nationalism and chauvinism, ego and pride. For Penfield, internationalism and interdisciplinarity emerged as related, overlapping scientific virtues.[59] At the same time, Penfield's conception of interdisciplinarity was profoundly personal. Interdisciplinarity had first to be embodied in the self by surveying and mastering multiple skills.

By doing so, Penfield aimed to transform his scientific and medical self while simultaneously improving his professional position within the competitive medical world of New York.

Returning to New York, Penfield embarked on the creation of a new laboratory at the Presbyterian Hospital where he could enact this new scientific self and train others in his practice. Within a year, he had proposed the creation of a laboratory of neurocytology "for the study of the central nervous system, particularly by means of staining methods used by certain Spanish investigators." He went on: "There remains [a] large group of mental and nervous ailments about the causes of which little or nothing is known. It is evident that the complaints of the patients are not imaginary and yet microscopic examination of the brain and nerves of such cases has not yet yielded the secret that must be hidden there. Examples of the commoner of such uncontrolled infirmities are epilepsy, neuritis, insanity and possibly certain types of so-called hysteria."[60] Penfield's plan was ambitious. He and his associates gathered pathological samples from the Presbyterian Hospital and the Columbia College of Physicians and Surgeons surgical laboratory and animal tissue samples from Columbia University. He even gained the cooperation of his old rival Frederick Tilney at the NYNI. From across New York, brain tissue found its way to Penfield's lab, where it would be analyzed by Penfield and his colleagues: a young doctor, William V. Cone, who became his close friend and surgical partner, and Edward Dockrill, a moody and nomadic Cockney who had initially lied about his qualifications but proved himself an able laboratory worker. A small but energetic circle began to form around the laboratory of neurocytology.[61]

As the activities of the laboratory expanded, and as, in consequence, his understanding of interdisciplinarity changed, Penfield became increasingly convinced that the task was simply too large for one mind to encompass. By 1927, he began to contemplate writing a book to describe the new Spanish methods and their use in studying neurological diseases. The size of the undertaking was overwhelming. How to cover the entirety of the nervous system? As Penfield later put it in the preface to *Cytology and Cellular Pathology of the Nervous System* (1932): "[A]uthoritative description of the cytology and cellular pathology of the nervous system is an impossible task for one man or, indeed, for the microscopists of one country." He concluded: "It has . . . been necessary to appeal to men working in different parts of the world to contribute separate chapters. . . . The work is therefore an international, and not in any sense an individual, contribution to science."[62]

By 1928, Penfield had come to embody Cushing's vision of the interdisciplinary neurosurgeon that he had articulated at the beginning of the century;

he was a surgeon and a neurologist, a clinician and a scientist, and his physiological approach to surgery drew the laboratory and the operating theater
into a tighter feedback loop. Yet, simultaneously, his experiences in Spain
and the quickly expanding neurocytology project shifted his approach. To
encapsulate all knowledge of the nervous system was too great a goal for one
scientist, no matter how broadly trained; a team approach was necessary,
one that would draw on the strengths of different national and scientific cultures. Despite his robust ego and sense of mission, Penfield's emerging vision
for neuroscience was an intrinsically collaborative one. In a letter to Isabel
Rockefeller, an early philanthropic supporter of his laboratory, he spelled this
out: "To be able to make a group attack on the problems of neurological surgery with really adequate facilities will make these three years of the laboratory of neurocytology memorable ones for me. . . . To do this sort of research
is a joy for me. It is also a sort of expression of my ~~religion~~ philosophy of
life."[63] If interdisciplinary research was Penfield's religion, it was a religion
that needed a congregation.

Mapping Brains and Surveying Science

In 1927, with the laboratory of neurocytology in full swing, Penfield received
a letter from Edward Archibald, the chief of general surgery at Montreal's
Royal Victoria Hospital. Archibald was Canada's most well-known and successful surgeon at the time, and he shared Penfield's scientific turn of mind.
He hoped to replicate at McGill University what Ward Halstead had done
at Johns Hopkins University a generation earlier: modernize all of surgery
and hand off neurosurgery to a capable young operator who could specialize
in the practice. In Baltimore, this had been Harvey Cushing. In Montreal,
it would, he hoped, be Penfield.[64] "The field is open," Archibald exclaimed.
"[T]here is nobody else doing it in Montreal."[65]

For Penfield, Montreal was a blue ocean. Since his earliest days in New
York, he had struggled to maintain a clinical practice in the city's competitive
medical marketplace. Years earlier, he had groused to his mother: "My ability
to build up a practice is very doubtful."[66] At the same time, he had turned
down a thriving surgical practice in Detroit because it offered no possibility
for research. Montreal, by contrast, was an environment conducive to medical research, particularly of a patient-centered kind. The country's largest
city at the time and its manufacturing and financial hub, Montreal had developed a French- and English-speaking population that rested in a delicate
truce. Medical education at McGill University was arguably the best in North
America in the late nineteenth century, and the legacy of its most famous

alumnus, William Osler, had made the school a prototype for the reform of medical education—didactic lecturing had been phased out in favor of laboratory work, dissection, and ward rounds. In 1920, the city's most advanced hospital, the Royal Victoria, was made a research hospital for McGill, and, in 1924, Johnathan Meakins, a Montreal native who had trained at McGill and taught at the University of Edinburgh, was named professor and chair of medicine at McGill, physician-in-chief at Royal Victoria Hospital, and director of the McGill University Clinic. Under his leadership, the Royal Victoria Hospital became a clinical research hub in the 1920s.[67]

Archibald's offer was greeted with a list of demands. Penfield insisted that "Neurosurgery must go hand-in-hand with Neurology" and that "there must be adequate provision for research as well as the development of clinical and operative teamwork." This collaborative enterprise required segregated surgical beds, a neurosurgical intern, a research fellow, academic standing for the neurosurgeon "equal to that of the neurologist," an association with all the city's hospitals to ensure enough patients, and, most importantly, a neuropathology laboratory.[68] Penfield put it more bluntly in a letter to his mother: "[T]his lab would be the heart of the Clinic. It is here that men should be attracted to come from away. And it is here where most of the investigation should be done."[69] Following a visit to Penfield in New York, Archibald replied: "If [you come to Montreal], I expect that ten years from now the Hub of surgical neurology, in this continent, will be transferred from Boston to Montreal."[70] Penfield accepted the offer from Archibald in January 1928, noting to his mother: "[I]t boils itself down to the fact that Montreal offers a chance to build a complete clinic, to be a personality, i.e.—To have a place in a community that does not drown you."[71] However, before diving into a new medical community, Penfield first needed to survey another one.

In March 1928, Penfield boarded the SS America for Europe with his wife and growing family of four children. His goal was twofold. First, he had become aware of the German neurologist Otfrid Foerster, who had developed a new operation for patients suffering from focal epilepsy. Working at his clinic in Breslau, Foerster began to operate on veterans of World War I whose brain wounds had rendered them epileptic. To treat these patients, he developed an innovative surgical procedure. He exposed the brains of his patients while they were under local anesthetic and applied a gentle electric probe to the brain's surface. The patients were conscious, felt no pain (the brain has no pain receptors), and could respond to questions and provide feedback. The use of electric probes to map brain function had formed a key plank of laboratory neurophysiology since the 1870s, when Gustav Fritsch and Eduard Hitzig used such probes to map the motor cortex of dogs. Foerster now used a similar technique on patients to locate the focal point of their seizures. He then attempted to remove the lesion causing the focal epilepsy (typically scar tissue) (fig. 1.4). In theory, this would reduce the frequency and severity of his patients' epileptic attacks. It also allowed the surgeon to map functional areas of the cortex in a conscious patient. This combination of physiological research tools and surgical intervention proved deeply intriguing to Penfield. If he could observe and master this technique, he might transplant it to North America.[72] "Our headquarters shall be established in Breslau if Prof. Foerster proves to be the sort I suspect," Penfield commented. "I shall study German very hard, and later take side trips."[73]

The "side trips" were the other goal of Penfield's second European sabbatical. As excited as he was about Foerster's brain maps, Penfield was also keen to survey the terrain of the modern brain sciences. Over several months, he traveled to over a dozen neurological clinics and laboratories in Germany, France, the Netherlands, and Belgium and eventually submitted a detailed report of his observations to the RF. That report was revealing, both of neurology in Europe and of his own developing views. "At the present moment," Penfield wrote, "there is no other branch of medicine in which *clinic organization* is of greater importance than in the study and treatment of nervous and mental disease." Tying disciplinary and national cultures together, he went on to note: "If I presume to criticize [different clinics] then it is against the background of my familiarity with English Neurology, American Neurosurgery and Spanish Neurohistology, and not from any personal feeling of superiority."[74]

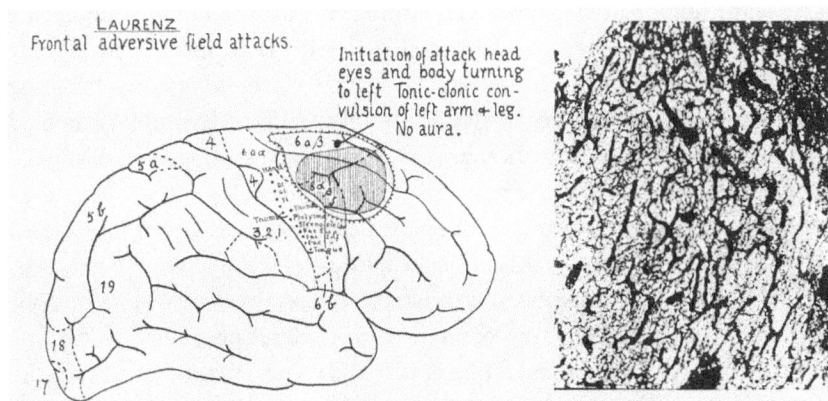

FIGURE 1.4. A case of focal epilepsy on which Otfrid Foerster operated. On the left, the brain of the patient, a World War I veteran with a cerebral scar from a shrapnel wound. On the right, Penfield's stained section of the scar tissue, removed during the operation. Otfrid Foerster and Wilder Penfield, "The Structural Basis of Traumatic Epilepsy and Results of Radical Operation," *Brain* 53 (July 1930): 109–10.

Despite his stated humility, Penfield criticized several clinics and laboratories. Of Viktor von Weizäcker in Berlin, Penfield reported: "There seemed
to be little enthusiasm and no original points of view in this clinic. The mere
creation of a department of Neurology does not seem to have improved upon
the condition of affairs found in Berlin." Paris came in for particular scorn: "It
is obvious that the successors of Marie and Dejerine (Guillain and Crouzon)
are not the teachers that their masters were."[75] A theme emerged from
Penfield's complaints: even in clinics he admired, there was a lack of integration with the laboratory and between neurology and neurosurgery, anatomy
and physiology. Even the clinic of Bernardus Brouwer in Amsterdam, which
Penfield described as having "the most complete neurological organization
of any in Europe," was found to be inadequate. Yes, "ward and laboratory
are interrelated in such a way that all assistants have some activity in both
places." But "the laboratory is an anatomical rather than a pathological one."
And "Brouwer, though a splendid anatomist, was deficient as a clinician."[76]
Again and again, Penfield found much to admire in each city and nation but
no clinic that could facilitate the kind of team approach that he now believed
to be necessary.

Penfield concluded with his own proposal for the future of neurology,
neurosurgery, and psychiatry. His agenda was both ecumenical and ambitious. Classical neurology was becoming increasingly impotent as other medical disciplines chipped away at its territory: "It is chiefly those diseases for
which adequate therapy is lacking which are left undisputed in the province of
Neurology." Tumors needed the specialized diagnostic knowledge of neurologists, but, as more surgeons mastered neurological diagnosis, this too would
end. Meanwhile, many "functional neurological cases" had come under the
purview of psychoanalysts and psychiatrists with little interest in organic
neurology.[77] Everywhere, the traditional territory of neurology was shrinking
as other specialties with effective (or seemingly effective) treatments claimed
more of its disorders. Penfield's solution? "Give to psychiatry neurohistology
and the means of mental therapy and [give] to neurology neuropathology
and complete facility for surgical therapy. Let them be as closely associated
with medicine and surgery, as is possible in hospital life, but allow them complete facility for experimental and therapeutic study of their own problems.
Then we can hope for solution of some of the riddles presented by the sufferers from nervous and mental diseases."[78] His recommendation was for the
neurosurgeon to incorporate the territory of the neurologist gradually, adding more effective surgical therapies for different neurological and psychiatric
disorders. At the same time, the psychiatrist would take over the functional or
mental disorders, with both groups united in their use of laboratory research.

This combined approach could happen only in an institute that allowed for physical proximity of both primary specialties and shared laboratory space. Using the power of the laboratory, psychiatry and neurosurgery would plumb the borderland regions of disorders that occupied both groups.

The Borderland

Penfield submitted his report to the RF in 1928, where it came to the attention of a crucial reader: Alan Gregg, the American doctor who would spend the next twenty years as the director of the RF's Medical Sciences Division and do more to direct the development of American medicine than perhaps any other figure. As the center of gravity for medical innovation and training shifted from Europe to the United States, few areas—from eugenics to the emergence of antibiotics—would escape Gregg's pervasive influence as the RF's key medical moneyman. From 1919 until 1951, Gregg and his mentor Richard Pearce oversaw the distribution of millions of RF dollars to doctors, scientists, universities, and other institutions in the name of medical research and training.[79] On his retirement in 1956, *Time* magazine declared: "No man alive has had a wider or deeper influence on both the practice and teaching of medicine than Dr. Gregg."[80] Indeed, Gregg would provide crucial support for Penfield and his interdisciplinary vision and shape the emergence of neuroscience as much as any physician or scientist.

Modernizing American psychiatry proved one of Gregg's key areas of interest. As a student at Harvard College, the young Gregg entered the circle of the neurologist James Jackson Putnam, the founder of the Harvard Medical School Department of Neurology. Putnam also introduced Gregg to Sigmund Freud, Carl Jung, and Sandor Ferenczi at a retreat in the Adirondack Mountains in 1909, an event that solidified his interest in all things psychiatric.[81] At the same time, Gregg also became enamored with the laboratory, doing influential work with the physiologist Alexander Forbes on the nerve impulse.[82] Gregg found his way from Massachusetts General Hospital to the RF in 1919 and, in 1922, was offered a job as an assistant to Richard Pearce, the RF's director of medical education. In 1924, he became the associate director of medical education in charge of the RF's activities in Europe, and, in 1925, he conducted an extensive survey of the medical schools of Italy, where psychiatry formed an important plank of medical education. His tenure at the RF also accompanied a transition in its funding priorities. By the late 1920s, the RF showed more interest in funding the behavioral and social sciences, and, by 1926, its president, George Vincent, noted in his annual report that money would now be devoted to studying "at least the borders of the fields of biology and psychology, as these have bearing on medicine and public health."[83] In 1928, the RF was reorganized, and the Department of Medical Education was folded into the new Medical Sciences Division. Instead of transforming medical education, the RF would now fund medical science directly via a series of grants in aid of research. By the time Gregg took over from an ailing Pearce in 1929, he was in a position to reorient American psychiatry toward laboratory research.[84]

Gregg's efforts were also part of a larger transformation in American psychiatry that had profound effects. Following the lead of the émigré Swiss neurologist Adolf Meyer, American psychiatry emerged in the 1920s as an expansive medical specialty that would no longer be contained within the isolated and stigmatized asylums of the state mental hospital system. According to Meyer's monistic philosophy of psychobiology, mind and body were inseparable, and neurological and psychiatric disorders lay on a continuum. His psychobiological vision also demanded interdisciplinary collaboration between the clinic and the laboratory. World War I expanded the reach of his viewpoint as Thomas Salmon, a Meyerian psychiatrist, was appointed to direct the army's program in mental medicine. Salmon eventually introduced the United States to the hybrid discipline of neuropsychiatry, a medical specialty that embraced those disorders that sat in the borderland between the world of the asylum and the private practice neurologist; these included the milder neuroses, stuttering, alcoholism, and, crucially, epilepsy.[85] As Penfield's close friend the Harvard neuropsychiatrist Stanley Cobb later put it: "Epilepsy is [a] borderland, claimed and disclaimed by neurologists, psychiatrists and internists" (see fig. 1.5).[86] Gregg was a prominent convert to Meyer's viewpoint and a proponent of linking laboratory research to university medical departments. The vision for unifying neurology, psychiatry, and neurosurgery that Penfield outlined in his 1928 report expressed almost perfectly Gregg's overall vision for scientific psychiatry, and, by targeting a neuropsychiatric condition like epilepsy, Penfield's proposal sat in the middle of a crucial borderland that Gregg hoped to seed with money.[87]

Penfield moved to Montreal in 1928 and settled into the professional rhythms of clinical practice and university teaching. Having ameliorated the city's French and English neurologists, by the end of 1928 he felt ready to approach the RF to secure a grant for an institute that would "provide a center for neurological thought which would serve the whole continent, [where] we could work effectively upon the unsolved problems in neurology unhampered by the artificial division between medicine and surgery." The institute would also "form a common meeting ground for Neurology, Neurosurgery and Psychiatry."[88] Notably, Penfield's initial requests for funding in 1929 were

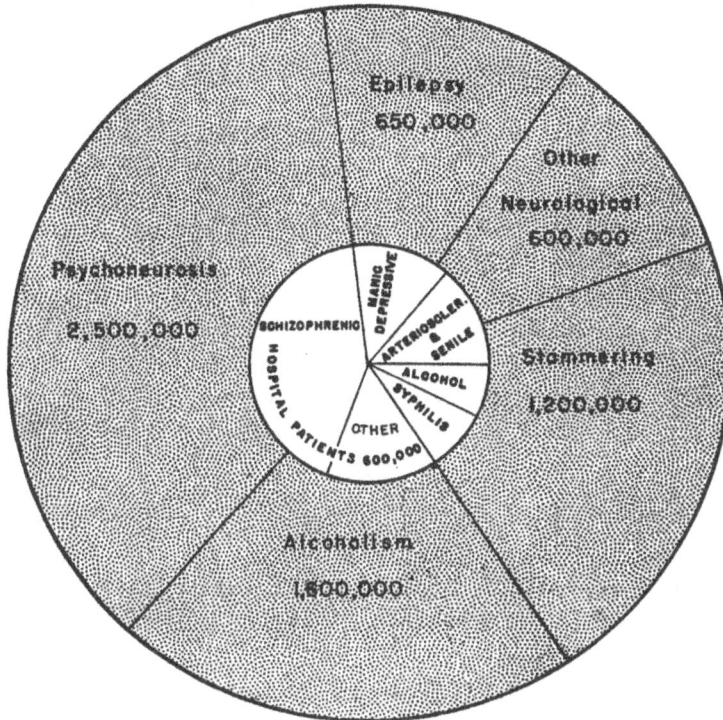

FIGURE 1.5. Stanley Cobb's diagram of the "borderlands" of psychiatry (numbers represent estimated numbers of cases in the United States). Note the prominent position of epilepsy among these borderland conditions. From Stanley Cobb, *Borderlands of Psychiatry* (Harvard University Press, 1943). Copyright © 1943 by the President and Fellows of Harvard College. Copyright renewed © 1970 by Stanley Cobb. Used by permission. All rights reserved.

refused; Richard Pearce, still head of the RF's Medical Sciences Division, was not a convert to Meyerian psychobiology.[89]

Alan Gregg provided the decisive factor. From 1928 to 1931, Penfield worked to establish a surgical practice in Montreal and consolidate his emerging approach to epilepsy. By 1931, Gregg was now in control of the RF's Medical Sciences Division and in a position to help. Gregg met with Penfield in March and October 1931. He had clearly absorbed Penfield's 1928 report and proved infinitely more receptive than Pearce had been. During a decisive meeting in Montreal, Penfield made his plans clear: a neurological institute that combined the clinic and the laboratory and ultimately aimed to unite neurosurgery, neurology, and, eventually, psychiatry into a single medical specialty. Penfield specifically referenced earlier conversations with Adolf Meyer about the eventual unification of neurosurgery and psychiatry. Gregg required little convincing and offered to help Penfield prepare a new proposal for the RF. In proposing to unify neurological medicine and science by way of an institute that would unite both clinic and laboratory and make epilepsy its chief area of interest, Penfield occupied precisely the kind of borderland that Gregg felt was so important.[90]

With Gregg's support, Penfield's second proposal to the RF was successful. The dollar value of the grant represented the single largest outlay of RF money for medical research to date: $1.25 million, which constituted one-third of Gregg's expenditure on psychiatry and 16 percent of the total money spent by the Medical Sciences Division between 1929 and 1939.[91] Among the projects funded under the RF's turn to psychiatry, Penfield's would prove the most enduring and successful; similar initiatives to study psychosomatic medicine and neurology at the University of Chicago floundered over internal divisions and conflicts between basic and clinical research.[92] Penfield's institute was the product of precisely the right confluence of social, scientific, and medical developments, and it was in the borderland between these developments that it would operate. As Gregg himself put it in the 1934 RF annual report, during a time of considerable financial strain at the height of the Great Depression, the foundation had chosen to place its emphasis on "the group of studies [that emphasized] the function, derangements, and diseases of the nervous system or, speaking in broader terms, of that psychobiological knowledge of the behavior of man which might be in the possession of the ideal psychiatrist."[93] Penfield's institute was the first program funded under this rubric.

If epilepsy constituted a borderland—an area of overlap between neurology and psychiatry—then Montreal itself occupied a crucial borderland as well, a place where disciplines and national scientific styles could blend and re-form. As negotiations with the RF developed, Penfield was offered a professorship of neurosurgery at the University of Pennsylvania as part of creating a new university hospital. The position provided a permanent budget for scientific research and several other attractive professional and personal accommodations for a young surgeon with a growing family. Penfield seriously considered the offer and received encouragement to take it when he visited the first International Neurological Congress in Bern, Switzerland, in 1931. However, he had come to see Montreal as having an advantageous strategic position, allowing for communication between European and North American medical and scientific cultures. As he recalled:

> I told [Gregg] what I had been thinking. Montreal is a quieter place for study, I explained, quieter than New York or Philadelphia or even Baltimore. Tradition and awareness link Montreal with Europe, especially Great Britain

and France, as well as with the United States. Our location here, above the American border and just off the main highroad to the great American university centers—might well prove to be the best place in which to be influenced by the work in other centers. It might be the ideal place in which to do constructive scientific work on the brain and the mind of man, work that might in time influence thinking in other centers.[94]

For Penfield, Montreal was a key borderland between medical and scientific cultures that might lead to productive work. In a pamphlet that accompanied the opening of the MNI in 1934, Penfield noted that Montreal "more than . . . any other city on this continent . . . is situated on a crossroads of scientific medicine."[95] In the coming years, the various travelers who would meet at that crossroads would transform the neurological sciences.

The MNI opened its doors in September 1934. The opening was attended by Harvey Cushing, whose keynote address, "Psychiatrists, Neurologists and the Neurosurgeon," detailed his own failure at creating a similar institute in the years after World War I.[96] Certainly, Penfield's success was as much a product of his context as it was of his character: the emergence of a physiological approach to surgery, the new laboratory and clinical tools from Spain and Germany, and, crucially, the new orientation of the RF and the unique environment of Montreal put Penfield in the right place at the right time. Simultaneously, his ecumenical style and willingness to embrace teamwork created an environment for interdisciplinary work that was a unique assembly of those influences. He had gone from trying to embody interdisciplinarity to institutionalizing it. In this respect, the MNI preceded by several years the interdisciplinary institutions and collaborations of the so-called big science that would characterize World War II and the postwar period.[97]

Commenting on the MNI, the *New York Times* described it as an example of "illocality": "[I]nstead of looking for our sanctions to the past . . . [the RF] shall increasingly look around the globe and take or encourage the best wherever it rises to view." It added that it hoped "the world will find its way to the laboratories and clinics and lecture rooms of McGill."[98] The newspaper writers were prophetic, but not in the way that Penfield had initially conceived. Inspired by the Spanish methods, Penfield believed that the breakthroughs from his institute would come from the pathology laboratory. Instead, the most exciting developments came from the operating room itself, which evolved into a neuroscientific laboratory. To study and understand what emerged from the operating room, Penfield would enlist the help of outsiders, whose different perspectives, techniques, and technologies crystallized the MNI's interdisciplinary environment.

In the rest of this book, I examine the many facets of interdisciplinary neuroscience that began in Penfield's operating room and follow the careers of the scientists who helped make sense of what transpired there. As we will see, what began at the MNI seldom stayed there. In chapter 2, I follow the careers of two English psychologists, Molly Harrower and Brenda Milner, who would legitimate the use of psychological methods in the study of surgical patients, resulting in one of neuroscience's critical discoveries about the nature of human memory. In chapter 3, I follow the career of Herbert Jasper, the physiologist and psychologist who would complete Penfield's localization project with the electroencephalograph and use the MNI's interdisciplinary ethos to organize the world's neuroscientists under the banner of the International Brain Research Organization. In chapter 4, I follow the career of Donald Hebb, who would translate developments at the MNI into the realm of theory, supplying a crucial source for the emergence of cognitive science after World War II. In chapter 5, I return to the 1940s and examine the case of Penfield's disastrous collaboration with the psychiatrist Ewen Cameron to understand, first, why some disciplines were incapable of participating in the MNI's interdisciplinary research community and, second, how the weak ties of Montreal could shape one of its most controversial afterlives. Finally, in chapter 6, I follow the career of the young physician David Hubel, who passed through weak ties from Montreal to Baltimore, Boston, and eventually to the receipt of a Nobel Prize in 1981.

The Montreal Method: Molly Harrower, Brenda Milner, and the Neuroscience of Memory

On December 12, 1928, six years before the opening of the Montreal Neurological Institute (MNI), Wilder Penfield could be found with his surgical partner William Cone and his laboratory technician Edward Dockrill combing through mounds of trash in Montreal's Rosemont landfill. The three men were searching for a barrel containing over three hundred brain tissue samples they had brought from the laboratory of neurocytology at the Presbyterian Hospital in New York that had been lost in the move to Montreal. As Penfield later recalled: "I thought of the bits of preserved human tissue in the vanished barrel. We needed them for comparison in future studies. I had removed some of those blocks of tissue at grueling operations. Others I had taken, at autopsy, from the patients I could not save. . . . I had made a promise to their loved ones. I thought of . . . the many abnormalities that had produced epilepsy . . . and the scars from the New York experimental animals. They were all there."[1] For Penfield, these histological samples had both scientific and moral value—they represented a research program as well as the sacrifices his patients had made. Always methodical, Penfield drew up a plan to survey the dump site despite the subzero temperatures. However, he was soon called away as his attentions were required for a different sort of emergency.[2]

Ruth Inglis had a serious problem. Born in 1885, she had suffered from incapacitating "fits" since childhood but had maintained something like an ordinary life. By age forty-three, however, these fits were much more common and severe; frequent headaches, vomiting, weakness, and paralysis convinced Ruth and her husband that something had to be done. Ruth did have one advantage; she had been born Ruth Penfield, and her younger brother was rapidly becoming one of the world's most respected neurosurgeons.

Ruth traveled with her mother from California to Montreal the same day Penfield surveyed the Rosemont landfill and was greeted by her brother at the train station. He performed a neurological exam and discovered that her optic nerve was swollen and distended, a common sign of a brain tumor. "My knees grew suddenly weak," Penfield recalled, "and for a moment I thought I might fall."[3]

What to do? After conferring with colleagues, Penfield decided to operate a few days later. Removing a portion of Ruth's skull, he confirmed the existence of a massive tumor in her right frontal lobe. Ruth was conscious during the operation, and Penfield communicated with her while he worked slowly and methodically to remove the tumor. Then, tragedy struck. Penfield discovered that the tumor extended back into the motor gyrus and encapsulated several cerebral veins. If any of those veins were cut, Ruth could quickly bleed to death. The tumor also extended into the left hemisphere, and, if Penfield continued his removal (already the most radical he had ever attempted), she might die or be permanently paralyzed. Never a man to relinquish control easily, Penfield felt incapable of continuing the operation and let his assistant take over while he regrouped.[4]

Ruth recovered from her surgery and lived for nearly two more years in California before dying in 1931 (fig. 2.1). In the intervening years, Penfield made a careful, if informal, study of the changes in his sister's behavior that signaled the possible effects of her frontal lobe operation. The resulting paper would become one of the most influential ever published on the functions of the frontal lobes and one of the most cited papers to emerge from Penfield's operating room.[5]

The transition embodied in the story related above—from searching for brain tissue samples in a local dump to studying the effects of surgery in a living patient—enacts in microcosm the critical transition that would make Montreal a hotbed of neurological knowledge production. Penfield had based his vision for his institute on the power of the laboratory to bring about new knowledge of the brain; indeed, he had sold his vision to the Rockefeller Foundation (RF) on these grounds. However, the most important scientific discoveries that emerged from his clinic came from the operations themselves. His approach to brain surgery—and epilepsy surgery in particular—transformed the surgical theater *into* a neuroscientific laboratory.[6]

The transformation of the surgical theater into a research space called for new types of collaborations, most notably with psychologists. Initially, Penfield saw these collaborations as flowing in one direction; the ability to stimulate the brains of conscious patients would, he hoped, produce new knowledge for the field of psychology, fulfilling his deeper goal to use physiological knowledge to unravel the workings of the mind. However, things would not play out precisely this way. Penfield's operations were, by his own

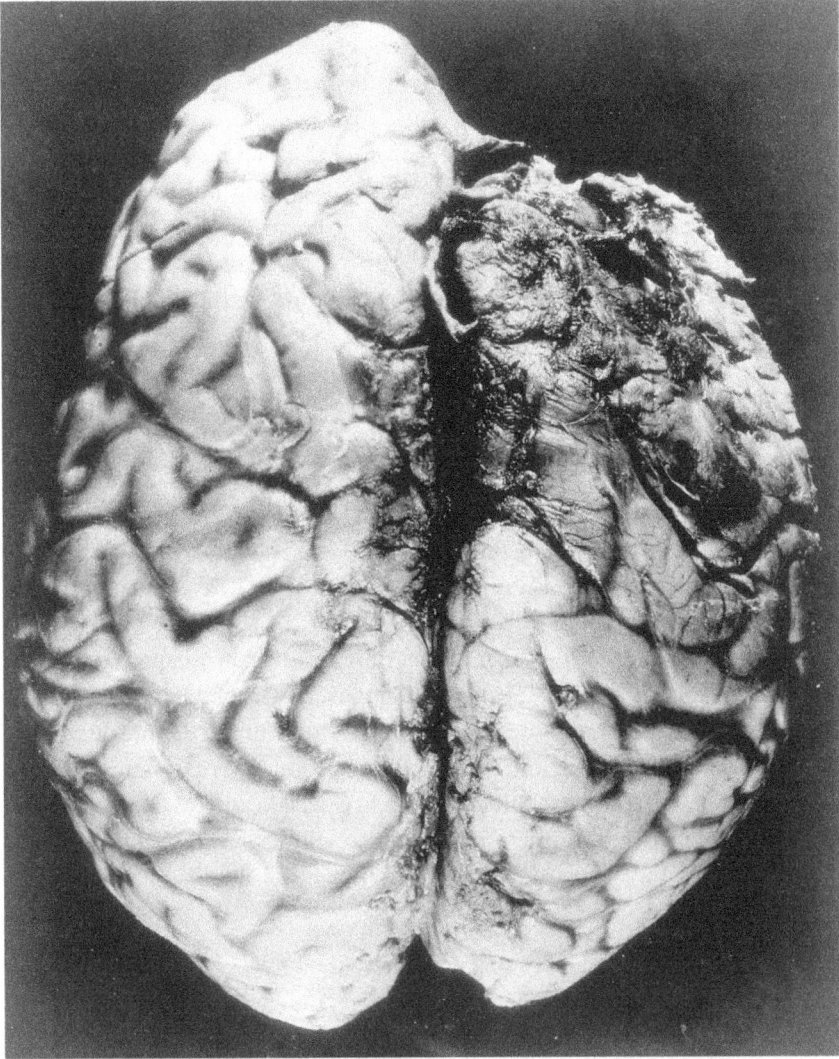

FIGURE 2.1. The brain of Ruth Inglis, Wilder Penfield's sister, removed at autopsy. Reproduced by permission of the Osler Library of the History of Medicine, McGill University.

admission, radical, and the need to understand and mitigate their possible ad-verse psychological side effects soon made collaboration with psychologists a necessity. It was this necessary collaboration that would eventually grow, by the 1950s, from a trading language into a fully fledged neuroscience. To understand this evolution, I follow the careers of two female psychologists— Molly Harrower, who was at the MNI in the late 1930s, and Brenda Milner, who arrived in the early 1950s and never left. Doing so allows us to observe not only how they legitimated the field of psychology within the MNI and forged a genuinely interdisciplinary neuroscience but also how they forged unique career paths within the heavily gendered mid-twentieth-century world of psychology. Beginning as a helping profession within the world of clinical medicine, not all that different from a glorified nurse, the role of the clinical psychologist would evolve under Harrower and Milner into a robust research field, and Milner would produce one of neuroscience's most iconic discoveries. In 1953, she took the ethos of the MNI and passed it through weak ties to Hartford, Connecticut, where her work with the famous patient H.M. transformed our understanding of human memory.

Radical Treatment

After receiving his RF grant in 1932, Penfield was consumed with planning and building the MNI. Exercising something close to dictatorial control over even the finest construction details, he collaborated closely with the Montreal architects Robert Henry Macdonald and George Allan Ross to turn his pencil sketch of the MNI into a brick-and-mortar reality (fig. 2.2). Nothing, from the layout of the surgical theater to the number of microscopes in the laboratories to the dimensions of the broom closets, escaped his eye. From its inception, the institute was designed as much to facilitate the production of new knowledge as to treat and heal; operating rooms were designed both to keep patients safe and to facilitate research. Special cameras, windows, and mirrors were installed in operating rooms to allow for precise photography

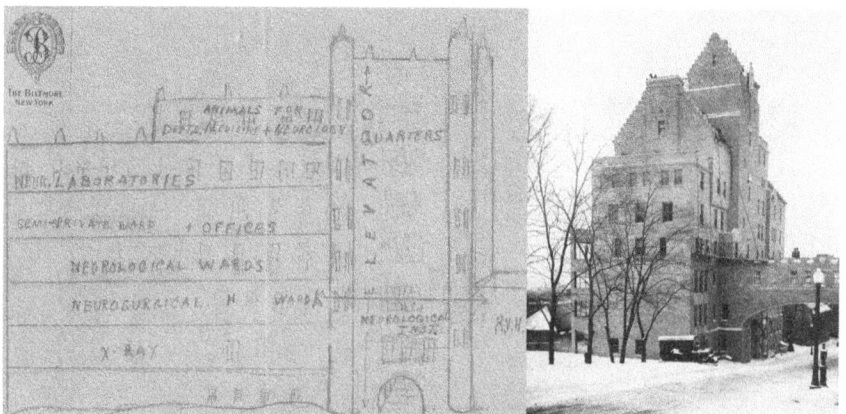

FIGURE 2.2. Wilder Penfield's original sketch for the Montreal Neurological Institute (MNI), ca. 1929 (*left*). The MNI nearing completion in 1933 (*right*). Reproduced by permission of the Osler Library of the History of Medicine, McGill University.

of operations and for secretaries, nurses, and surgical assistants to take meticulous notes (the emphasis on precision and recordkeeping was likely a hold-
over from Penfield's laboratory training).[7]

Ironically, given Penfield's emphasis on the microscope and the new
Spanish histology techniques, the MNI laboratories initially produced few
scientific breakthroughs. Excitement about the new methods attracted a small
community of scientists to Montreal, but, despite this enthusiasm, the early
successes of this reductionistic approach were modest.[8] Research emerging
from the MNI's "Madrid cubicles" (as the pathology labs were called) was
conventional rather than revolutionary: classifications of tumors and studies
of brain scar tissue. Even the most promising initial collaboration between
the clinic and the laboratory—a theory that posttraumatic epilepsy might be
caused by an involuntary vasomotor spasm of cerebral arteries—proved to
be a bust.[9]

What ultimately shifted the MNI's practice was the operations themselves.
This was for two reasons. First, Penfield's radical treatment for epilepsy created a unique environment for exploring the functions of the brain in a living
patient. Patients could report their experiences under local anesthesia while
the surface of their cortex was stimulated with an electric probe; the reports
were carefully recorded, allowing correlation of movement and sensory experience with the spot on the brain that was stimulated. As Penfield noted, these
operations allowed for the extension and revival of one of the oldest research
paradigms in neurology—the localization of physiological and psychological
processes to specific brain areas.[10]

Other neurosurgeons had used cortical stimulation in brain surgery
and had recorded their patients' sensory experiences (most notably Otfrid
Foerster, from whom Penfield learned the technique). What distinguished
Penfield's approach was volume and documentation. By 1937, Penfield and
his colleague Edwin Boldrey had accumulated over 160 patients whose brains
were stimulated during operation; a stenographer carefully recorded all patient responses, and sterile square tickets were placed on the brain's surface
to mark the place that elicited the response. Penfield and Boldrey published
a summary of their stimulation research in 1937.[11] Building on the work of
Foerster, they mapped the motor cortex in the posterior of the fissure of
Rolando—correlating the stimulated portions of the cortex to the parts of the
body that moved in response (e.g., lips, limbs, eyes, toes, and fingers) and
also the bodily sensations produced by stimulation of the pre-Rolandic cortex. These operations produced a map of the somatosensory cortex and the
differential representation of body regions. To illustrate this, Penfield and
Boldrey produced an image that has become an icon of neuroscience—the

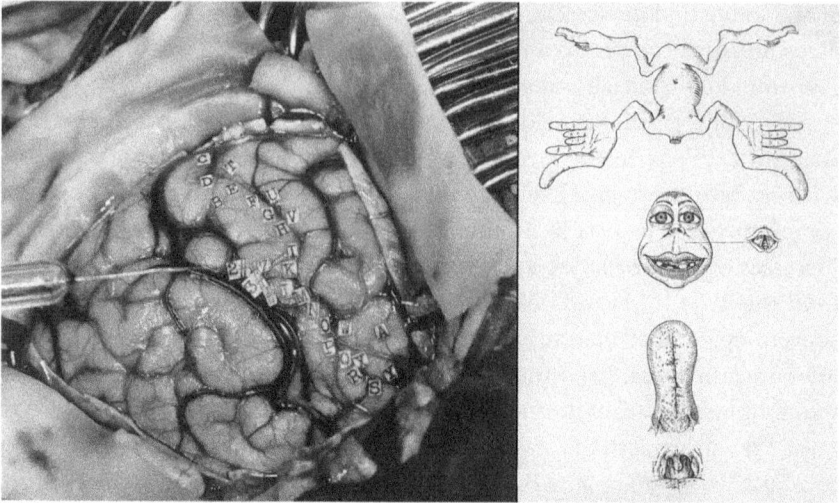

FIGURE 2.3. Wilder Penfield's stimulation procedure (note the sterile tickets) and the resulting sensory homunculus. Reproduced by permission of the Osler Library of the History of Medicine, McGill University.

sensorimotor homunculus, a humanoid figure whose features are drawn in proportion to the space taken up by their representation in the cortex (fig. 2.3). Fundamentally dependent on his patients' input and participation, Penfield transformed his operating theater into the equivalent of Sherrington's physiology laboratory, a fact that was not lost on the older man, who commented to Penfield: "It must be great fun to have the 'physiological preparation' actually speak to you."[12]

At the same time, Penfield's operating room also became a psychological laboratory. "A neurosurgeon has a unique opportunity for psychologic study when he exposes the brain of a conscious patient," Penfield noted. "No doubt it is his duty to give account of such observations on the brain to those more familiar with the mind. He may find it difficult to speak the language of psychology, but it is hoped that material of value to psychologists may be presented, the application being left to them. It seems quite proper that neurologists should push their investigations into the neurologic mechanism associated with consciousness and should inquire closely into the localization of that mechanism without apology and without undertaking responsibility for the theory of consciousness."[13] In the late 1930s, Penfield sought potential colleagues who could "speak the language of psychology."

There was, however, another reason why Penfield sought the outside advice of psychologists. His radical treatment involved quite extensive removal of brain tissue, an aggressive approach that distinguished his surgical practices from the more conservative ones of Cushing and his students. This always entailed potential loss of function, and Penfield was aware that his aggressive operations could lead to paralysis, aphasia, or other debilitating deficits. Much of his surgical practice involved mitigating this risk. For instance, Foerster's electric probing technique helped locate the offending tissue, but it also helped map the cortex to identify no-go areas that subserved speech or motor functions, the loss of which would be an unacceptable trade-off for relief of epileptic seizures.

The problem was that not all brain areas revealed their function to Penfield's probe. Large portions of the cortex—particularly the frontal lobes—produced no interpretable response in Penfield's patients, even when conscious. Moreover, postoperative examination did not always make it clear what the effects of the operation had been. An example makes this problem clear. In 1927, shortly before moving to Montreal, Penfield operated on a seventeen-year-old boy named William Hamilton who developed epilepsy after having been struck on the head by a falling brick. Penfield's operation was certainly radical; he removed nearly half of the boy's right frontal lobe (the first successful lobectomy). When Hamilton recovered from the operation, little seemed to have changed in his personality or intelligence: "The patient is in excellent condition, working and showing no mental symptoms or neurological signs as a result of the procedure."[14]

The clinical evaluation of postoperative patients in American neurosurgery was, in 1927, more art than science, resting typically on a surgeon's subjective judgments. As one of Penfield's colleagues, Percival Bailey, put it regarding the removal of cancerous brain tissue: "It is always necessary to weigh the result for the patient. I am not one of those who would remove half of the brain without regard to the fact that I might leave the patient without his intellect or the means of making it effective." At the same time, Bailey also captured the subjective nature of these judgments (often colored by class and gender stereotypes): "I hesitate before amputating a frontal lobe. This procedure is always followed by a more or less great alteration in character and defects in judgment. In a washerwoman these results may be of little concern, but when the patient is a professional business man, who must make decisions affecting many people, these results may be disastrous."[15]

Penfield had little choice but to operate on Ruth in 1928. Immediately following the operation, he and his colleagues tried to determine how she had fared. Remarkably well was their initial conclusion. Ruth seemed reasonably normal, so much so that one of Penfield's colleagues noted: "[S]he expressed her gratitude so nicely that one could not help wondering how much the frontal lobe had to do with the higher association processes."[16] However, the patient being Penfield's sister, the situation presented a unique opportunity for follow-up study. Between 1929 and 1931, Penfield closely studied Ruth's condition as she readjusted to life in California. He concluded that, although much of her personality remained intact, she had lost something that psychological tests could not measure. A natural setting, however, was more revealing:

> [Ruth's] own home provided in some ways a better background for study than the consulting room of the psychologist. The following test, though not sanctioned by psychological usage, may illustrate her shortcomings. One day about fifteen months after operation she had planned to get a simple supper for one guest (W.P.) and four members of her own family. She looked forward to it with pleasure and had the whole day for preparation. This was a thing she could have done with ease ten years before. When the appointed hour arrived she was in the kitchen, the food was all there, one or two things were on the stove, but the salad was not ready, the meat had not been started and she was distressed and confused by her long continued effort alone. It seemed evident that she would never be able to get everything ready at once. With help the task of preparation was quickly completed and the occasion went off successfully with the patient talking and laughing in an altogether normal way. Although physical examination was negative and there was no change in personality or capacity for insight, nevertheless the loss of the right frontal lobe had resulted in an important defect. The defect produced was a lack of capacity for planned administration. Perhaps the element which made such administration almost impossible was the loss of power of initiative. If we express it as she did with no attempt at analysis: she could not "think well enough," was a little "slow," a little "incapable."[17]

Penfield presented his sister's case, along with two comparable removals, at a meeting of the Association for Research in Nervous and Mental Diseases in 1932.[18] This presentation prompted the Yale physiologist John F. Fulton to write to Penfield that he considered it "the most crucial case of frontal area extirpation that will probably ever appear in neurological literature."[19] Penfield's subsequent 1935 "The Frontal Lobe in Man: A Clinical Study of Maximum Removals" became one of the most influential papers on the frontal lobes of the period (an issue that will become crucial in an examination in chap. 5 of the complex relationship between Penfield and psychosurgery).[20]

By the mid-1930s, then, two developments were converging at the MNI. The operations for epilepsy were producing new information about brain function that would, Penfield felt, be interesting for psychology. At the same time, they were producing psychological deficits that bore watching.

However, Penfield felt that the method for evaluating these deficits was common sense; the salient psychological details could be observed with the naked eye. In short, he thought that he could be his own psychologist, much as he had earlier thought that he could function as his own pathologist and neurologist. In the early 1930s, he felt that his operations had something to offer the world of psychology but that psychology had little to offer the surgeon.

Given the wealth of potentially interesting psychological data produced in his surgical theater, Penfield became increasingly interested in obtaining trained psychologists to join the staff at the MNI by 1935. He saw potential value in bringing in outsiders, much in the same way that he enlisted scientists from other nations to help in the completion of his neurocytology project (see chap. 1), and he now searched for psychologists who could help him investigate and interpret the effects of his brain operations.

The search could not have come at a worse time. Psychology as a discipline in the 1930s had very little interest in the brain or the nervous system. The growing dominance of behaviorism—either of the classical Watsonian kind or the neobehaviorism championed by figures like Clark Hull and B. F. Skinner—eschewed the study of the neurological in favor of constructing a bloodless science of behavior. Behaviorists and their rat mazes increasingly dominated American psychology departments.[21] One exception was the Gestalt psychologists then fleeing the rise of Nazism in Germany.[22] In 1935, Penfield invited Wolfgang Köhler, the Gestalt psychologist recently arrived in the United States following his ouster from Germany, to join the MNI as a resident psychologist. "From a psychological point of view," he wrote Köhler, "there is an important piece of work to be done on these patients. As I told you the other day, I would like very much to be able to get you to come here for a year or two or permanently, if we could induce you to do so."[23] Köhler had undoubtedly been intrigued by the research possibilities of Penfield's operations but was unwilling to leave Swarthmore College, which had graciously taken him in. In 1937, Penfield informed the McGill dean of faculty of his frustrations: "We are at present planning to secure one or two psychologists to work at the Neurological Institute."[24] A month later, however, he could still complain that "there are not many psychologists who are fitted for this kind of work."[25] Within the month, however, he would find the right candidates.

Thursday's Child

Three figures arrived in Montreal in 1937–38: two psychologists, Molly Harrower and Donald Hebb, and a psychologist-turned-physiologist, Herbert Jasper. In different ways, each of these actors would transform activity at the MNI in a relatively short time by bringing critical new techniques and technologies. I examine the contributions of Jasper and Hebb in chapters 3 and 4. Here, I first need to lay out how psychological expertise became institutionalized at the MNI, and this involves an examination of the career of the nearly forgotten Molly Harrower (fig. 2.4). Today mainly

FIGURE 2.4. Molly Harrower, around 1930. Reproduced by permission of the Drs. Nicholas and Dorothy Cummins Center for the History of Psychology at the University of Akron.

remembered (when she is remembered at all) for creating the first large-scale Rorschach test for evaluating military recruits,[26] Harrower's less well-known efforts at the MNI were, in many respects, more consequential. In a short period of time, she created a functional trading zone between surgery and psychology at the MNI, inaugurating the first psychology department in a neurosurgical hospital, and began the process of wiring together an assembly of scientific actors.

Born in South Africa in 1906 to Scottish parents, Harrower grew up in the small city of Cheam, twelve miles south of London. The child of a depressive father and an intelligent but subordinated mother, she was shaped by years in a brutal English education system featuring considerable hazing and isolation. These experiences left her with "an empathy for the underdog," along with an inner resolve and independence.[27] In her unpublished autobiography, "Thursday's Child," she recalled that the first time she heard the word *psychology* was in a church sermon while on vacation in Switzerland; the minister stressed that the modern sciences of mind made it possible for one to control one's destiny, a prospect that appealed to her independent spirit.[28]

Inspired by this search for self-mastery, Harrower eventually undertook training in psychology at the University of London. Her education in psychology was notably heterodox, a reflection of the unstable position of that discipline in the British academy in the 1910s. Courses drew on the German laboratory tradition of Wilhelm Wundt, while Harrower wrote course papers with titles like "The Unscientific Nature of Behaviourism," "An Attempt to Understand Consciousness," "In Search of a Clear Understanding of Time," and "A Study of the Act of Thinking." Her interest in the more personal side of psychology—questions of mind, consciousness, and the inner life—eventually drew her into the orbit of C. K. Ogden, the eccentric English scholar who combined studies of literature with his editorship of the psychological journal *Psyche*. Ogden introduced Harrower to the work of the Gestalt psychologists Wolfgang Köhler and Kurt Koffka and ultimately secured her a place to study with Koffka in the United States.[29]

In 1928, Harrower arrived at Smith College, where she worked in the recently opened Neilson Laboratory for psychology, which was "100% international," with students from China, Russia, Western Europe, and the United States working on a diverse array of projects.[30] For her and others, Gestalt psychology, stripped of its many German cultural associations, served as a scientific counterculture for those who thought that psychology was still a science of mind rather than behavior.[31] Harrower's research generated a doctorate and a minor publication, tellingly titled "Organization in Higher Mental Processes."[32] While at Smith, she also began a romantic relationship with the then-married Koffka that was to continue, in different forms, for nearly two decades. Their relationship was as much a romance of minds as it was one of the flesh, and Harrower, a prolific letter writer, carried on a voluminous correspondence with Koffka. The love letters they exchanged (over two thousand from Harrower alone) constitute a remarkably detailed and intimate source for historians.[33]

Indeed, Harrower's voluminous correspondence allows us to date fairly precisely the pivot in her career that would bring her into Wilder Penfield's orbit. In December 1933, her physician allowed her to observe the operations at a hospital and put her in contact with the RF. "I want to make clinical studies of patients recovering from brain operations," Harrower informed Koffka, "sort of Gelb and Goldstein detailed experiments. I am so utterly and crazily thrilled over the prospect that I can hardly contain myself."[34] Her excitement was matched only by her sheer lack of qualifications for the position—she had received no clinical training whatsoever at that point. Nevertheless, four months later, she sent a proposal to the RF outlining her plans: "I would like to continue the line of approach suggested by Lashley in his 'Brain Mechanisms and Intelligence,' but using, instead of animals who have been operated on, human patients who have suffered brain injuries."[35] Through his connections with the RF, Penfield became aware of Harrower's application and wrote her an enthusiastic letter: "We have a good deal of available material of the type in which you are interested, and I think perhaps that our brain excisions are so carried out as to make a careful correlation between the anatomy of the brain and psychological study possible."[36]

How much precedent was there for Harrower's plan to apply experimental psychology to clinical work? As psychology matured into an experimental science in the late nineteenth century, there had been limited attempts to apply it to clinical issues. By the 1920s, the intelligence tests developed by Alfred Binet in France had been adapted for extensive use with American soldiers during World War I as well as in the classification of mental deficiency, mental retardation, and "feeble-mindedness" by the American psychologist Henry H. Goddard; few, however, applied psychological tests within clinics or hospitals.[37] In Weimar Germany, the work of Kurt Goldstein and Adhemar Gelb with brain-injured veterans had given birth to the field of neuropsychology, but few outside Germany had followed their lead.[38]

Harrower's proposal dovetailed closely with the interests of the RF and with Penfield's plans. Despite this obvious meeting of interests, Harrower would not arrive in Montreal for an additional three years. Why the delay? Harrower would later explain her conversion to clinical work with a charming story about wanting to understand the psychological changes suffered by her close friend Mabel Roys, the dean of Wells College, who emerged from a "drastic surgical procedure as a seemingly different person": "What . . . was evoked by obvious traumatic events, physically, such that behavioral changes were noticeable? How could one study such a happening? How, perhaps even to restore people to their former, rightful selves? All of a sudden these questions seemed to become the province of psychology that I wanted to explore. I wanted to deal with the individual person, not make studies with his retina."[39] The charm of this later retelling notwithstanding, Harrower's letters reveal a more personal reason for the delay, one indicative of the conflict between professional and personal lives that was likely true for many female scientists of the era. Harrower initially turned down offers of employment as director of students at the New Jersey College for Women primarily because she thought that Koffka was on the verge of divorcing his current wife so he could finally marry her. Two years later, with no divorce in sight, Harrower resumed her discussions with the RF and Penfield about coming to Montreal. Penfield was still keen to hire her:

> I was glad to hear from you after two years of silence. There is no psychologist working with us, I am sorry to say. There is opportunity for different types of study here. . . . I have begun to try to think for myself without psychological blessing or guidance. I am sending this along because it may illustrate to you the many lines of attack which are possible. The thing which intrigues me most is the problem of various types of disturbance in consciousness. . . . If you find this work challenging I should be glad to have you come on here.[40]

After a meeting in New York, Harrower wrote enthusiastically to Koffka about Penfield's plans while also alluding to the different disciplinary languages that would have to be bridged if the collaboration were to be successful: "His idea is for psychology to contribute to his 'round table' of different approaches. . . . [Penfield] has the ingrained notion that he cannot understand any psychological terms. Telling him what 'gestalt' meant almost ruined my delicious [meal] for me. On the other hand, on his first explanation, I could not understand the essential difference between removing a tumor and the scar of epilepsy. . . . Well, beloved: what do you think? Shall I accept the challenge and go to the cold north?"[41]

Notably, Penfield's was not the only offer Harrower contemplated in the winter of 1936–37. She also received offers from Stanley Cobb at Harvard and the neurologist Walter Freeman, then developing his lobotomy procedure in Washington, DC. What linked these three sites was their RF funding and commitment to the paradigm of psychobiology. Despite her numerous options, Harrower preferred Penfield, whose plans and institute impressed her, as did his character: "[T]here is a quality of 'gentleness-in-the-midst-of-strength' in that man that makes me suddenly want to cry."[42]

Harrower's chances were nearly scuttled by Penfield's immediate decision to hire Donald Hebb (about whom we will learn more in chap. 4), who already had training in brain physiology (which Harrower lacked), but Harrower appealed to Penfield's emerging sense of the value of interdisciplinary collaboration:

> I am more than ever convinced that psychology has a real contribution to make in the fields of neurology and neurosurgery. In saying this I am not thinking of myself, or any one psychologist, but of the contribution that can ultimately be made by the interaction of these fields of knowledge. Thus any psychologist who goes to the Institute has a tremendous responsibility, for he can not only content himself with working on specific and isolated problems that will be of interest to his fellow psychologists, but must also attempt to bridge a gap, and justify psychology's inclusion among that constellation of subjects which are being brought to bear on nervous diseases.

As a specific example of where psychologists could contribute, Harrower made a prophetic prediction:

> I believe, for instance, that very different kinds of memory disturbances would come to light, for after all the term "memory" is but a convenient short cut for many different psychological activities, and a comparison of these in the light of the different physiological and operative findings would be valuable. Then I believe it would be possible to get a little nearer to an exact formulation of what this lack of initiative, and lack of concentration, means. Such studies would be of interest to the theoretical psychologist and such defects, if properly understood, might prove amenable to some form of psychotherapy.[43]

Alongside her psychological expertise, Harrower also offered to undergo training as a clinical psychologist. This new training represented a deviation from her previous career path as a laboratory-based experimental psychologist, but for a woman this was not an unusual transition. By the 1920s, the field of psychology was, in contrast to many areas of science, relatively well populated with women, who held one in five advanced degrees. However, the field retained gender differences in other areas; graduate education for women was much more restricted, with few American universities offering them doctoral education (Koffka's group at Smith was an exception). Simultaneously, a dual labor market had evolved in psychology; male psychologists tended to occupy university research departments, while women tended to move into service-oriented positions in hospitals, clinics, courts, and schools.[44] In this respect, Harrower conformed to these trends while evincing different reasons for doing so; although she had attained graduate training as a laboratory experimentalist, she had abandoned that path, seeking more exciting research possibilities in the clinic.

Before arriving at Montreal, Harrower undertook a brief apprenticeship with Kurt Goldstein, who had fled the Nazi regime in 1933 and arrived at the Montefiore Hospital in the Bronx. Harrower's time with Goldstein was educational in three ways. First, she learned "almost immediately that [she] could, by being [her] natural self, relate to any ill person whom [she] met or worked with."[45] By shadowing Goldstein in his ward rounds with neurological patients, she quickly developed a natural clinical style and imbibed much of the older man's holistic philosophy and emphasis on therapy and recovery following brain injury.

Second, Harrower added a series of new technologies to her arsenal. She had been experimenting with aspects of Gestalt perceptual psychology for some time and, in 1936, had developed a variation on the classic Rubin-vase figure modified "so as to have a series in which both the Profile and the Vase were made more and more explicitly the figure" (fig. 2.5). Administering the test to brain-injured patients, she "found significant differences [between the patients and control subjects]": "Patients were often unable to make anything out of the completely ambiguous figure. They would also show marked perseveration of the first response, despite the increased articulation of the

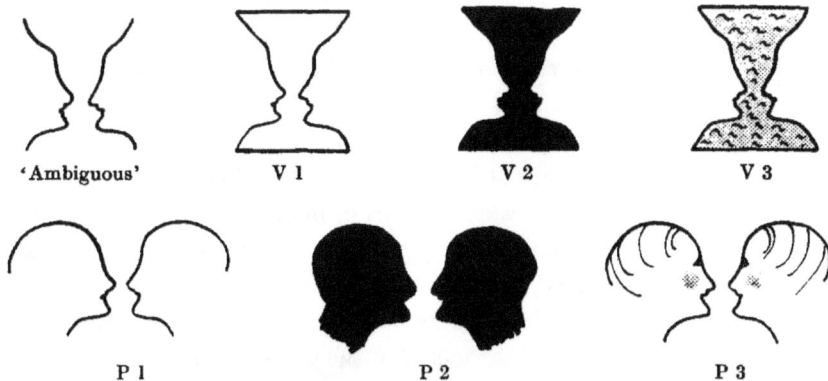

FIGURE 2.5. Molly Harrower's modification of the classic Rubin ambiguous figures used to test patients with brain damage. From Molly R. Harrower, "Changes in Figure-Ground Perception in Patients with Cortical Lesions," *British Journal of Psychology* 30, no. 1 (1939): 48.

alternate part of the field, that is, seeing the profiles rather than the vase to begin with, would continue to see them despite additions to make the vase more visible. They tended to introduce concrete, familiar objects from their own immediate experience. For example, a farmer called the vase area a 'mound of hay.'"[46] This finding was in line with her mentor Goldstein's emphasis on the loss of "abstract attitude" often experienced by patients with frontal lobe damage. More significantly, this early success, however small, indicated to her the possibility of using Gestalt and other psychological principles as the basis for tests that might reveal the changes to mental life brought about by damage to the brain. For Harrower, different types of tests might provide a more subtle method of probing the effects of brain damage (or surgical intervention) than quantitative measures of intelligence.[47]

In addition to her profile-vase test, Harrower also acquired a new tool set that would eventually alter her life course. She was introduced to the Rorschach by none other than its most vocal proponent, Bruno Klopfer, who paid occasional visits to Goldstein's clinic. Her enthusiasm for the Rorschach was colored by her personal experience with it; she underwent Rorschach testing by Zygmunt Piotrowski, one of Klopfer's disciples, and found the experience "extraordinarily convincing": "By that I mean I recognized myself in his summary as in no other test."[48] Setting aside the scientific validity of this type of characterological cold reading, she also grasped the possibilities of the Rorschach within a clinical setting: "I was immediately struck by the research and clinical possibilities of the method. . . . Before long Piotrowski and I had projects which we planned to interrelate in the use of the Rorschach with organic, or brain damaged patients. We decided then and there to make comparisons between cortical involvement and non-cortical involvement, between localized tumors and general diffused tumors, and between frontal lobe tumors and other locations."[49]

Finally, Harrower received an education in the crass sexism that characterized much of medicine in the 1930s: "Almost immediately I ran into a (to me) quite unexpected problem, a confrontation about the fact that I was a woman. How come I did not know that the world had two levels of citizenship? And that hospitals were bastions of male supremacy?" Up to that point, she had spent much of her career in "situations that were virtually devoid of sex distinctions" and with men for whom "discrimination along those lines . . . would have been unthinkable." On her first day at Montefiore Hospital, however, she was informed that she could not eat in the doctors' dining room. "Because I was a psychologist? No, the other psychologists in Goldstein's lab were permitted to go there. The implication was clear: doctors were men!"[50]

The Psychologist at Work

Following three months of training with Goldstein, Harrower arrived in Montreal alongside Herbert Jasper and Donald Hebb. "Nothing could have been more startling than the difference in atmosphere between the Montefiore Hospital and the Neurological Institute," she recalled, "for the whole staff at the Institute was unified, dedicated, and untiring. One got caught up in the esprit de corps, the likes of which I have never again encountered."[51]

Despite the "esprit de corps," Harrower's first introduction to the MNI operating room was shocking. In her autobiography and other accounts of the institute, the operating room often appears as a tranquil space of humane medical intervention and investigation. Surgery, rendered painless by anesthesia, proceeds cautiously and expertly, with the gentle electricity of Penfield's probe revealing the brain's hidden secrets and ridding sufferers of their tumors and epilepsy. Harrower's private letters, by contrast, at times reveal a more chaotic side to the early days of brain exploration. In one early instance, Penfield and his newly arrived electroencephalographer, Herbert Jasper, disagreed on the focal point of an epileptic attack. In an attempt to localize the seizure, Penfield used his probe to provoke four violent epileptic attacks before he had to abandon the procedure, which Harrower recalled as "a very nerve-wracking experience": "I felt absolutely dead."[52] In another instance, she raised doubts about the effectiveness of Penfield's epilepsy operations following the death of a young boy owing to an infection contracted during surgery:

> [A]n epileptic I was in on, in the OR, has developed a terrific infection, such a marvelous boy. I just can't bear it. Tumor people one doesn't mind quite so much about, but this boy could have lived happily for years if this operation had not been thought of. It seems very unlikely that it does any good. [Theodore Erickson, Penfield's surgical apprentice, whom Harrower would later marry,] is very pessimistic as he is checking over the results of five years. It's a colossal task these people set themselves [as] it's the most disappointing and dangerous of all procedures in medicine.[53]

Perhaps most frustrating for Harrower was that Penfield's aggressive operative style initially seemed to sideline the patient. "Penfield is kind," she lamented to Koffka, "but he is the most unimaginative great person along these lines. He has no conception of what the things he says to patients do to them. I would not want to leave here until I had made a slight dent in this direction, but it will be some job."[54]

Harrower, too, felt herself initially uneasy with the all-male surgical environment. To Koffka, she confided that McGill University was "lousy with its male prejudice." Of the MNI, she related: "Penfield has very severe notions about women and work, no Christian names allowed, no girl allowed to smoke . . . because such breaking down of barriers might lead to 'flirting,' and no girl who flirts during business hours is a lady." She also lamented that she had hoped to play with the male fellows on the MNI's one luxury installation—a squash court on the eighth floor. The matter was left "under consideration."[55]

Over time, however, Harrower became integrated into the operating environment. Penfield initially assigned her the task of communicating with the patient during brain operations: "[C]loseted under the sterile green drapes with the unanesthetized patient for 8 hours at a time while stimulation of the cortex proceeded, my task was to give tests, ask questions, record answers." The purpose?

> [To] observe what actually went on in the patient's experience, in his mind or emotions. . . . I had to improvise, to choose my questions to elicit what would be most helpful to the surgeon. I might have to ask a patient a question which, if he could not answer it, might indicate that the removal of a particular area was dangerous to his mental achievements. Or, if a patient was describing an experience (evoked as if from nowhere, by the external electrical stimulus), I had to ask questions which would enable the patient to contrast this immediate experience with comparable experiences from everyday life. If the patient said, for example that she suddenly heard the voices of children, I would ask her were they her children? Did they say anything they had said before? Was an old experience happening again, or was it something new?[56]

It would be easy, at this point in her work at the MNI, to confuse Harrower's role with that of the anesthetic nurse: create a sense of ease in the patient to facilitate the surgical examination. Nevertheless, Harrower's improvised examination style brought a greater psychological sophistication to Penfield's approach. Soon, Penfield was soliciting her opinion on the more psychologically rich phenomena he observed in the operating room. As his probe began to call forth "dreams and hallucinations and evoked memories," he called on Harrower to "prevent him from getting in wrong and spoiling the work in the eyes of some people by the too laymen-like use of psychological terms."[57] The intellectual and physical effort expended during these operations left Harrower physically drained; after her first operation (over eight hours long), she returned home, ran a bath, and promptly fell asleep, awakening to a flooded apartment.[58]

Even more important than her activity during operations, however, was Harrower's expanding use of psychological tests both before and after brain operations, and it was here that a meaningful trading zone began to develop. Harrower completed her study using the Rubin profile-vase figure with surgical candidates at the MNI and determined that patients with lesions in areas of the frontal lobe had significant difficulty transitioning from perceiving the vase to perceiving the face (or vice versa). "The value of such findings," she noted, "would seem to lie in the fact that they may give us clues to certain qualitative differences in the experiences and behaviour of persons suffering from cerebral lesions, differences which might well leave unaltered a quantitative estimate of the intelligence level, and might well pass unnoticed in ordinary daily life."[59] Her use of pen-and-paper psychological technologies made possible a more subtle understanding of the effects of surgery and of brain function.

What ultimately wired together surgeons and psychologists at the MNI, however, was Harrower's use of psychological tests that could aid in diagnosis. The MNI surgeons, including Penfield, were initially skeptical about the value of psychological testing. Harrower convinced them of its utility. The success or failure of a brain operation rode on the ability of the surgeon to locate and remove a discrete lesion. Since the earliest days of brain surgery, the ideal procedure had been for a neurologist to localize the site of a possible tumor or other lesion on the basis of known neurological signs and then for the surgeon to operate to remove the lesion at the suspected site. In practice, this was more easily said than done, particularly in cases where the lesion might be in a neurologically "silent" area such as the frontal lobes or where the symptoms might be confused with a psychiatric or so-called functional condition. In a time when technologies for peering inside the skull were few and their reliability questionable, Harrower's test provided a crucial window into brain function that could spare Penfield and other surgeons from making a devastating mistake. In this respect, the flow of knowledge that Penfield had predicted between his operating theater and psychology was reversed; while he had anticipated the operations producing knowledge for psychology, what truly cemented the place of psychology at the MNI was when Harrower began providing valuable knowledge for the surgeons.

Between 1938 and 1941, Harrower began to employ different tests, both pre- and postoperatively. Patients who were rendered aphasic by their neurological conditions, for instance, were tested pre- and postoperatively to determine the extent of their language recovery. Most importantly, Harrower began to employ the Rorschach test to distinguish between organic cases, such as tumors and epilepsy, and psychiatric or functional conditions unamenable to surgical treatment. Her tests proved to be particularly valuable as divining rods within the border region between neurology and psychiatry that the MNI occupied as they could distinguish between organic/neurological cases and mental/psychiatric ones, and this established their trading value. In a crucial instance in January 1939, Harrower administered a Rorschach diagnosis on a "very exciting case, 99% psychotic from the clinical or neurological point of view, but with 8 out of the 10 Rorschach organic signs." The next day, the same patient was revealed, during an exploratory operation, to have "a huge tumor of the right frontal lobe."[60] As Harrower later recalled the episode: "One by one, the [research] Fellows found meaning in the psychological tests for themselves, which had seemed so alien at first. . . . I shall never forget the occasion when the tests showed this cerebral pattern [of an organic lesion] in a patient diagnosed as psychotic. Exploratory operation revealed a large frontal lobe tumor . . . in a silent area. Penfield announced this achievement at ward rounds. I must have blushed scarlet, for one of the interns whispered, 'Can you turn any other color?'"[61]

This triumph established the utility of psychology at the MNI and facilitated its institutionalization there. In February 1939, Penfield wrote Alan Gregg "to express our gratitude to the [Rockefeller] Foundation for the help which they have given us at the Montreal Neurological Institute in psychology." According to Penfield, Harrower and Hebb were "carrying out work of real scientific value and somewhat unexpectedly, we have come to lean upon them for help in our clinical problems." In addition to differentiating organic from functional cases, Harrower's tests "seem to have real localizing value in the study of cerebral lesions."[62] By September of that year, Harrower would write Koffka: "I feel at long last that [Penfield] has 'passed' me, something I have never felt before. I delight in the idea of building up even a small department."[63] By November, she could remark: "I [finally] came to grips with the Chief [Penfield] instead of skating along paths of completely different thought."[64]

By establishing the utility of her tests for localizing brain lesions, Harrower established their trading value in a language that the surgeons could understand. In doing so, she had built a trading language between psychology and neurosurgery and institutionalized a trading zone in the first psychology department to be integrated into a neurosurgical hospital. By 1940, she was collaborating with Penfield and his surgical colleague Theodore Erickson on their major publication, *Epilepsy and Cerebral Localization* (1941), applying the Rorschach technique to evaluate the claim, popular at the time among psychiatrists, that there existed an "epileptic personality." Her studies provided strong evidence that this entity was a fiction.[65] The inclusion of her efforts was a major validation of psychological expertise as a part of modern neurosurgery.[66]

In many respects, her time at the MNI legitimized a professional role that was encapsulated in the title (if not the content) of a book Harrower had begun writing years earlier while working with Koffka—*The Psychologist at Work* (1938).[67] While the book itself served as an introduction to the psychological laboratory and experimental psychology, Harrower had instead forged a professional identity not in the male-dominated spaces of the psychology laboratory but in the underexploited space of the clinic by rolling up her sleeves and working with patients. In doing so, she normalized the presence of a psychologist in the most elite of medical clinics. In that connection, her success did much to validate the presence of psychology at the MNI and of female scientists there as well. As she described the weekly fellows' meetings: "Once a week one of our number talked of his specialty. The occasion when 'she' [Harrower] first talked about 'her' specialty was a milestone, the focus of great anxiety, but with comparable relief and even triumph afterwards!"[68] Reflecting on her time at the MNI, Harrower noted: "I do not feel that I as a person was in any way handicapped in anything that concerned my major objective—namely, to bring psychological understanding to areas where it did not exist. My various mentors responded to clearly expressed requirements if it was within their power. Perhaps at some point in my hard-to-describe-or-formulate career, it may even have been helpful to be a woman rather than part of the larger masculine gestalt."[69]

Wartime Disruptions

Canada being drawn into war preparations earlier than its southern neighbor, its mobilization for World War II brought considerable disruption to the MNI and its staff. Convinced of the need to aid the war effort, Penfield retooled the MNI's research agenda for war work and established a forward neurosurgical hospital in England, Neurosurgical Hospital 1.[70] The war also brought important changes for Harrower. Increasingly convinced of the value of her testing work, Penfield endorsed her application to the Canada Research Council to develop Rorschach tests to "weed . . . out the neurotics and incipient neurotics" from military service.[71] This study evolved into the first large-scale Rorschach testing technique for screening military recruits. The resulting tests were eventually adopted by the US Navy and played a significant role in armed forces recruitment procedures when the United States entered the war in 1941.[72] That same year also saw personal changes for Harrower. Her professional relationship with the neurosurgeon Theodore Erickson had blossomed into romance, and she and Erickson were married at Penfield's home in 1938. In 1941, Erickson was offered a position at the University of Wisconsin–Madison, and the couple left Montreal in 1942.[73]

Despite her departure, Harrower's legacy at the MNI proved crucial to its future direction. Having established the first psychology department in a hospital, Harrower fundamentally shifted the disciplinary makeup of the institute. Even more than her colleague Donald Hebb (who will be discussed in chap. 4), with her institutionalization of psychological expertise she inaugurated a tradition of disciplinary trading and coordination that proved crucial in the years after World War II. Reflecting on her work, Harrower invoked a linguistic metaphor to explain how to function in an interdisciplinary environment: "One must become accustomed to a new language and new concepts, so that as far as possible the household words of the 'foreign environment' become one's own. One must acquire enough perspective to be able to use and adapt one's own professional expertise in the new field."[74] In the years after World War II, another English psychologist would expand the MNI trading zone into something considerably more robust.

Temporal Lobes

As the war ended in 1945, Penfield and the MNI returned to one of their most promising lines of work. Surgery for epilepsy had continued during the war and followed the original paradigm established by Foerster: develop a tentative localization of the epileptogenic lesion on the basis of clinical history, expose the brain, and, if a lesion is present, determine whether it is the likely cause of the seizures, map the area with the electric probe, and remove the lesion if possible. Penfield attempted to determine the effectiveness of these operations in 1935 on the basis of a crude analysis of postoperative success, and the results seemed encouraging. In 1941, he again tried to assess the effectiveness of his interventions with a more sophisticated analysis and was discouraged to see that only 43 percent of patients had a satisfactory outcome (i.e., a decrease in seizure frequency by 50 percent or more).[75] More disturbing, nearly one-quarter of patients who had only an exploratory craniotomy (meaning no removal of tissue) experienced some reduction in seizures, raising the possibility of a kind of surgical placebo effect. Although discouraging, the data clarified something important: in patients in whom there had been a clear lesion identified (tumor or scar tissue), the results of the surgery had been much more positive. This put an emphasis in Penfield's operations on identifying a discrete lesion.[76]

At the same time, Penfield was aided in his quest to improve surgical outcomes by a man who brought a revolutionary technology to the MNI and soon became his closest collaborator. Arriving at the same time as Harrower, Herbert Jasper fundamentally altered the institute's work and the modern understanding of epilepsy. The electroencephalograph (EEG), which registered a typical electric discharge pattern in cases of epilepsy, could be used before and during operations to confirm that a particular lesion was the cause of the epilepsy. Much like how Penfield had used the electric probe and psychological testing to increase the safety of his operations, he now employed Jasper's skill with the EEG to improve his surgical accuracy and efficacy.[77] Jasper's work with the EEG will be detailed more thoroughly in chapter 4. For now, however, it is crucial to understand that his skill had two effects. First, over time, Penfield came to trust Jasper's EEG to localize epileptogenic portions of brain that *did not* always present an obvious lesion (such as a scar or a tumor). Second, Jasper's EEG increasingly pointed to sources of epilepsy within the temporal lobes, where operative success up to this point had been disappointing.

Jasper's presence led Penfield into uncharted territory. Early surgeries for temporal lobe seizures tended to remove tissue from only the lateral surface or tips of the temporal lobes. Penfield, like other neurosurgeons of the time, was reluctant to move deeper into the temporal lobes, toward the mesial structures such as the uncus, the amygdala, and the hippocampus, because the functions of these structures were largely unknown. Moreover, experiments conducted with monkeys by the German neurophysiologist Heinrich Klüver had indicated that removal or damage of these structures might cause irreparable *psychic blindness* (now termed *visual agnosia*) and a host of behavioral side effects, including "hypersexuality."[78] The details of the so-called Klüver-Bucy syndrome were complex, but the possibility of leaving a patient with profound disabilities after an elective operation left Penfield reluctant to venture deeper into the brain.

Nevertheless, Jasper and his EEG beckoned in this direction. Moreover, the temporal lobes were a tempting target for Penfield. Since John Hughlings Jackson first identified them in 1888, the forms of epilepsy that produced the so-called dreamy state auras were the most intriguing for neurologists. These seizures were usually preceded by experiences of déjà vu, auditory hallucinations, bizarre doublings of consciousness, and, most alarmingly, periods of automatism in which epileptics behaved seemingly normally but later had no memory of their actions. Jackson's autopsies of epileptic patients indicated the likely source of these seizures to be in the deeper, mesial structures of the temporal lobe, but clinicoanatomical correlation provided only limited evidence.[79] The dreamy state seizures were rechristened *psychomotor seizures* in 1938 by the Harvard epileptologists Erna and Frederic Gibbs, whose EEG studies also pointed to deeper structures in the temporal lobes.[80]

Between the time of his arrival in 1938 and the early 1950s, Jasper and his colleagues came to persuade Penfield that the source of temporal lobe seizures might be in these deeper brain structures. If Penfield and his colleagues could locate the source of these seizures, which had so frequently in the past been thought of as idiopathic, then the symptoms of epilepsy might be brought under a single Jacksonian paradigm. At the same time, the possibilities for research were impressive as "[n]ew knowledge [might be] gained of the function of the human temporal cortex," including its role in brain functions much more complex than the sensory and motor functions that Penfield had already mapped.[81] By the early 1950s, Jasper's EEG, along with stimulation and recording electrodes that could be inserted more deeply into the brain, had been correlated with pathological findings and clinical histories to indicate that the source of temporal lobe seizures was almost certainly within these deeper structures. Moreover, these findings corresponded with pathological findings of tissue excised from the deeper portions of the temporal lobes that was frequently atrophied and infested with astrocytes. It increasingly appeared to Penfield and his colleagues that psychomotor epilepsy was caused by damage done to the mesial structures of the temporal lobes, possibly during birth through compression of the skull and herniation of the mesial structures.[82]

This new approach quickly bore fruit. Patients on the operating table who had their deeper brain structures stimulated by depth electrodes experienced the kind of emotional, mental, and physical symptoms that typically accompanied the auras of their seizures. When combined with a new operative technique developed by Penfield and the young surgical intern Maitland Baldwin that left the surface of the temporal lobes intact while removing a smaller portion of brain tissue in the mesial temporal lobe, the newly christened "Montreal method" of temporal lobe surgery began to show real promise. Operative success rates climbed quickly from 50 to 75 percent, and news of the new procedure's success brought more patients to Montreal seeking treatment.[83]

Penfield's success with the temporal lobes could not have come at a better time for the MNI. Despite its initial success with epilepsy surgery and impressive endowment by the RF, the MNI ran into considerable financial trouble immediately after World War II. A funding crisis consumed the institute from 1947 to 1949; demand for operations was high, but the existing clinical facilities were inadequate. Patients who traveled from Europe and the

United States to Montreal for operations were still housed in a temporary wooden annex built for military patients during the war, and Penfield petitioned anyone who would listen about the need for a new wing to house the incoming patients. By 1949, he could state publicly: "[E]ither the public must support voluntary hospitals, or medicine must be socialized." He continued: "[T]he decision was made to build [the institute in Montreal] because of the hospitality of the citizens of Montreal, the City Council, and the Provincial Government. . . . If there is no Canadian, or group of Canadians, ready to make permanent its organization—then let the doors of its hospital close."[84]

Privately, Penfield used the emerging promise of the temporal lobe operations to secure the funding he needed. In a letter to John Bassett, the publisher of the *Montreal Gazette*, he related the story of Madame Poinso-Chapuis, the wife of France's minister of health, who had brought her son to Montreal to receive the temporal lobe procedure that, as Penfield put it, he could "only get here."[85] Bassett evidently leaned on several connections because Penfield was eventually accorded an audience with the premier of Quebec, Maurice Duplessis, who secured the funding necessary to construct a new wing of the MNI to handle the increasing volume of temporal lobe cases. Only a year later, in 1950, Penfield could report: "The Montreal Neurological Institute may now fulfill its destiny as a provincial hospital and a national institute, and its doors will never be closed."[86]

The temporal lobe operations were the most exciting development at the MNI following World War II. Exploring these areas with an electric probe and the EEG produced some of the most exciting findings in the postwar brain sciences. The result of several hundred such operations was an elegant hypothetical brain system that Penfield and Jasper referred to as the *centrencephalic system*: "a neurone system centrally placed within the brain and equally connected with the two hemispheres" (fig. 2.6). Because epileptic seizures that originated within this region invariably produced unconsciousness, the men proposed that the system acted as "the indispensable substratum of consciousness." Patients on Penfield's operating table who had their temporal lobes or deeper structures stimulated would frequently recount fragments of songs, long-forgotten experiences of childhood, and other vivid recollections. While Penfield initially dismissed these experiences as hallucinations, he and Jasper came to believe that they were actual recollections and that his team was charting the anatomical structures responsible for human memory. According to Penfield, the centrencephalic system served as a director of attention, determining what experiences in the stream of consciousness were laid down in the neuronal pathways of the temporal lobes: "Whenever a normal person is paying conscious attention to something, he is simultaneously

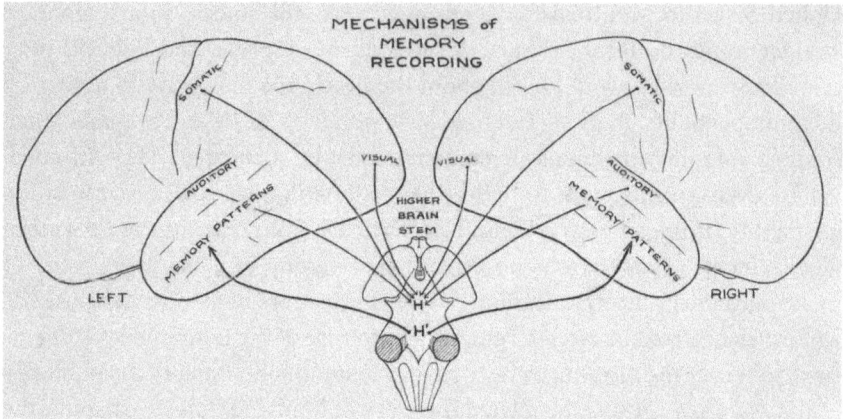

MECHANISMS of MEMORY RECORDING

LEFT   RIGHT

FIGURE 2.6. Wilder Penfield's "centrencephalic" integrating system and his model of memory formation in the temporal lobes. Reproduced by permission of the Osler Library of the History of Medicine, McGill University.

recording it in the temporal cortex of each hemisphere." During recall, the centrencephalic system accessed neuronal patterns in the temporal lobes; on the operating table, Penfield's electric probe evoked a similar reaction by forcing a memory, long hardened by epileptic activity, into consciousness. According to Penfield: "[T]hese results . . . prove the existence of memory patterns; but how are they formed there, and how are they used? The answers to those questions are of great psychological importance."[87]

Penfield and Jasper's centrencephalic system also encroached on a long-running debate over the possibility of localizing memory. For the first half of the twentieth century, the dominant voice in that debate was the American neuropsychologist Karl Lashley. Since the 1920s, Lashley had undertaken a quasi-behaviorist experimental program in which he trained rats to run mazes and then surgically destroyed portions of their brains to determine the location of so-called *engrams*—the physical sites of memory. His findings, presented in his 1929 *Brain Mechanisms and Intelligence* and defended in his classic 1950 "In Search of the Engram," suggested that no such location existed; memory was spread throughout the cortex and could not be localized to any brain region or system (even one as diffuse and complex as Penfield's). Lashley confronted Penfield directly at the 1951 meeting of the American Neurological Association: "Dr. Penfield's observations on the effects of stimulating the temporal lobe raise many problems, but I do not believe that they justify the conclusion that memories are stored in that region."[88] At the center of a critical scientific debate, the Montreal method of temporal lobe operations was the key not only to Penfield's medical and scientific agenda but to the very survival of the MNI itself. By the early 1950s, the volume of temporal lobe operations had increased dramatically, and those operations formed the core of the research program that was producing the most exciting publications from the institute since its founding in 1934.

This made it even more imperative to find a solution when things began to go wrong. According to Penfield's initial theories and prevailing ideas about brain structure, patients who had portions of their temporal lobes (including the hippocampus) removed should be minimally affected since only one side of the structure would be removed and previous unilateral excisions had produced no significant impairments. It came as quite a shock, then, when a young engineer, P.B., emerged from a temporal lobe operation in September 1951 with anterograde amnesia, a profound deficit that prevented him from forming new memories. Penfield initially dismissed this result as a tragic anomaly, but just over a year later, in October 1952, it happened again, this time to a young glove cutter identified as F.C. These surgical accidents demanded an explanation. Fortunately, although Harrower had left the institute before the war, she had created an interdisciplinary culture at the MNI that encouraged collaboration between neurosurgeons and psychologists. Help would come in the form of a young English psychologist recently arrived in Montreal.

"We need you"

Brenda Langford (1918–) was born into a Manchester home that valued art and education, music and language. Her father, a music critic for the *Manchester Guardian*, and her mother, a music teacher at a local elementary school, nurtured their precocious daughter's intellectual interests despite their bitter disappointment at her total lack of musical talent. A robust education in languages, notably conversational French, combined with a passion for algebra and mathematics, led Brenda to Cambridge University. Her hopes for a mathematics career were dashed by discovering her limitations and the tight competition for limited scholarships available to women in the 1930s. As a daughter of England's struggling middle class, she would have to find more practical training.[89]

Experimental psychology hardly seemed an appropriate solution to the problem of financial stability. Still, the subject appealed to Langford, and, in the worst-case scenario, she could always move to London and become a factory inspector, the only vocation for which psychology seemed applicable. She began training in the Psychology Department at a perspicuous time and with two crucial advisers. The first, F. C. Bartlett, had reinvigorated the relatively small community of Cambridge psychologists with his experimental studies of memory. His 1932 *Remembering: A Study in Experimental and Social Psychology* was a minor success in England but remained relatively unknown elsewhere. Bartlett had argued, contrary to the prevalent notion of memory as an indelible trace left on the brain, that remembering was, in fact, a reconstructive enterprise in which memories were altered and edited during the act of recall.[90] Langford's other major influence was the psychologist Oliver Zangwill, later the prime mover for the development of neuropsychology in Britain during World War II. Significantly, through Zangwill, Langford developed her interest in organic brain damage: "It was he who first taught me the value of studying the behavioral effects of brain lesions, because he believed that through an analysis of disordered function one could gain insights into the functioning of the normal brain."[91]

The outbreak of World War II altered Langford's career trajectory. In 1939, C. P. Snow recruited her as an experimental officer for the Ministry of Supply, the sole woman in her unit. As the Battle of Britain raged in the air, Langford was transferred to a research establishment at Malvern, where she began collaborating with a young electrical engineer named Peter Milner on creating control mechanisms for radar and antiaircraft defense. Over the next two years, their professional relationship became romantic. When John Cockcroft asked Peter to move to Montreal after the war to help initiate the Canadian Atomic Energy project, Langford decided the relationship was worth preserving. In late 1944, Peter and Brenda (the latter hereafter referred to as Brenda Milner; fig. 2.7) were married and two weeks later set sail for Boston with a party of war brides in a converted troopship, zigzagging across the Atlantic to avoid German submarines.[92]

Peter and Brenda arrived in Montreal in late 1944. With Peter at the nuclear facilities being constructed at Chalk River, Brenda was left at home for much of the time with little to occupy her. "I couldn't be a kept woman," Milner recalled. So she sought work in her new home.[93] Her facility with French made her an attractive applicant for the French-language Université de Montréal, on the northern side of Mount Royal (the mountain served as the demarcation line between Montreal's English- and French-speaking populations), and the emerging Psychology Department hired her as a lecturer. Milner realized that, without a PhD, her career would not likely advance further, and she eventually decided to undertake a study with Donald Hebb, the psychologist who had joined Harrower at the MNI in 1938. Following an American sojourn, Hebb returned to McGill University in 1947, bringing with him the manuscript of what would become his magnum opus, *The Organization of Behavior*, a theoretical treatise on the nervous system that

FIGURE 2.7. Brenda Milner, mid-1950s. © Montreal Neurological Institute. All rights reserved.

he had developed as part of his earlier experiences at the MNI. (Hebb's work will be explored more thoroughly in chap. 4.) Milner was invited to attend the evening seminars at McGill led by Hebb: "[D]iscussion after the seminars often continued late into the night. It was an exciting time and hastened my decision to do a Ph.D. at McGill."[94]

Milner began her PhD with Hebb in 1949, intending to work on tactile form perception in the congenitally blind, but once again, circumstances intervened. On returning to Montreal in 1947, Hebb had secured a promise from Penfield that he could send one graduate student to the MNI to study his patients. Milner was the perfect candidate, given her previous training with Zangwill. In June 1950, she began investigating perceptual deficits in patients who had undergone the relatively novel procedure of temporal lobe resection for epilepsy. She would commute from her day job on the French side of Mount Royal by streetcar, trying to guess which patient Penfield planned to operate on so that she might run a few perceptual tests before and after the operation. Milner finished her PhD in early 1952, having found some small but intriguing visual deficits in the temporal lobe patients. Despite the difficulties of those early days, she became enchanted by her encounters with surgical patients and hoped to stay on at the MNI. At this point, Penfield began to encounter the startling memory deficits in his two temporal lobe patients, P.B. and F.C., and it was this emergency that led Penfield to revive the interdisciplinary trading zone fostered by Harrower a decade earlier. "I'll get you an office in the building," Milner recalled Penfield saying. "We need you."[95]

The Sleeping Beauty of the Brain

Milner was incorporated into an existing research program that Penfield had inaugurated to study his temporal lobe patients. As Penfield wrote Robert Morrison, the new director of medical research at the RF, in 1952: "[E]xciting information is at hand on memory, dreams, interpretation of perception, deviations of normal behavior and psychoses." He continued: "[T]hese facts open a field for a psychological-psychiatric-neurological study which can only be undertaken adequately by a group working together. We have the material, we have the cases, we have a group of relatively young people ready now to tackle the job." Despite this optimism, he added an ominous detail: "In some cases removal of one whole temporal lobe has reduced the capacity for memory; in other cases there has been no interference whatever."[96] Much like the earlier frontal lobe work, Penfield's current work often involved removing brain structures whose functions were poorly understood. He was, in effect, partially flying blind. Yet the risks seemed worth it—a surgical treatment for temporal lobe epilepsy seemed close at hand, and it was largely taken for granted that, if one removed one half of a brain structure (such as the hippocampus), the other half could compensate.

This understanding of hemispheric redundancy was shattered by the surgical patient P.B. An American engineer, P.B. underwent a temporal lobectomy that initially involved surface removal of the portions of the temporal lobe, following Penfield's initial understanding of temporal seizures. Unsatisfactory results led to a second operation in which Penfield removed more of the temporal lobe's deeper mesial structures, including the hippocampus's left horn. The results were disastrous. Emerging from his postoperative recovery, P.B. apparently said to Penfield in an accusatory tone: "[W]hat have you people done to my memory!"[97] He had been stricken with anterograde amnesia, a debilitating condition that rendered him incapable of holding on to experiences for more than a minute or two. According to Milner: "He did not remember what he'd had for breakfast, he did not remember whether his wife had been to see him that day, . . . he did not recognize me."[98] Milner tested P.B. extensively, using many of the intelligence tests bequeathed her by Hebb when she had begun work at the MNI, and found that his intelligence was unaffected. Penfield consulted with Milner and Jasper on this peculiar case; Jasper argued that the cause was probably an unpredictable anomaly relating to the patient's brain stem and was unlikely to be repeated.[99]

Thirteen months later, it happened again. This time, the patient (F.C.), a young glove cutter, had the same left temporal lobectomy and removal of the left hippocampus and experienced precisely the same memory deficit. "This really stopped Penfield in his tracks," Milner recalled. Indeed, the possibility that temporal lobe operations would need to end altogether stood in the way of Penfield's surgical agenda.[100]

In 1953, with a cloud hanging over the future of temporal lobe operations, Milner began work on what was coming to be called the Temporal Lobe Research Project. Meeting minutes record that she could quickly correlate her psychological testing of the patients with Jasper's review of their EEG abnormalities. Because her neuropsychological testing had become routine at this point, she was also able to confirm the clinical impressions of P.B. and F.C.—they had no ability to remember new events, but their memory for the past before the surgery and their general intelligence remained intact. A tentative hypothesis emerged that the memory loss resulted from a previously undiagnosed impairment of the hippocampal structures on the opposite, unoperated-on side; in effect, removing one side of the hippocampus left only an atrophied, impaired hippocampal horn, effectively depriving the patient of the entire structure. This was the most important clue as to the nature of the memory dysfunction; if the patient had an undiagnosed abnormality on the remaining horn of the hippocampus, the surgical procedure would have been effectively bilateral, leaving the patient with no compensating structure.[101] Such a conclusion was possible only because of the coordination of surgical, psychological, and electrophysiological data that was the hallmark of the

FIGURE 2.8. The regular EEG conferences at the Montreal Neurological Institute became a key site for trading between disciplines. This example includes, from left to right, the head of radiology Donald McRae, Herbert Jasper, Wilder Penfield, Brenda Milner, and the neurosurgeons Jake Hansberry and Theodore Rasmussen. Reproduced by permission of the Osler Library of the History of Medicine, McGill University.

MNI's clinical EEG meetings, which had become the heart of the institute's interdisciplinary trading zone (fig. 2.8).

One notable feature of the MNI trading zone between surgery and psychology was the extent to which coordination and trading could occur among disciplines that did not share theoretical presuppositions. In contrast to attempts at scientific unity that stressed working out the meaning of underlying concepts (e.g., the postwar unity-of-science movement that grew from logical positivism), those at the MNI were capable of coordinated work despite disagreement on several theoretical commitments.[102] This can be seen clearly in Milner and Penfield's joint work on memory. Since his early explorations of the temporal lobe, Penfield had advanced a theory of memory formation that grew from his experiences at the operating table. However, despite Harrower's interventions, that theory was only slightly more theoretically sophisticated than folk psychology. Penfield argued that the temporal lobes contained a permanent record of the stream of consciousness, resulting in a memory trace that was unchanged from birth to death (fig. 2.9). Milner, steeped in the Bartlett tradition, which challenged the notion of a permanent, unalterable memory trace, disagreed. Yet, in addition to creating a space where psychological expertise could be called on, Harrower had left Penfield and his MNI as a place where psychological insight would be taken seriously.

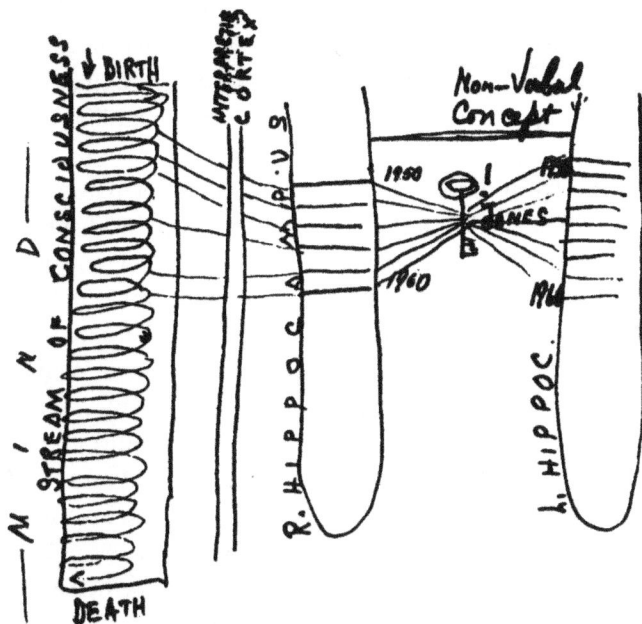

FIGURE 2.9. A sketch by Wilder Penfield illustrating his hypothetical mechanism of memory formation sent to Brenda Milner in the 1970s. Milner herself noted of the diagram: "This formulation will seem to most psychologists little more than a picturesque metaphor." See Brenda Milner, "Memory Mechanisms," *Canadian Medical Association Journal* 116, no. (1977): 1376.

In writing up the cases of P.B. and F.C., Milner and Penfield negotiated their disagreements over memory in exacting detail. For instance, Penfield argued that removals from the hippocampal zone indicated that memory and recall were localizable to these regions; Milner objected that "recall" ought not be included because "so much can be recalled by our patients from the past, despite . . . some retrograde amnesia."[103] Crucially, while Milner and Penfield disagreed on central elements of the nature of memory, they could still conclude that, in contrast to the theories of Lashley, which stressed whole brain functioning in memory, the hippocampus and mesial structures were vital to memory formation. These subtle distinctions, which survived into the published work, were evidence that psychology had become an established part of the MNI community once again and that the Montreal method was more than an operative procedure; it was a formula for combining disciplines and knowledge.[104]

Milner's investigation of P.B. and F.C. ultimately prepared her for a further encounter that would allow her to test her hypothesis that their memory disruptions resulted from bilateral disturbance of the hippocampus. It would also establish her reputation, and it would establish the reputation of neuropsychology as a fully fledged scientific field rather than merely an adjunct helping profession. In 1953, the American neurosurgeon William Beecher Scoville at the Institute for Living in Hartford, Connecticut, contacted Penfield to report a similar case of anterograde amnesia. Inspired by Penfield's operations, Scoville had adapted his own psychosurgical procedure—orbital undercutting—to treat a twenty-three-year-old epileptic man named Henry Molaison. However, unlike Penfield, who refused to operate bilaterally (to remove both halves of a brain structure), Scoville performed a more radical procedure—complete bilateral removal of the hippocampus. After the operation, Molaison, who would soon be known to the world only by his initials, H.M., displayed the kind of profound anterograde amnesia experienced by P.B. and F.C. Scoville became aware of Milner's work with Penfield and invited her to examine H.M. Commuting by night train from Montreal to the Institute of Living in Hartford, Milner confirmed the loss of his capacity to form new memories—precisely like the patients in Montreal. She also made a surprising discovery: using a simple mirror-star drawing test, she discovered that H.M. could still learn new motor skills, skills he had no memory of learning. Milner's work with H.M. not only confirmed the role of a specific neuronal system for memory formation (contra Lashley) but also established the existence of separate memory mechanisms within the brain, a discovery that considerably revitalized the experimental study of memory in the ensuing decades. The impact of Milner's work was perhaps best summarized by the Russian psychologist Alexander Luria, who wrote her on the publication of her work with H.M.: "Memory was the sleeping beauty of the brain—and now she is awake."[105]

Conclusion: The Montreal Method

Luria's pronouncement proved accurate. Milner's work with H.M. became one of the most iconic discoveries of the brain sciences in the 1950s and 1960s and presaged an explosion of interest in the neurobiology of memory.

In subsequent decades, scientific reputations would be made in the field of memory research, and H.M. would become the single most studied individual of the twentieth century (more will be said about his fate and his work with Milner's student, Suzanne Corkin, in the conclusion).[106] Perhaps just as important, however, Milner's work with H.M. served as a proof of concept for an interdisciplinary approach to studying the brain, one that was the product of trading between surgery and psychology over several decades at the MNI. Harrower's validation of neuropsychology as a diagnostic service formed a tentative trading language in the 1930s, and the synchronized activity of the surgeons and psychologists created a strong assembly of actors at the MNI; in the 1950s, Milner could easily join this already-formed assembly and grow this interdisciplinary work into a fully fledged neuroscience of memory.[107] Pursuing career paths that were shaped at least partially by the gendered nature of clinical psychology, this pair of English women had forged unconventional careers that, over time, elevated neuropsychology from an adjunct profession to a fully fledged research specialty, effectively making them the most prominent mothers of neuroscience.

By the mid-1950s, the staff at the MNI were putting the capstone on nearly two decades of scientific investigation. Moreover, the studies of memory and memory loss in temporal lobe patients provided a crucial demonstration of the value of their interdisciplinary approach; these scientific achievements could be reduced not to the contributions of surgery, histology, physiology, pathology, or psychology but rather to a successful integration of these disciplines. In the next chapter, we will see how one scientist, Herbert Jasper, would use the MNI trading zone as a launching pad to bring this kind of interdisciplinary brain research to the world. We will also see how the technology of which Jasper was a key user—the EEG—served as a binding object to unify disparate approaches to the study of the brain. At the same time, we will also track a competing approach to unifying the brain and mind sciences, one that would rely almost totally on the reductionist approach that had proved so unsuccessful in Montreal. The amalgamation of these conflicting approaches in the 1960s, with each having to pivot from their original intentions, would conclude in the establishment of neuroscience as a stable scientific field.

A Tale of Two Sciences: Herbert Jasper, Francis Schmitt, and the Origins of Neuroscience

In 1938, another newly appointed fellow would transform the interdisciplinary environment of the Montreal Neurological Institute (MNI) and also do more than any other to spread its growing interdisciplinary ethos to the world. Indeed, in the career of the psychologist-turned-physiologist Herbert Jasper, we can see two different yet overlapping roles: one as a laboratory scientist whose interdisciplinary work in the clinic transformed neurosurgery and neurological knowledge; the other as a scientific leader whose organizing efforts allowed the disparate scientific disciplines that studied the nervous system to cohere into a new scientific community.

What united both roles was a key technology—the electroencephalograph (EEG). First developed by the German psychiatrist Hans Berger in 1924, the EEG would in Jasper's hands transform knowledge of epilepsy and neurosurgical practice. Beyond its value as a clinical tool, however, it would also create the social ties that brought together the world's brain researchers into a self-consciously interdisciplinary scientific community (in aspiration, if not always in practice). While the community at the MNI formed strong interdisciplinary assemblies of actors through its coordinated work, it was through Jasper's weak ties to the global EEG community that this ethos eventually cohered in the form of the Interdisciplinary Brain Research Organization (later becoming the International Brain Research Organization [IBRO]), which served to organize the world's brain scientists into a self-consciously interdisciplinary scientific community by the late 1960s.

Jasper's career also brings us into close contact with an alternative origin story for modern neuroscience, one that revolves around the biophysicist Francis O. Schmitt of the Massachusetts Institute of Technology (MIT). Schmitt, a key figure in the emergence of biophysics and molecular biology in the middle of the century, was in many ways Jasper's mirror: both were young men raised in religious homes who turned to science to find purpose and meaning in their lives. However, their paths quickly diverged, leading each to embrace different unifying visions of the brain sciences. Jasper took his cue from the MNI, where scientists applied their different approaches to the problem of brain function, hammering out unified knowledge where possible. Integration was Jasper's goal, and the EEG was the binding object that allowed this integration, first in Montreal, then, he hoped, the world.

By contrast, Schmitt's goal was not integration but transcendence. Schmitt fixed his view rigidly on the molecular and sought to transcend the differences between scientific disciplines with a single discovery—a memory molecule—that he hoped would unify the brain and mind sciences in one fell swoop. As we shall see, the organization that Schmitt founded—the Neurosciences Research Program (NRP)—failed to find this memory molecule but paradoxically succeeded in its public relations effort to popularize the concept of neuroscience. If the IBRO community of brain researchers was neuroscience in all but name, then the NRP was neuroscience in name only, composed of scientists who knew very little of brain research or neurology. While IBRO ultimately undertook the difficult groundwork of organizing the global brain science community into the Society for Neuroscience (SFN) by 1970, its historical role was eclipsed by the superior public relations of the NRP—ironically just as it became clear that its organizing concept (the memory molecule) was a chimera.

"Restless genes"

Like his future collaborator Wilder Penfield, Herbert Jasper was also born in the Pacific Northwest, albeit fifteen years later, in La Grande, Oregon, in 1906 (fig. 3.1). Jasper later attributed a lifelong wanderlust to the "restless genes" he had inherited from his parents; his father was a Methodist minister and brilliant religious scholar who was also a talented mathematician and committed pacifist, while his mother was a descendant of French Huguenots who escaped persecution in France, fleeing first to Switzerland and later the United States. Growing up in Oregon, Jasper, encouraged by his father, cultivated an intellectual curiosity that led him to study comparative religions and, later, modern philosophy. At Willamette University, he was "saturated" in the works of classic European philosophy (Descartes, Bacon, Kant, Hegel, Spinoza, Berkeley, Locke, Spencer, Hume, and Bergson) as well as contemporary American pragmatism. An early professor also introduced him to the world of experimental psychology in the form of William James's *Principles of Psychology* (1890), and the young Jasper became convinced that this exciting

FIGURE 3.1. Herbert Jasper, ca. 1937. University of British Columbia Archives, UBC 5.1/1491.

new approach to the mind might help him resolve the "philosophical problems that had been bothering [him]."[1]

Beyond his interest in abstract philosophical questions, two more immediate experiences pushed Jasper toward the study of the mind and the brain. The first was a dalliance with recreational drugs. While a college student in the 1920s, Jasper took mescaline with his roommates and was "astounded by the profound effects of a few drops of [the drug]": "The whole world changed.

I was disoriented completely, had hallucinations and delusions, and sensations of floating in air. . . . I have never forgotten the dramatic effect of such
a small amount of a chemical substance upon the mind."[2] Second, the suicide of a close friend led him to take an interest in psychiatric problems. He
channeled his grief into scientific work, completing an undergraduate thesis
that used a variety of psychological tests to predict student success in college.
His studies represented one of the earliest attempts to construct an objective measure of depression. In doing so, Jasper was typical of a generation of
American students who saw in the science of experimental psychology the
tools necessary to understand and manage thorny psychosocial problems.[3]

Psychiatric problems continued to intrigue Jasper as he began graduate
study in the late 1920s, leading him to the new neuropsychiatry that became
popular in the United States after World War I (see chap. 1). Alongside epilepsy and psychoneurosis, stammering or stuttering constituted one of the
significant borderland problems that lay between neurology and psychiatry, and Jasper began an experimental program with Lee Edward Travis, the
founder of American speech pathology, at the University of Iowa to determine whether stuttering was caused by interference between the left and the
right brain hemispheres. This seemingly strange experimental program introduced Jasper to some of the most technically sophisticated instruments
in neurophysiology; Travis and his engineer had developed a device for the
electrical measurement of the nerve action potential, the electric signal that
passed along and across nerve fibers.[4] Thus, it was through Travis's neuropsychiatric research that Jasper was introduced to the emerging world of the
electric brain.

During the mid-nineteenth century, electricity emerged as an increasingly important tool for investigating and understanding the functions of
the nervous system. In the 1840s, the German organic physicist Emil du
Bois-Reymond used a highly precise galvanometer to show that the nervous
impulse—the signal sent by nerves to peripheral muscles—was electric. This
promising research program stalled at the turn of the century owing to a lack
of instruments sensitive enough to dissect the nerve impulse further. Nerves
were simply too small and their electric properties too fleeting to be observed
by existing galvanometers. The fundamental unit of nerve activity—the impulse that coursed along individual nerve axons—remained secluded within
bundles of nerve fibers whose summed activity amounted to no more than
a brief electric blip. The discovery in the 1910s by the English physiologist
Edgar Douglas Adrian that the nerve impulse propagated in an all-or-none
fashion put the capstone on classical neurophysiology, but little else was possible with existing laboratory instruments. As a problem for reductionistic

laboratory science, the electric properties of the nervous system remained tantalizingly out of reach.[5]

Technological developments in the radio industry following World War I transformed the situation. In the 1920s, as Jasper entered the world of neuropsychiatry, the physiologists Alexander Forbes and Herbert Gasser at Washington University, St. Louis, absorbed technologies from the radio industry (vacuum tubes and cathode-ray tubes) into their laboratories to amplify the nerve signal from individual nerve axons; the nerve impulse could finally be seen, and Adrian eventually used the same technologies to observe his all-or-none principle in naturally occurring nerve impulses.[6] The triumph of this electrical technique fascinated young physiologists like Jasper, who had seen early attempts to do just such a thing in Travis's speech laboratory. In the winter of 1929, Jasper was introduced, through the St. Louis group, to the French physiologists Alexandre and Andrée Monier. The Moniers recruited him to study with them at the Sorbonne in the laboratory of Louis Lapicque, who was investigating the actions of individual nerves. With a fellowship from the Rockefeller Foundation (RF), Jasper set sail for Paris in September 1931.[7]

Jasper's time in Paris was crucial for his later career. Within the American school of electrophysiology at St. Louis, reductionism reigned. The experimental program of Gasser and his associates focused almost exclusively on breaking the nervous system down into smaller and smaller components that could be studied with the increasingly sophisticated laboratory instruments at their disposal; christened the "axonologists," these men were dismissive of attempts to study the electric activity of the brain itself.[8] As Gasser himself informed Jasper: "[I]t would be a waste of time to try to record electric activity from the brain since it would be a composite of the action potentials from millions of simultaneously active nerve cells, impossible to interpret."[9] Yet, seemingly eschewing this advice, Jasper carefully conducted animal experiments on the electric excitability of peripheral nerves and related their functions to that of the entire cortex. It was very likely this continued interest in the whole brain, carried over from his early interest in neuropsychiatry, that allowed him to appreciate the importance of a new medicoscientific instrument that had recently emerged in Germany—the electroencephalograph (EEG).

Brain Waves

The EEG emerged from a radically different medicoscientific culture than that of the axonologists. First developed in 1924 by the German psychiatrist Hans Berger while he was serving as professor of psychiatry at the Jena

Psychiatric Clinic, the EEG grew from Berger's eccentric interest in the possibility of discovering "psychic energy." At the same time, Berger drew on more mainstream research traditions of the physiology laboratories of Europe, particularly the work of Jules-Etienne Marey, the French physiologist whose so-called graphic method aimed to render physiological functions of the body such as pulse, breathing rhythm, heartbeat, and changes in blood volume as the oscillating curves of the kymograph's rotating drum and stylus. In his decades-long quest to discover the dynamics of psychic energy, Berger moved from tracking brain and blood volume as a measure of mental activity to eventually transforming brain-stimulating electrodes into *recording* electrodes, allowing him to register the electric activity of the *whole* brain. Beginning with patients at the Jena Psychiatric Clinic, and culminating in his observation of his own brain waves, Berger documented the existence not only of regular brain waves but also of their relationship to simple psychological processes. Electroencephalograms revealed that an idle brain produced a stable ten-hertz basic rhythm—what he christened the *alpha* rhythm. They also revealed that sensory stimulation, calculation, and other mental activity consistently blocked or disrupted this rhythm. In his isolation from the main currents of American and European neurophysiology, Berger seemed to have discovered a kind of holistic "brainscript" that might unravel the relationship between mind and matter.[10]

Attempts to measure the electric activity of the whole brain would have made little sense to the American axonologists or European physiologists more generally. When Berger published his initial findings in 1929, they were largely ignored; indeed, it was not until 1934 that the phenomenon of brain waves was validated as real by Edgar Douglas Adrian, whose credibility as a careful experimenter (and his recent Nobel Prize) ensured that the phenomenon would be taken seriously.[11]

One notable exception to the deafening silence that initially greeted the EEG, however, was Jasper, who had recently returned to the United States in 1933. Jasper split his time between Woods Hole Marine Biological Laboratory, where he was completing his research on crustacean neuromuscular systems, and Brown University's Bradley Psychiatric Hospital for Children, where he was setting up an electrophysiology laboratory. Jasper, with one foot in reductionist laboratory physiology and the other in the more holistic psychobiology of the 1930s, had heard of the EEG from a colleague and quickly confirmed the existence of Berger's alpha rhythm. Notably, he did so *before* he heard of the simultaneous efforts to confirm Berger's results by Adrian in England. His internationalism and methodological pluralism led him to entertain the possibility that the EEG was a real and valuable tool and understand its technical limitations and possibilities more profoundly than others did.[12]

An example here will illustrate the difference between Jasper and his colleagues. Simultaneously with Jasper's work with the EEG at Brown, fifty miles north in Cambridge the epileptologists William Lennox, and Frederic and Erna Gibbs joined forces with their neurophysiological colleague Hallowell Davis and, in 1934, recorded a brain wave pattern from their epileptic secretary—a distinctive three-per-second spike pattern they realized could reliably distinguish epilepsy from other neurological and psychiatric conditions. This discovery proved the value of the EEG as a diagnostic tool in clinical neurology, a fact that Jasper had predicted: "It may well be that the electroencephalograms . . . may prove significant in psychology and clinical neurology. It is even possible that this technique may provide information in regard to brain action which will be comparable in significance to the information in regard to heart function which is provided by the electrocardiograph."[13]

However, because Jasper had already spent considerable time working with electrophysiological devices in the United States and France, he understood the technical limitations and possibilities of the EEG better than his Harvard colleagues did. This led him and the Harvard group to considerably different views of epilepsy and whether it could be localized to different areas of the cortex. Lennox and the Gibbses argued, against the prevailing notion of localizable Jacksonian epilepsy, that epilepsy was, in fact, a disorder of whole brain function. Generalized, whole body seizures (grand mal) and lapses of consciousness (petit mal) epilepsies that seemed to have no clear focus were to be renamed *paroxysmal cerebral dysrhythmia* or *disordered rhythm of brain potentials*. Drawing a direct comparison to the heart, the Harvard group argued that the brain had its own pacemaker and that epilepsy resulted from a failure of this central control system. Employing an automotive metaphor, they argued:

> [T]he brain may be likened to a motor car which is being driven along a road with a ditch on the right called the too slow ditch and a ditch on the left called the too fast ditch. The rate-regulating mechanisms of the brain normally steer along the centre [sic] of this road, but when these mechanisms are out of order they steer the brain first to one side and then to the other. The more defective the steering (rate-regulating) mechanism the less quickly are deviations from normal corrected. Petit mal is an example of a minor disorder of rate-regulation in which the brain swerves quickly back and forth between too fast and too slow. In grand mal the rate regulators are so completely defective that the brain plunges far over to the too fast side, stays there for a considerable time, then goes over to the too slow side and remains there for a considerable time.[14]

According to the Harvard group, epilepsy was, thus, a disorder of the whole brain and not susceptible to the kind of localizing that was characteristic of Jacksonian seizures.

Jasper, by contrast, had a more sophisticated view of the EEG and its value for diagnosis because of his more robust technical education in neurophysiology and his exposure to other research traditions in Europe. In their 1936 paper detailing their theory of cerebral dysrhythmia, the Harvard group noted: "[T]he method [for observing epileptic electroencephalogram patterns] is exceedingly simple."[15] To this, Jasper responded: "[I]t is important to point out that the method is actually exceedingly complicated."[16] Jasper, with his laboratory experience, was able to distinguish epileptic patterns from errors and artifacts but also able to localize epileptic discharges to specific areas of the cortex. For the Harvard group, epilepsies were distinguished by their different temporal patterns, easily visible to anyone trained to see them. For Jasper, epilepsy was still a localizable functional disturbance in specific brain tissue that might be amenable to surgical intervention in the way that Jacksonian epilepsy had been.

Localization of function, then, was the medicoscientific trading language that both Jasper and Wilder Penfield could speak. In 1937, Jasper and his collaborator Leonard Carmichael invited Penfield to Brown to discuss his recently published work with Edwin Boldrey on the motor homunculus and his operations on conscious patients (see chap. 2). Jasper informed Penfield that he thought he might be able to use the EEG to localize the focus of epilepsy in at least *some* patients. Penfield was initially skeptical, but, after a demonstration on a twelve-year-old patient from the Bradley Psychiatric Hospital, he was intrigued enough to agree to operate on the boy. The operation and another like it were successful, and, for the next several months, Jasper and Penfield undertook an almost "unthinkable commuters research project." Piling his EEG equipment into the front seat of his car, Jasper made a weekly trip from Rhode Island to Montreal for nearly a year and with Penfield's assistance succeeded in having his RF grant transferred. According to Penfield: "It was as though far-away Rhode Island were a suburb of Montreal."[17]

The success of the Penfield-Jasper collaboration was evident almost immediately, and the commuter project became one of permanent residence. Jasper moved to Montreal in 1938 and opened a dedicated EEG laboratory in the basement of the MNI in February 1939 with an additional $50,000 from the RF.[18] The opening of the EEG laboratory was attended by stories in the local press announcing a new "brain laboratory" that would "bear much the same relation to the Montreal Neurological Institute that an X-ray department does to any hospital."[19] More than a diagnostic aid, however, under

Jasper's direction the EEG laboratory would become a focal point for a new vision of brain science.

Scientific Leader

Jasper's first years in Montreal were productive ones. The MNI had become the most important location for epilepsy surgery in North America, and its prominence ensured that a nearly endless supply of patients would pass through the EEG laboratory.[20] This wealth of patients allowed Jasper to complete an extensive electroencephalographic study of epilepsy, resulting in a more definitive classification based on localization (in contrast to the Harvard group's cerebral dysrhythmia concept).[21] His grand ambition to study all mental life with the EEG may not have panned out, but, through the study of epileptic patients in the clinic, he was now producing knowledge of the brain that was medically and scientifically valuable.

The outbreak of World War II disrupted Jasper and Penfield's productive start on temporal lobe epilepsy (see chap. 2) but enriched the MNI's interdisciplinary environment in other ways. Jasper engaged in wartime research on nerve injury and the phenomenon of pilot blackout at high altitudes. Now married to a Canadian nurse, Margaret Goldie, Jasper obtained an MD and spent a stint in the Canadian army as a captain. In this respect, the experiences of the MNI were like those of the rest of the biomedical sciences and contrasted with those of physics. While physics was transformed by its engagement with wartime projects like the Manhattan Project and radar, productive work in the biomedical sciences was put on hold as doctors and biological scientists were conscripted to help with the more immediate problem of wartime medicine.

The years following the war saw more profound transformations. The MNI, Montreal, and Canada generally experienced significant postwar growth. As Jasper recalled: "Canada . . . benefited by the war. She had grown in self-confidence and in industrial and political independence from Great Britain. The work of the [MNI] had become well-known worldwide so that there were . . . postdoctoral students from many countries applying to come to Montreal for a period of research and training."[22] Jasper, in particular, had become the MNI'S main draw. From around the world, young researchers and doctors came to learn from the master of the EEG (fig. 3.2). In a 1951 letter to the dean of medicine recommending Jasper's promotion to full professor, Penfield noted that "about half of the letters that come to the Institute from abroad from graduate doctors seeking advanced training are requests to work with Jasper."[23] Simultaneously, Jasper had emerged

FIGURE 3.2. Herbert Jasper instructs fellows in the interpretation of the electroencephalogram. Reproduced by permission of the Osler Library of the History of Medicine, McGill University.

as the leader of the clinical profession of electroencephalographers, whose numbers were expanding rapidly. He founded the Eastern Association of Electroencephalographers in 1939 and served as the inaugural president of the American Electroencephalographic Society (AEEGS) in 1946 and the International Association of Electroencephalographers in 1947.

Far from being simply an expert clinical technician, akin to a radiologist, Jasper maintained a productive laboratory research program that embraced more and more approaches as the capacity for interdisciplinary work at the MNI expanded. An example will illustrate the point. In 1944, the MNI hired a young South African chemist to assist with a wartime project on brain edema (brain swelling following injury). K. A. C. Elliott, born in Kimberley, South Africa, in 1903, followed a circuitous route to medical chemistry, beginning with a job in a dynamite factory near Johannesburg before eventually being hired by the MNI.[24] This unorthodox career path reflected the underdeveloped state of research on brain chemistry, which barely existed in the 1940s. While Henry Dale and Otto Loewi had discovered chemical transmission between nerves and visceral organs, few neurophysiologists suspected that these *neurohumors* (later renamed *neurotransmitters*) played any role in the brain.[25] It was an act of great optimism when, on hiring Elliott in 1944, Penfield gave him the title *neurochemist*, a term he invented merely by adding the prefix *neuro-*.[26]

The mere creation of a term, of course, did not create a new scientific field. Instead, collaborative work on discrete problems brought scientific disciplines together at the MNI. Here again, epilepsy provided the impetus. Beginning in 1950, Elliott began investigating brain tissue removed by Penfield at surgery that Jasper's EEG confirmed was epileptogenic. Jasper and Elliot began to speculate that, as Jasper later put it, "the abnormality in epileptic tissue might be due to defective inhibition rather than to excessive excitation, the latter resulting from a release from normal inhibitory controls."[27] The American physiologist Eugene Roberts had demonstrated the existence of gamma-aminobutyric acid (GABA) in the brain in 1950, and the Australian physiologist Ernest Florey had demonstrated that a newly discovered substance, "Factor I," served an inhibitory role in the nervous systems of crayfish. Elliott and Jasper brought Florey to Montreal, where they coordinated what they knew about epilepsy with Florey's work. Elliott and Jasper knew that, in animals, vitamin B_6 deficiency produced seizures and that B_6 was also an inhibitor of GABA, suggesting that Florey's Factor I might be GABA. Elliott and his team confirmed this using infrared spectroscopy and chromatography. The significance of this discovery was considerable. By establishing the functional significance of GABA in the central nervous system through coordinated work between clinic and laboratory, Jasper and Elliott wired together two scientific fields and provided the new hybrid discipline of neurochemistry with a tentative research program.[28]

As the story related above reflects, Jasper's presence at the MNI and his stewardship of the professional network of electroencephalography established the EEG as a tool allowing clinicians and laboratory scientists to trade and promote fertile interdisciplinary collaborations. More than merely a device for confirming epilepsy or brain tumors, the EEG and the networks attached to it allowed the interdisciplinarity of the MNI to diffuse outward. Initial enthusiasm for the device—that it would resolve the mind/body problem, that its waves could decode the language of thought, that it could be used to construct thinking machines—eventually gave way to more restrained but more significant research in laboratories and clinics across the globe. For laboratory scientists, clinical researchers, and working physicians, the EEG functioned as a binding object that could link the clinic, the laboratory, and human and animal research. Indeed, under Jasper's direction, the house journal of the International Association of Electroencephalographers, the *Journal of Electroencephalography and Clinical Neurology*, became an essential place for the publication of clinical *and* basic research in the brain sciences.[29] Reflecting on the fate of the AEEGS later in its existence, one of Jasper's colleagues remarked to him: "[U]ntil the early 1970s we were *the* neuroscience society."[30]

The Anatomy of Consciousness

By the early 1950s, Jasper, Penfield, and the MNI were firing on all cylinders. The interdisciplinary environment of the MNI contrasted notably with the situation of the brain sciences in the United States in the immediate postwar years. In June 1945, the American National Research Council (NRC) set up a small committee on neurobiology consisting of, among others, the scientists Karl Lashley, John F. Fulton, Rafael Lorente de Nó, Detlev Bronk, Paul Weiss, and Louis Weed and the former Jasper associates Leonard Carmichael and F. O. Schmitt. The committee met periodically between 1945 and 1948 and produced a final report in 1952. The report painted a picture of a disunited and uncoordinated American brain science community. The report noted: "Research on the nervous system has followed the lines of specialized techniques, and therefore, has become highly departmentalized, with the various lines, such as physiology, psychology, anatomy, histology, embryology, clinical neurology, psychiatry, neurosurgery, and comparative biology, following, for the most part, independent courses." It continued: "Lack of cross correlation of information and the reliance upon excessive superstructures of speculation, built upon narrow and often shaky foundations of fact, have led to serious incongruities among the views prevailing in the various separate disciplines, all purportedly pertaining to the same nervous system." This lack of unification in the American brain sciences could be "blamed on the failure to evaluate, correlate and integrate properly the available knowledge so that different lines of research might utilize each other's sources of information, techniques and results to best mutual advantage."[31] The report delineated virtually everything that the MNI was and everything that the American brain science community was not. As the American physician Paul McLean noted to Penfield after a visit to the MNI in the same year that the NRC issued its report: "It is always a thrill to visit the Neurological Institute. . . . I have frequently wondered why we have nothing comparable to the Institute in the United States. As some of us were saying the other day, it is certainly not because of lack of means."[32]

At the time of McLean's comment, Penfield and Jasper were putting the capstone on a decade's worth of work. Their studies of temporal lobe seizures had elucidated the role of subcortical structures in memory formation (see chap. 2), and Jasper had refined his electrophysiological classification of the epilepsies. He had also come to occupy a leadership role that placed him at the center of the global brain science community, which had grown since World War II.

In 1954, Penfield and Jasper published their valedictory statement on epilepsy, *Epilepsy and the Functional Anatomy of the Human Brain*. In it, they formalized a theory of the role of the brain stem in consciousness that they had been developing since their earliest collaborations. Before World War II, the prevailing wisdom in neurology was that consciousness was the product of activity in the entire cortex, with an emphasis on the activity of the frontal lobes. As a contemporary of Jasper's put it: "[M]any of us thought of the brain as an organ consisting of long fiber tracts, serving sensory and motor function. Stimuli travelled along the fiber paths from the periphery to the cortex. At arrival in the cortex stimuli were transformed into conscious experience. How this was achieved, nobody could understand."[33]

However, Jasper's intraoperative EEG recordings, combined with Penfield's brain stimulation, had suggested otherwise. Patients whose seizures originated in deeper cortical structures like the diencephalon invariably experienced periods of lost consciousness during their seizures, a phenomenon that Penfield and Jasper could reproduce and record during operations. The result of several hundred such operations was an elegant hypothetical brain system that they referred to as the *centrencephalic system* (see chap. 2), which Penfield described as "a neurone system centrally placed within the brain and equally connected with the two hemispheres." The proposed centrencephalic system "acted as the indispensable substratum of consciousness" and served as a director of attention for incoming sensory stimuli and a coordinator for outgoing motor response.[34] Far from being a product of the entire brain's holistic integration, consciousness now had a distinct, decidedly prosaic anatomy. No longer the product of humanity's distinctive neocortex, it was demoted from the so-called new brain of the cortex to the old structure of the brain stem.

Penfield and Jasper's centrencephalic system was controversial among neurologists and psychologists. However, it was consistent with the discoveries of the Italian neurophysiologist Giuseppe Moruzzi and the American physician Horace Magoun. In 1949, building partially on Jasper's earliest work, Moruzzi and Magoun published the results of their EEG experiments in cats, showing that specific brain stem structures were responsible for sleep, wakefulness, and alertness.[35] Less grand than Penfield and Jasper's proposal, Moruzzi and Magoun's "ascending reticular activating system" still provided experimental support for the centrencephalic system, and the pair of theories constituted perhaps the most exciting development in neurology in decades (fig. 3.3).

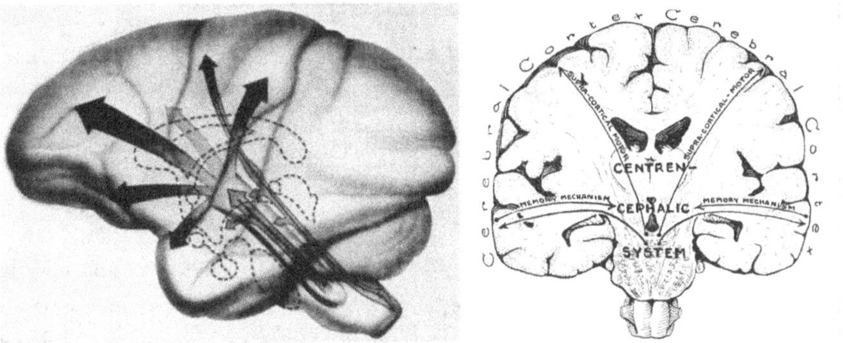

FIGURE 3.3. On the left, Giuseppe Moruzzi and Horace Magoun's "reticular activating system." On the right, Wilder Penfield and Herbert Jasper's "centrencephalic system." From, respectively, Herbert H. Jasper and Frederic Bremer, eds., *Brain Mechanisms and Consciousness* (Charles C. Thomas, 1954), 13; and Wilder Penfield and Theodore C. Erickson, *Epilepsy and Cerebral Localization: A Study of the Mechanism, Treatment and Prevention of Epileptic Seizures* (Charles C. Thomas, 1941), 475.

The centrencephalic system, the product of nearly two decades of inter-disciplinary work at the MNI, proved crucial to what happened next. In 1953, scientists from around the globe gathered in Montreal for the International Congress of Physiology. At a satellite conference in the Laurentian Mountains, the site of the skiing conferences of the international EEG community since Jasper's arrival in Montreal, the world's leading brain researchers met to dis-cuss the nature of human consciousness. The conference was a veritable who's who of postwar brain sciences, and, despite the different disciplinary back-grounds represented, it seemed clear that these differences were being knit together by the MNI's careful coordination of clinical and laboratory studies. As Horace Magoun put it:

> Dr. Penfield, in private discussion yesterday, used the simile that his work was like a hand reaching down from the cortex to grasp another hand reaching up from the brain stem. . . . It seems to me that this simile could be broadened to a hand reaching down from clinical studies to grasp another reaching up from the experimental laboratory. The idea of the centrencephalic system reaching up into the cortex arose in Montreal and now the hand reaching down from the cortex to the centrencephalic system is beginning to emerge from studies in Dr. Penfield's Institute and elsewhere.[36]

Indeed, the Laurentian conference, later published as *Brain Mechanisms and Consciousness* (1954), represented a summary statement both of the MNI's interdisciplinary style and of a new conception of the brain: not as a series of distinctly localizable faculties or as a formless equipotential mass but rather as a series of interacting and interlocking systems.[37]

The concept of the centrencephalic system did not sit well with one group of attendees at the 1953 conference: representatives from the Soviet Union. The death of Joseph Stalin in March 1953 led to a thaw in the political control of culture and science in the Soviet Union, allowing several Russian physiologists to travel to Montreal. At the height of McCarthyism in the United States, Canada and Montreal were considered relatively safe ground by Russian scientists.[38] Declassified CIA documents indicate that the Soviet delegate G. D. Smirnov felt that he was among friends: "We were greatly impressed by our visit to the [MNI]. . . . Clinical work and thorough experimental work are well coordinated at this institute." Moreover, the Soviet physiologists appreciated that Penfield and his associates understood the importance of Pavlovian theory: in constructing the MNI, Penfield had included "the Name of Pavlov, together with several of the world's greatest neurologists, . . . on the wall in the main entrance hall of the institute." Penfield also told the Soviets: "Pavlov's deep penetration into the physiological mechanism still guides the thinking of clinicians."[39]

The gesture to Pavlov was not one of mere hospitality. Since his earliest work at the MNI, Penfield had invoked the notion of Pavlovian conditioning alongside his ideas about the centrencephalic system; indeed, the two theories fit together. In Penfield's view, while much of the brain's activity consisted of unconditioned reflexes present from birth, there also had to be a system of conditioning possible that would allow the brain to learn. He argued that because his probe called forth from his patients experiences more complex than simple movement or sensation—hallucinations, memories, and other epileptic auras—the temporal lobes constituted a portion of "uncommitted cortex" onto which experience could be written. Language was laid down in the dominant left temporal lobe, and other experiential phenomena were laid down in the right temporal lobe. The centrencephalic system tied the arrangement together, directing conscious attention and laying down the record of the stream of consciousness in the temporal lobes. The temporal lobes were the location where Pavlovian conditioning occurred.[40]

Despite this, Soviet physiologists and electroencephalographers viewed Penfield and Jasper's work with ideological suspicion. At the time of the Laurentian conference, they still saw the centrencephalic and reticular systems as anathema to Pavlovian physiology, which sat at the center of Russian physiology and psychology. To Soviet physiologists and psychologists, these central, directing systems in the brain were "a foundation on which physiological idealism was being rebuilt."[41]

Jasper provided a solution to this impasse. Inspired in part by Milner's work on memory (see chap. 2), he began experiments with a new technology—the microelectrode—that he hoped would clarify the relationship between gross surface EEG phenomena and the discharge of individual nerve cells. This technology allowed for so-called single-unit recording of neural activity—in essence, recording individual neurons' electric activity, pierced by a remarkably thin micro glass pipette or metal needle. With the aid of the Chinese MNI fellow Choh-Luh Li, who had arrived in 1950, and his pupil and technician David Hubel, Jasper used these glass and tungsten microelectrodes to tunnel down to deeper levels of functional organization within the brain stems of cats and monkeys. Beginning with studies of the thalamocortical system of anesthetized cats, Jasper and his MNI colleagues began to clarify the relationship between the observed slow-wave oscillations of the large-scale EEG recordings and the actions of single cells within the brain. Jasper was also inspired by his relationship with the MNI veteran Donald Hebb (see chap. 5) to examine the relationship between electric activity in the brain, the brain stem reticular system, and the Pavlovian conditioned response. As Jasper and Hebb's departments began to trade theory and technique, data accumulated that expanded classical learning theories by suggesting that conditioned responses, observed with microelectrode methods, could be enhanced, inhibited, or otherwise altered on the basis of additional stimuli affecting the experimental animal's attention.[42] "We had a beautiful picture of what the cells do in the cortex during conditioned reflexes," Jasper recalled (fig. 3.4).[43] Although it was certainly not meant to mediate between capitalist and Communist science, Jasper's research program did create an ideological bridge to his Russian counterparts by reconciling the EEG with Pavlovian ideas.

Meanwhile, Jasper's global EEG community built connections with Russian scientists. Breaking the ice at the 1953 conference paved the way for future contacts between the global EEG community and Soviet neurophysiologists. That conference was followed by a colloquium in Marseilles in 1955 and eventually the Moscow Colloquium on Electroencephalography of Higher Nervous Activity in 1958. This meeting, facilitated by the global network of electroencephalographers that Jasper had been organizing since the end of World War II, brought the MNI's brand of interdisciplinary neuroscience to the world stage.[44]

Moscow and IBRO

Jasper departed for Russia on October 3, 1958, his pockets heavily laden with lantern slides for presentations to be given in Moscow. He recorded a feeling of genuine trepidation aboard the DC-7 that took him on the final leg of his

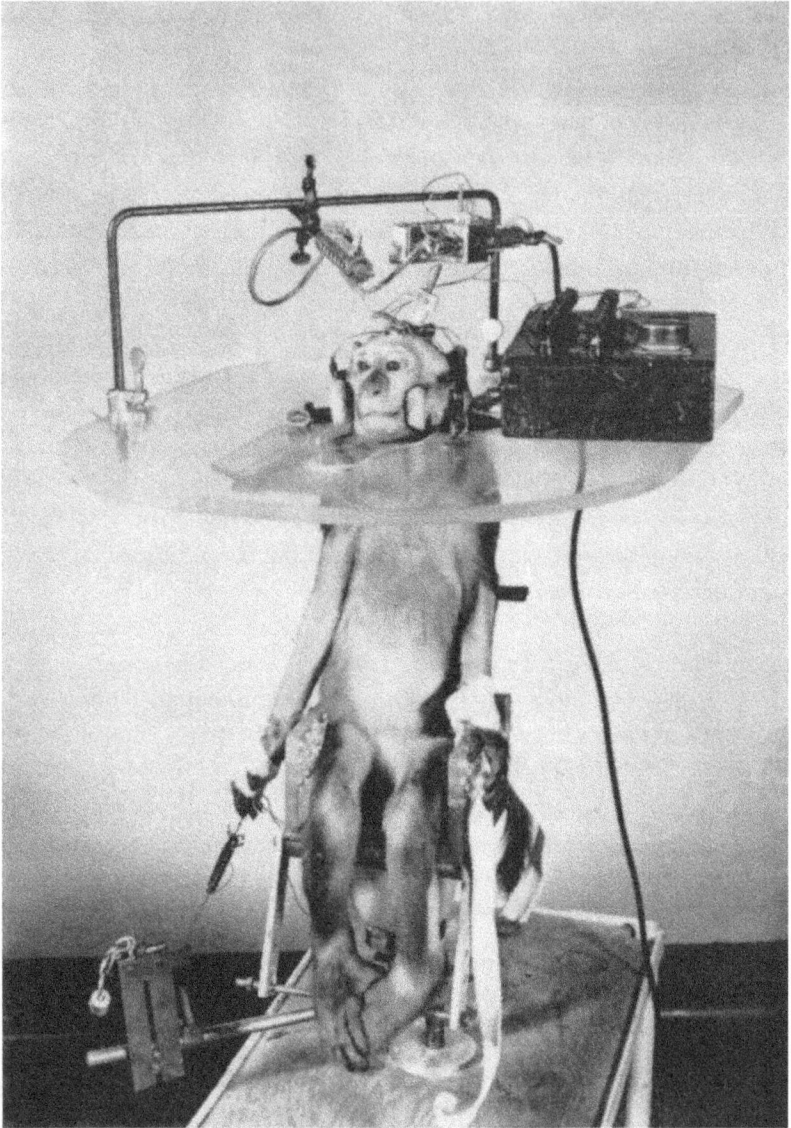

FIGURE 3.4. A monkey in Herbert Jasper's experimental program to reconcile large-scale electro-encephalograph phenomena with microelectrode studies of individual cell activity in conditioning. H. H. Jasper and G. D. Smirnov, eds., *The Moscow Colloquium on Electroencephalography of Higher Nervous Activity* (EEG Journal, 1960), 139.

journey from Prague to Moscow, noting that the airplane stewardesses occasionally whispered to each other "as though not wanting to be heard": "All of this was very mysterious and I felt that finally I had entered the atmosphere behind the Iron Curtain."[45]

Despite his trepidation, Jasper was greeted by friendly faces. Electroencephalography had grown in the Soviet Union in places like the Institute of Higher Nervous Activity in Moscow, and the Russian side of the colloquium was almost entirely a product of the Russian EEG community. Vladimir S. Rusinov and Georgiy Smirnov, the key Russian organizers of the colloquium, had first connected with the global EEG community through one of Jasper's colleagues, the French electroencephalographer Henri Gastaut, who had performed considerable work to organize neurophysiology and electroencephalography in France.[46] The colloquium began on October 6, 1958, and was held in a building that was "obviously a residence of a former wealthy Muscovite taken over for this purpose [and] equipped with simultaneous translation rooms."[47] In attendance were twenty-four delegates from Western nations and twenty-four delegates from Soviet satellite nations, representing seventeen countries overall. The main topics discussed during the sessions read like a recitation of Jasper's research efforts since the early 1950s: the location of electric changes in the brain during conditioning and the activities of single neurons. Jasper's paper, which reported his microelectrode work in Montreal, was perhaps the most well-received of the colloquium as it made the most explicit links between Pavlovian methods and the EEG.[48]

At the close of the colloquium on October 11, Jasper, Rusinov, Gastaut, and the Russian physiologist Ivan Solomonovich Beritashvili proposed the formation of an international brain research year, patterned after the International Geophysical Year, which had just ended in 1957. They also proposed the creation of an organization that would "facilitate the systematic exchange of scientific workers in brain research between the leading centers in various countries throughout the world."[49] After the colloquium, Jasper and his colleague Alfred Fessard traveled to Paris to propose the creation of an interdisciplinary brain research organization under the auspices of UNESCO. UNESCO eventually agreed, with the provision that the name be the *International* Brain Research Organization to reflect its multinational makeup and to further UNESCO's overall mission.[50] Over the next two years, Jasper and Penfield played an intimate role in organizing the structure of IBRO, a fact that Jasper frequently pointed out: "My inspiration came at first from working with Dr. Penfield and his team at the Montreal Neurological Institute which was a very good early example of the neurosciences at work, including many of the basic disciplines together with clinical sciences. . . . The first financial contributions were also made by Canadians, by the Medical Research Council of Canada, by the Canadian National Commission for UNESCO, by the Montreal Neurological Institute, and by generous individual donors."[51] Explicitly internationalist and interdisciplinary in its orientation,

IBRO represented the first serious attempt to organize and coordinate the world's brain researchers since before World War I.[52] And, although it contained actors from many laboratories and clinics, Jasper's proven track record of interdisciplinary work—most notably in his recent microelectrode studies—made him an obvious choice to run the new organization. IBRO was incorporated by a Canadian Act of Parliament in Ottawa in September 1961. Jasper would run its operations in Paris during a yearlong leave of absence from the MNI. Those who participated in the founding of IBRO referred to it as the global birth moment for "neuroscience,"[53] but, given Jasper's crucial involvement, it would be better to think of 1961 as the moment when the MNI's approach became the organizing principle for neuroscience on the global stage.

The founding of IBRO represented a rare moment of scientific internationalism at the height of the Cold War. However, neuroscience followed a very different course in Cold War America. Simultaneously, and paralleling the events in Montreal and Moscow, another independent actor also claimed the title *father of neuroscience*. The remainder of this chapter tells the parallel story of Francis O. Schmitt, his scientific career, and his work to discover a memory molecule. In pursuit of this goal, Schmitt established a wholly separate organization—the NRP at MIT—that was, in many respects, a perfect inversion of the interdisciplinary brain research advocated by Jasper and IBRO.

"Schmitty Vereins"

Francis O. Schmitt's upbringing mirrored many aspects of Herbert Jasper's; both men were born to minister fathers, were talented scientists, and were absorbed by the possibility of using laboratory science to unravel complex philosophical and even theological questions about the human mind. Yet the scientific and medical worlds they passed through would lead them to very different views of how to unify the brain sciences.[54]

Born in St. Louis, Missouri, in 1906 to a strongly Lutheran family, Schmitt was destined for a scientific career from an early age. A talented student and an actual Boy Scout, he absorbed a certain stubbornness and tenacity from his father. In an incident he recalled vividly seventy years later, he decided to bicycle the 150-mile trip to Jefferson City at age fifteen. On the journey, he encountered torrential downpours, "trusty" prisoners on work release, and poisonous copperhead snakes. He called his father several days into the trip asking to be picked up and was informed that he should try pedaling just a bit further to the next town. Schmitt eventually made it home.[55]

Schmitt was pressured by his father into a medical education, which later led him to study physiology. His training at the Marine Biological Laboratory at Woods Hole, Massachusetts, oriented him toward the microscopic structures and chemical constituents of cells (fig. 3.5). Cytology and cellular chemistry became his passion and inculcated in him a lifelong orientation toward the microworld. An early lecture he organized in 1920 by the physicist Arthur Holly Compton introduced him to the wondrous new laboratory technique of X-ray crystallography, by which the structure of molecules, including organic molecules, could be uncovered. The technique would eventually allow Rosalind Franklin, James Watson, and Francis Crick to uncover the molecular structure of DNA, and Schmitt decided to pursue "a career focused on the investigation of the functions of cells and tissues as nearly as possible at the molecular or macromolecular level."[56] This reductionist orientation continued during his graduate education. Immersed in the emerging world of molecular biology, he completed a PhD on the electric activity of heart muscle and postgraduate work at the University of California, Berkeley, and the Kaiser Wilhelm Institute of Biology, where he studied with the pioneering biochemist Hans Krebs.[57]

Returning to Washington University in 1929, Schmitt was on the cutting edge of the new technologically inflected laboratory biology. This placed him at an important position to take advantage of a critical transition in American science funding. The RF also played a crucial role in his career, although in a different way than it did for Penfield and the MNI. The Washington University Medical School and Zoology Department were creations of the RF, part of the foundation's efforts to develop scientific medicine in the American South. Schmitt's time there was defined by the patronage of Alan Gregg's replacement, Warren Weaver, who shifted the RF's priorities away from the holistic psychobiology of Gregg and toward a reductionist paradigm. Weaver's "molecular biology" would use "delicate modern techniques . . . to investigate ever more minute life processes."[58] Schmitt's research program at Washington University, which focused on the molecular constituents of the nerve cell, was a perfect fit for Weaver's agenda.

At Washington University, Schmitt developed a particular style of collaboration that characterized much of his career. He has been described as an *über*scientist, "equally the theoretician and experimentalist, model builder and biologist."[59] Much of his research revolved around new instruments, often constructed by his polymath brother Otto. However, much of the actual benchwork was done by subordinates. Schmitt played the role of intellectual leader, which took the form of somewhat dictatorial control of discussion groups referred to by participants as "Schmitty Vereins" (from *Vereine*,

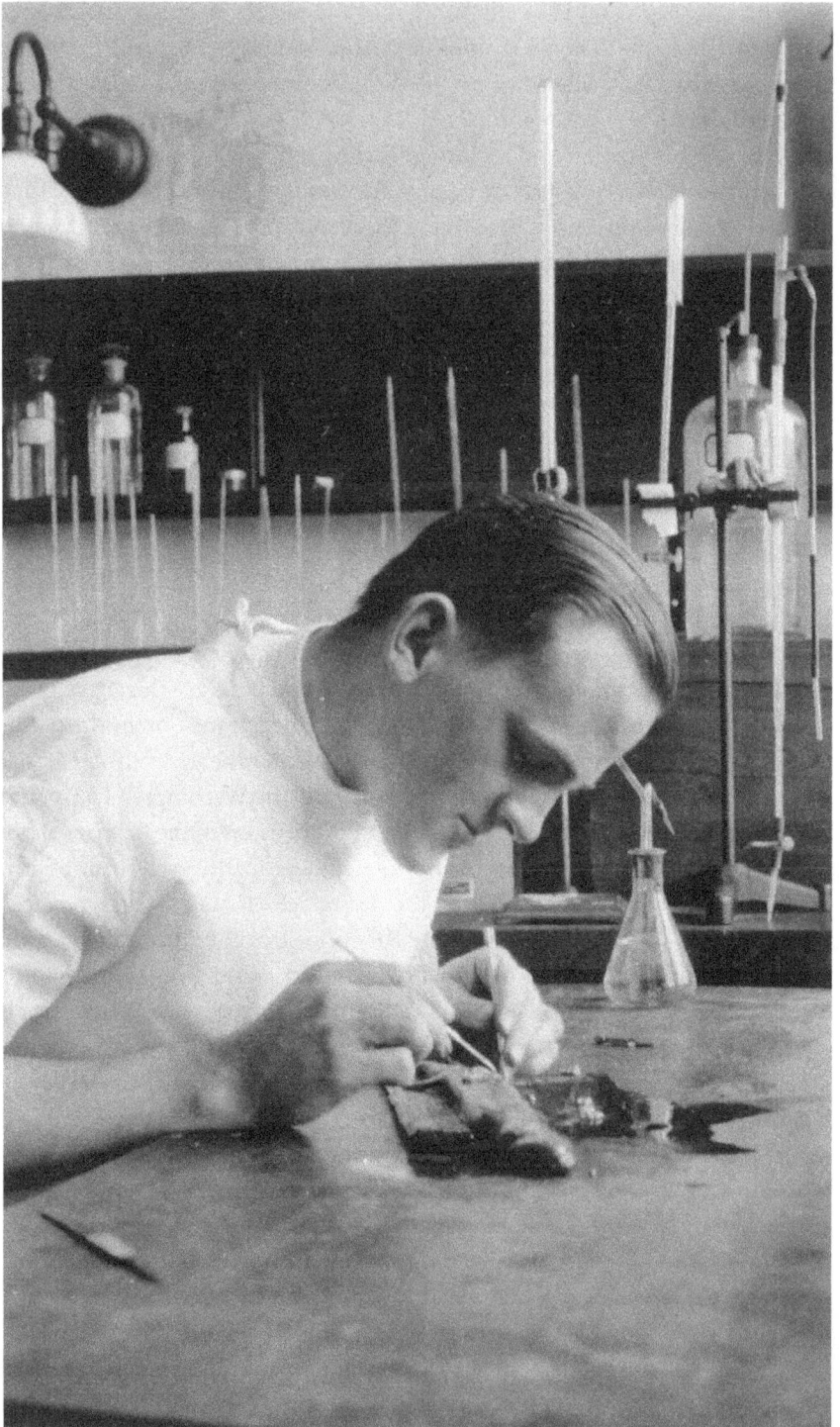

FIGURE 3.5. Francis O. Schmitt performing a dissection in a laboratory, Marine Biological Station, Woods Hole, Massachusetts, 1924. Courtesy Marine Biological Laboratory Woods Hole Library.

German for *clubs*). The contrast with Jasper's role at the MNI during roughly the same period (the late 1930s) is stark. Jasper and Penfield collaborated through shared work in the operating room, and their goal was to bring different techniques and approaches to a shared problem—epilepsy and what it could reveal about brain function. This was disciplinary hybridization. By contrast, Schmitt sought disciplinary transcendence; in his club, the disciplines of, say, embryology and zoology would be transcended by focusing on shared underlying biophysical processes.[60]

In 1940, with Weaver's backing, Schmitt was recruited by Karl T. Compton and the soon-to-be czar of American wartime research Vannevar Bush to rejuvenate and revitalize the Biology Department at MIT. He arrived in Cambridge six months before the attack on Pearl Harbor drew the United States into World War II.[61] Much as for Penfield and his team at the MNI, the war was more disruptive to biological research at MIT than it was for disciplines like physics. "The transition from peacetime to war-oriented research was quick and relatively easy for some units such as the radiation laboratory," Schmitt recalled. "However, . . . applications of basic biology and molecular biology to the wartime effort were not immediately obvious, nor had we been officially asked to attack particular research problems."[62]

Much as with the MNI, the real transformations for Schmitt would come at war's end. MIT emerged as a crucial hub of postwar technoscience—a role that only intensified as the Cold War dawned. While Jasper was consolidating his postwar role as the leader of the EEG community and an organizer of global brain research, Schmitt took on a very different role as a leader in the new hybrid field of biophysics. From this very different disciplinary style and practice, he would eventually develop a highly speculative hypothesis for molecular memory storage that ultimately formed the core of his hopes for a transdisciplinary neuroscience.

From Biophysics to Molecules of Memory

In the same decade that the MNI's interdisciplinary style cohered around memory in surgical patients, Schmitt's disciplinary style also crystallized at MIT, mainly around the progressively more complex tools of molecular biology such as X-ray diffraction and electron microscopy. Much as in his prewar years at Washington University, Schmitt served primarily as "the intellectual leader of a group of people who [did] a lot of good things," according to one laboratory member.[63] For instance, he encouraged the young scientist Betty Geren to combine X-ray analysis and electron microscopy of the myelin sheath of nerves, culminating in her crucial "jelly-roll" model of nerve myelination. This kind of laboratory triumph was typical of Schmitt's scientific style: provide intellectual encouragement for work at or near the molecular level of biology, spotting potential connections that others could then work out. As his laboratory at MIT found its feet, the resources and technological base for such work expanded. Alongside state-of-the-art X-ray equipment and electron microscopes, Schmitt built an infrastructural pipeline to fishermen in Rhode Island and eventually Chile to feed the nearly insatiable appetite of his laboratory for giant squid, perhaps the crucial model organism for experimental neurophysiology in the 1950s because of its giant axons.[64]

At the same time, Schmitt emerged as a linchpin figure in postwar science funding and management. In the early 1950s, the National Institutes of Health (NIH) tapped him to develop criteria for judging the increasing number of grant applications that seemed to fall between biology and physics. His efforts culminated in a four-week-long workshop in Boulder, Colorado, in the spring of 1958 that would set the terms of reference for the new hybrid field of biophysics and resulted in the publication of *Biophysical Science: A Study Program*.[65] "The general objective," he wrote, "was to aid and encourage the further blending of concepts and methods of physical science with those of life science in the investigation of biological problems."[66] Despite the rhetoric of collaboration on borderland problems in biology, the study program consisted almost exclusively of lectures summarizing key areas of research—little cooperative work was done. Regardless of its limited success in terms of research output, *Biophysical Science* was a publishing success, and it put Schmitt's stamp on biophysics. Hot off this success, he turned his attention to a new, grander project.

To understand the emergence of Schmitt's notion of neuroscience and how it differed from the interdisciplinary brain research at the MNI and IBRO, we must first understand his most idiosyncratic *idée fixe*. For Schmitt, the landscape of postwar biophysics was divided into two camps—the "wets" and the "dries." The wets, in his view, were biophysicists who studied the molecular constituents of the cell in its natural, aqueous environment. The dries were everyone else: those who approached biology with mathematical tools, such as cyberneticians and information theorists, but also physiologists who studied the electric properties of the nervous system. "The 'wets' were the good guys," as Schmitt's close associate Theodore Melnechuk put it. "Anybody that was interested in neurons and their molecules was a 'wet.' A 'dry' was everybody else, which was a heterogeneous bag of anatomists and physiologists, developmental guys, and certainly the computer guys. . . . It was like the Battle of Agincourt. Henry the V with a few heroic wet knights against the massive . . . dry French army."[67]

The brain was in the 1950s still primarily the property of the dries, but, if Schmitt could provide a better understanding of a "higher" function—like memory—by refocusing investigation on the internal components of the nerve cell (his métier since his days in St. Louis), it would constitute a significant victory against the dries. "It is probable," Schmitt emphasized in 1958, "that certain aspects of nerve function may involve chiefly the nerve cell or axoplasm without obvious relationship to the axon surface film. Such processes would probably not lend themselves to study or detection by bioelectric methods."[68] Indeed, the electrophysiology of memory remained mysterious. The most plausible electrical theory of memory formation was that of the neurohistologist Rafael Lorente de Nó, who thought that memory might consist of "recurrent brain circuits"—self-activating loops of neurons in the cortex.[69] There were objections (e.g., memories seemed to survive disruptive electric disturbances to the brain, such as electroconvulsive therapy), but no one had provided a more plausible suggestion. In an era before the discovery of brain neurotransmitters, the dry electrophysiologists held the high ground.[70] However, if Schmitt could proffer a more plausible alternative—one that placed memory in the neuron itself rather than in the interactions of nets of neurons—he could recolonize the nervous system for wet biophysics. His goal was to transcend biology and psychology by reducing a vital property of the mind to a complex property of macromolecules and, in so doing, inaugurate a new field that he initially called *mental biophysics*.[71]

The genesis of Schmitt's memory molecule hypothesis displayed his transdisciplinary practice at work. Schmitt possessed a fertile (even fanciful) scientific imagination and was a prodigious borrower of ideas. However, he limited his borrowing to those biophysicists and neurophysiologists who might provide evidence for his molecular memory hypothesis; in his quest to transcend disciplines, he would appropriate any fact or idea that might be useful. For example, one possible objection was that, at the time, no known chemical reaction was fast enough to be the analogue of a mental process—especially the lightning-quick process of on-demand recall. Schmitt neutralized this objection in 1959 when he met the physical chemist Manfred Eigen, whose coauthored paper "Self-Dissociation and Protonic Charge Transport in Water and Ice" seemed to show the possibility of fast proton transfer reactions in aqueous systems—the wet environments of Schmitt's wet biophysics.[72] Eigen's paper stimulated Schmitt "to a speculation . . . that neurofilaments, which course the entire length of nerve axons, might be the substratum of information processing by fast transport of elementary-charge carriers": "The still very vague idea was entertained that to account for processes like retrieval of memory . . . would, of necessity, require fast reactions."[73]

Schmitt also made prodigious use of metaphors of information storage in nucleic acids, prompted by efforts to crack the genetic code. However, his ideas were frequently borrowed from scientists considerably outside the mainstream of both genetics and neurophysiology. He became entranced by the Swedish neurocytologist Holger Hydén, whose investigations seemed to show that training and learning (in short, memory) led to alterations in RNA base pair frequency in neurons and glial cells. Also featured prominently was the controversial work of the American psychologist J. V. McConnell, who in 1955 claimed to have transferred learned memories between planarian worms through an act of cannibalism: planaria trained on specific tasks were ground up and fed to naive planaria, who then appeared to show improved performance on the same tasks.[74]

The fruit of Schmitt's transdisciplinary speculation and communication was a seminar series he organized in 1961, "Macromolecular Specificity and Biological Memory," in which he declared that "the necessary diversity and specificity demanded in fundamental life processes, particularly for encoding of experience, seems realizable only in macromolecular polymers."[75] Although it was the product of a relatively insular network of biophysicists, the book that resulted from the conference was a publishing success, convincing Schmitt that a more ambitious program was possible.[76] In early 1962, he began to make proposals to both the NIH and NASA. In his NASA proposal, he made the goal of his program clear: "Revolutionary discoveries in molecular genetics and molecular immunology demonstrate that information may be stored, processed, and read out in coded form in giant macromolecular polymers of DNA and RNA and that the genetic memory of the race and the chemical memory of the individual rest on such macromolecules."[77] Indeed, Schmitt was buoyant about the possibility of unifying the brain and mind sciences through molecular biology. "The great success story of modern life science is the proof that our heredity, the genetic memory of the species, is conveyed in a molecular code in the nucleic acid macromolecular polymer DNA," he noted. "Would it not now be profitable to inquire whether our personal memory and other components of cognitive function may also be coded in and upon the cells and networks of the brain? . . . In short, are we not ready as molecular scientists to pass from molecular genetics and molecular immunology to molecular neurology?"[78] Biology and psychology could now be transcended through a molecular approach. Schmitt's enthusiasm was persuasive, and his project, now the NRP, received a five-year grant from the NIH in 1962 (fig. 3.6).

In contrast to the activities in Montreal, which were driven by medical imperatives, Schmitt seemed most animated about the philosophical and even theological implications of his new neuroscience. Raised in a strongly

FIGURE 3.6. Francis O. Schmitt on the cover of *Modern Medicine* in 1964, in front of a diagram explaining the operations of the Neuroscience Research Program. Courtesy National Library of Medicine Digital Collection.

Lutheran family, and a practicing Christian all his life, he was preoccupied with the theological issues raised by materialism. In his 1961 MIT baccalaureate address, "Life, Science, and Inner Commitment" (delivered while organizing the NRP), he began by outlining his search for the macromolecular basis of memory. He noted, however, that this did not imply a materialistic understanding of consciousness. He sought to oppose the "malignant materialism" of the Soviet Union through a renewed commitment to Christian principles—a commitment that would be supplied by an inner spiritual sense, not dependent on any material basis, and therefore inexplicable by science.[79]

Privately, however, Schmitt agonized over the need to reconcile materialism with his belief in an immaterial spirit. In a remarkable private note written in February 1963, he made it clear that the larger importance of his molecular memory hypothesis was the very status of the human soul. He began by reviewing recent work on macromolecular assemblies. He then speculated that neurons might synthesize an antibody-like protein representing specific engrams. The recall of these engrams might involve fast transfers of protons along macromolecular structures. Schmitt continued speculating that the mind, soul, or spirit might consist in the continuing hum of these fast protonic transfers among neurons. In his estimation, the process of fast transfers that constituted the soul could exist "independent of the molecular substrate which originated or nucleated its formation." Once detached from their material substrate, these "bionic systems" might participate in a "universal, super-bionic matrix" of similar systems by virtue of their wavelike effects—a bionic "heaven" residing in a "celestial brain, giving rise to a cosmic mind consciousness": "A seeming duality (matter-energy: . . . soul or spirit) is resolved in a grand unity which may symbolize the system through which God works." For Schmitt, neuroscience would begin by unifying the brain and mind sciences through molecular memory and end by unifying the material with the spiritual, providing a scientific grounding for his inner Christian commitment: "If these speculations be vain, the evanescent product of a hyperactive scientist's mind, they will evaporate or be lost in the vigorous metabolism of science. If they have substance, they will couple with similar ideas of others and . . . reach threshold of scientific acceptance."[80] Neuroscience would transcend the brain and mind sciences and, eventually, the gap between body and soul.

A Tale of Two Neurosciences: IBRO and the NRP in Practice

By 1962, then, two wholly distinct organizations dedicated to unifying the brain sciences existed—IBRO and the NRP. For both organizations, the problem of memory was foundational. For IBRO, memory was the test case that showed the possibilities of interdisciplinary brain research; for the NRP, it was the object that would lead to a new transdisciplinary neuroscience. How would these disciplinary styles compare when they were put into practice?

The MNI's interdisciplinarity had cohered around studies of memory and memory loss, but this style was also based on close collaboration at a central site. With Jasper serving as the first IBRO secretary-general, one might imagine that this would lead to an international laboratory or clinic; this, however, was not to be. UNESCO was unreceptive to forming any kind of brick-and-mortar laboratory or institute.[81] As early as 1958, it was clear that IBRO would have to work differently. As Jasper wrote a colleague: "I do not think it advisable for us to consider establishing an international brain institute. . . . [M]ost careful workers consider it better to have such institutes under national supervision. . . . The function of our international organization . . . [would] be better for the coordination of the work of such institutes."[82] Nevertheless, IBRO retained the desire to scale up the close interdisciplinary coordination achieved at the MNI to a global level. Its initial agenda included several primary objectives, such as a fellowship and exchange program (to encourage international collaboration) and programs to provide material assistance to laboratories. To promote practical interdisciplinarity, IBRO established six panels covering different but still related disciplines—behavioral sciences, neuroanatomy, neurochemistry, neuroendocrinology, neuropharmacology, and, under a single panel, neurocommunication and biophysics. Most importantly, it facilitated the creation of special research teams in which scientists were encouraged to "go from their own laboratory to another laboratory to work for a short or longer period of time in a temporary team on a special research project." This approach was deemed "peculiarly suitable to the purpose of IBRO for such teams would usually involve a scientist from one discipline seeking the collaboration of scientists specialized in another discipline working on a common problem of brain function or structure."[83]

Despite IBRO's ambitious agenda for international collaboration, there was a notable tension in its vision—one noticed only in hindsight by an early participant: "[T]here was . . . a little confusion in the aims of the organization, between the objectives of promoting scientific internationalism and of encouraging practical interdisciplinism [sic]."[84] The special research teams experienced some success in IBRO's first decade, but the primary outcome was local interdisciplinary projects at individual labs.

Although it had been inspired by the MNI's approach, with no central clinical object to give its interdisciplinary program form and substance, scientific internationalism became IBRO's raison d'être. At the height of Cold War tensions in 1961, IBRO commenced the immense "World Survey of Resources and Needs in brain research," a survey program intended to map the world's brain research sites, surveying over forty countries, and publishing the results between 1964 and 1968. While the organization had been founded to promote interdisciplinary collaboration on an international scale, the IBRO surveys instead functioned to map the diverse social world of brain researchers.[85] Without a central object of study like memory, internationalism, rather than interdisciplinarity, became IBRO's organizing concept.

What did Schmitt's NRP look like in practice? Like IBRO, although for a different reason, the NRP did not generate a laboratory or a clinical site. According to Schmitt: "[A]nother research laboratory . . . though multidisciplinary . . . would not necessarily introduce a novel scheme of scientific communication and interaction."[86] Indeed, communication, rather than collaboration, was his primary concern. One of the earliest NRP hires was not a scientist but a journalist. Theodore Melnechuk, the associate editor of the magazine *International Science and Technology*, was brought on as the NRP's communications director in 1963 and exercised an outsized influence on the new organization. Early in the NRP's life, he suggested that the organization's primary activity should be to conduct extended meetings of top researchers at its headquarters at Brandegee House in Brookline, Massachusetts. These work sessions would then be recorded and printed as the *NRP Bulletin*. Melnechuk's journalistic connections ensured that the bulletin arrived at important institutions like the Salk Institute, and early editions were "snapped up like crazy."[87]

As Melnechuk later recalled, the purpose of Schmitt's communication strategy was less ecumenical than that of IBRO: "[T]hough Frank called it [the] 'Neurosciences Research Program,' the particular piece of progress he thought it was going to bring off in those days was the demonstration of what he passionately believed . . . that memory—psychological memory—would be found to be encoded on certain molecules of some sort, much as genetic memory was in DNA." The *NRP Bulletin* served as the mouthpiece for Schmitt's imperial goal: "The NRP, in its early days, was really a lobby for molecular neuroscience. . . . I believe Frank really thought that the mission of the NRP was to prove this belief in molecular memory."[88]

The difference in disciplinary style between the NRP and IBRO became apparent when the two groups finally encountered one another in the early 1960s. Jasper and Schmitt had met in the 1930s but had traveled along very different paths in the subsequent decades. However, they reconnected in 1961, and, in 1964, Jasper was invited to make the trek from Montreal to Cambridge to attend NRP meetings and present on a topic that had become a specialty of the MNI—memory. "It was a curious hodgepodge of people [Schmitt] brought to these workshops," Jasper later recalled. "They were sometimes not very successful because they couldn't understand each other's language. If you wanted to talk to a physicist about the functions of the hippocampus, first you had to tell him what the hippocampus is, what it does, where it's connected and so forth. We spent a lot of time, wasted a lot of time, some of us thought, teaching physicists about neuroanatomy and neurophysiology instead of getting on with our discussion."[89] Jasper was right to pick up on the disciplinary makeup of the NRP work sessions, which were dominated by physicists, molecular biologists, and biochemists.

Jasper presented the work of his colleague Brenda Milner (see chap. 2), and a discussion ensued that displayed the gulf between the interdisciplinary understanding of memory developed at the MNI and the transdisciplinary ambitions of Schmitt and the NRP. Schmitt inquired as to whether the hippocampus might be the storage space of memory. Jasper replied that Milner's work disproved that hypothesis since long-term memory seemed not to be affected by removal of the hippocampus—only the ability to consolidate new memories. One participant noted that Milner's findings made it "increasingly hard to consider memory to be a unitary mechanism": "[I]n fact, it seems to indicate a complex of essentially different phenomena." Schmitt replied: "[T]he reductionist approach, which has permitted geneticists to discover a basic mechanism underlying apparent diversity, may resolve the memory problem in the same manner."[90] For Jasper, Milner, and the MNI, memory had united their interdisciplinary enterprise, even if it had disunified the concept of memory itself; for Schmitt and the NRP, an underlying molecular mechanism might reunite memory (and unite neuroscience) in the same way that it had unified genetics and biochemistry.[91]

Despite the skepticism of more orthodox brain researchers, Schmitt and the NRP soldiered on between 1962 and 1966, intent on presenting a molecular vision around which neuroscience could cohere. Schmitt hoped to bring this vision to fruition in Colorado in 1966 with a second conference that might do for neuroscience what he had done for biophysics in 1958. According to Melnechuk: "[Schmitt] went to Boulder [in 1966] thinking it would be a one-time event like the one in 1958. . . . [T]his was going to do it for neuroscience and we were about to show the world that molecules encoded memory. And the major news at that Boulder [conference] was that . . . memory almost certainly doesn't work that way."[92] A fellow molecular biologist, Bernard Davis, finally disabused Schmitt of his memory molecule obsession. "It seems to me exceedingly implausible that in neural learning newly received information is encoded in new macromolecular sequences," Davis informed Schmitt and the rest of the Boulder participants. "I am convinced that the hope of a rapid shortcut . . . to get directly at the postulated informational macromolecules, is based on taking home the wrong message from the microbial and genetic experience."[93] Schmitt's transdisciplinary practice had failed to produce his memory molecule, and his transcendent neuroscience was dead on arrival.

IBRO, the NRP, and the SNF

By 1966, both IBRO and the NRP had failed to enact their more ambitious agendas. For IBRO, there was no obvious way to scale up the MNI's close-knit interdisciplinary environment, and there was no scientific object like memory to direct research toward. Global interdisciplinary brain research remained a chimera. For Schmitt and the NRP, the primary justification for the organization's existence—the memory molecule hypothesis—had evaporated. The brain would not be quickly unlocked with a transcendent molecular discovery. Surprisingly, however, the loss of their organizing concepts led both organizations to pivot in the second half of the 1960s in ways that would launch neuroscience onto the global stage.

Of the two groups, the NRP made the more dramatic adjustment. According to Melnechuk, Schmitt returned from Colorado dejected and disillusioned: "We got back and [Schmitt] was low. Because now his dream [had] been punctured. . . . He wanted to have a one-time Boulder, and wrap up neuroscience and then retire. . . . He would have done his life's work—solved how the brain works. But Boulder 1966 dashed that dream." Fortunately, Melnechuk's transdisciplinary practice of working groups and communication provided the solution: the NRP would transition away from being a lobby for molecular neuroscience and instead become a think tank for integrating and synthesizing approaches to brain research. Borrowing the concept of levels of analysis from Schmitt's longtime rival Ralph Gerard, Melnechuk convinced Schmitt to continue to host working groups and intensive study programs to bring together state-of-the-art work on different "levels" of brain research: "We should mark the domain of neuroscience in terms of levels, of which the molecular was one. There was the organism level, or behavior; and then there's the brain, and then the brain is made out of cells, and the cells use molecules." Melnechuk's publication savvy allowed for a relatively smooth transition; subsequent NRP study programs encompassed approaches to the brain that ranged from the molecular to the behavioral and enjoyed steady publishing success. If Schmitt's transcendent project had fizzled, Melnechuk's transdisciplinary practice had not; the NRP was especially effective at promulgating *neuroscience* as an umbrella term for the new perspective, particularly when it embraced the plural *neurosciences*: "We called it the neurosciences, because [we] had added a bunch of sciences

together—neuroanatomy, neurophysiology, neurochemistry, neurowhat-
ever, and so, [we] called it neurosciences. . . . We got a lot of grief for saying
'neuroscience.'"[94]

While the NRP transitioned to a think tank, IBRO continued the survey
work it had begun in the early 1960s; by 1968, it had surveyed the relevant
laboratories and clinics worldwide except for the United States. The survey of
American sites commenced in 1965 when Keith Cannan, the NRC's Division
of Medical Sciences chair, convened the Committee on Brain Sciences. Long
a booster and advocate for IBRO in the United States, Cannan arranged to
make the NRC Committee on Brain Sciences into the American arm of the
IBRO survey. The resulting survey detailed the activities of over four thousand
researchers and 370 laboratories and clinics. The results were published in
1968 as *IBRO Survey of Research Facilities and Manpower in Brain Sciences in
the United States*, the first comprehensive compilation of brain researchers
in the United States.[95] In addition to documenting the extent of the brain research community both in the United States and worldwide, the IBRO survey
"opened [the National Academy of Science's] eyes about the possibilities that
nobody had ever seen . . . the combination of all these different disciplines
into one neuroscience": "Out of this IBRO survey with the Academy was
born the Commission on Brain Research, . . . and from the Commission on
Brain Research was developed the Society for Neuroscience."[96] The adoption
of Schmitt's term *neuroscience* reflected the success of Melnechuk's proselytiz-
ing. Still, IBRO's organizing made the creation of the SFN possible. Founded
in 1969, the SFN rapidly grew into the world's largest neuroscience organization and, indeed, one of the largest scientific organizations of any kind.[97]

The SFN's first annual meeting in 1971 coincided with a moment of
reckoning for IBRO. An internal report submitted to UNESCO, "IBRO at
Crossroads," suggested that its very success in "promoting brain research at
the national or international level" might have created a situation "where
IBRO is no longer necessary."[98] In effect, IBRO successfully catalyzed a shared
sense of disciplinary identity among brain researchers, even without any unifying theory or discovery. Schmitt's neuroscience, which sought to seize the
brain from the diverse disciplines represented by IBRO, provided much of
the SFN's sense of mission and zeal but little of its actual practice or social
makeup. In that respect, the neuroscience embodied in the birth of the SFN
reflected an awkward amalgam of Schmitt's transdisciplinary ambitions (now
detached from his molecular goal) and IBRO's interdisciplinary organizing. If
modern interdisciplinary neuroscience was born in the crucible of the MNI's
operating room, then the founding of the SFN represented the field's matu-
ration. Neuroscience was now an established part of the modern scientific

landscape, yet it rested on an uneasy amalgamation of Jasper's interdisciplinary practice and organizing and Schmitt's armchair transdisciplinary vision.

The founding of the SFN represented the end of a genealogical inheritance of interdisciplinary brain science that began in the MNI's trading zone, passed through the global EEG community and IBRO, and eventually fused with the NRP's transdisciplinary ambitions. In this respect, the neuroscience that grew after the 1970s inherited both the MNI and the NRP's ethos and practices; however, like most children, neuroscience resembled both parents while not looking precisely like either. In the conclusion of this book, I examine the legacies of this genealogical inheritance for neuroscience and the disciplinary tensions that it contains.

First, however, I examine several remaining legacies of the MNI. Notably, the MNI's brand of interdisciplinary brain research transformed as it passed through several weak ties to the global scientific community. In part 2 of this book, we will see how the weak ties of the MNI wired neuroscience together with a range of related disciplines in the middle of the twentieth century by examining three additional actors who passed through Montreal. In some cases, these weak ties from the MNI to other areas deeply enriched disparate fields of science with new ideas and instruments. In other cases, the weak ties and the afterlives of the MNI ended disastrously. To begin this story, I first examine the career of a final psychologist who arrived at the MNI in 1937, Donald Hebb. As will be seen in the next chapter, Hebb's theory of wired-together cellular assemblies eventually carried the work of the MNI as far as possible, from very concrete clinical puzzles to the most abstract theoretical discussions of intelligence and behavior. More than any other participant, Hebb embedded the products of the MNI trading zone in psychological theory and, in so doing, helped create the new field of cognitive science.

PART II

Weak Ties and Afterlives

4

From Natural Intelligence to Artificial Intelligence:
D. O. Hebb, K.M., and the Cognitive Revolution

Alongside Molly Harrower and Herbert Jasper, a third figure arrived in Montreal in 1937–38, the psychologist Donald Olding Hebb. After an extended sojourn in the United States, Hebb, a Canadian by birth, returned to Montreal to study Penfield's surgical patients, who offered an unprecedented opportunity to observe the brain in action.

Hebb quickly became a crucial participant in the assembly of surgery, neurology, and psychology at the Montreal Neurological Institute (MNI). However, he would also chart a different path from Harrower and the other MNI fellows. Indeed, his influence would extend far beyond the walls of the MNI and shape the emergence of modern neuroscience and cognitive science as profoundly as the work of Penfield, Milner, or Jasper, although in a different way. His most significant contribution was to extrapolate the clinical findings of the MNI into the realm of theory. In doing so, he developed one of the most influential neuroscientific theories of the twentieth century and established his pervasive (if somewhat covert) influence on the emergence of neuroscience and cognitive science.

Today, Hebb is primarily remembered in connection with the eponymous "rule" or "postulate" of brain cell activity that he formulated in his 1949 masterwork *The Organization of Behavior*: "When an axon of cell A is near enough to excite a cell B and repeatedly or persistently takes part in firing it, some growth process or metabolic change takes place in one or both cells such that A's efficiency, as one of the cells firing B, is increased."[1] A hypothetical mechanism for learning at the level of the synapse—the microscopic gap between brain cells—Hebb speculated that this intercellular learning would eventually lead to *cell assemblies* of interconnected neurons that might themselves become associated with each other and form the physiological substrate of perceptions, concepts, and ideas. Cell assemblies might then trigger other cell assemblies in what Hebb termed a *phase sequence*, roughly corresponding to a train of thought. A foundational concept for the modern field of neural networks, Hebb's postulate is today a staple of textbooks in psychology, neuroscience, cognitive science, and artificial intelligence, and often summarized by the colloquial expression "neurons that fire together, wire together."

Admirers and former students have spoken of Hebb in rhapsodic terms, asserting that his work would one day be ranked alongside Darwin's theory of natural selection because of its elegance and explanatory power.[2] Indeed, he exercised a pervasive influence on the brain and mind sciences; a simple Google search reveals his book to be one of the most cited in the twentieth-century sciences (thirty-nine thousand citations according to Google Scholar as of 2023, surpassing Noam Chomsky's *Syntactic Structures* [1957], B. F. Skinner's *The Behavior of Organisms* [1938], and Clark Hull's *Principles of Behavior* [1943]). A 2002 ranking of the most influential psychologists of the last century places him in the top twenty, above many of his colleagues, mentors, influences, and rivals, including Hull, Chomsky, George Miller, Harry Harlow, Jerome Bruner, Stanley Milgram, and Ivan Pavlov.[3] At the same time, he remains a peripheral figure for historians, obscured by more charismatic and confrontational characters like Chomsky, Skinner, J. B. Watson, and others. An unassuming and modest man, he once described his work as a "crackpot theory" and remarked of his book that it had all the makings of a "classic," in that "classics" were "normally . . . cited but not read!"[4]

How could Hebb be acknowledged as of such pivotal importance for neuroscience and cognitive science yet simultaneously be so poorly understood historically? How do we reconcile two contradictory images of him: one of a towering genius who fundamentally changed our understanding of the brain, the other of an obscure toiler whose ideas were, by his own admission, neither wholly original nor unimpeachable? To answer these questions, we must place the man and his ideas in historical context and trace them to and from their origin point at the MNI. As we will see, his attempt to explain the intelligence deficits (or lack thereof) of Penfield's surgical patients drove him to synthesize a wide array of ideas and experimental findings from the worlds of experimental psychology and physiology into a general theory of brain function. However, it was through weak ties to other scientific communities that the MNI's clinical problems drove developments in both neuroscience and cognitive science. In the second part of this chapter, I trace how Hebb's students took his work out of Montreal and transmitted it to the rest of the scientific community. Through these weak ties, his theory escaped the clinic and exercised a pervasive influence on ideas about cognition, behavior, pleasure, pain, and intelligence, both natural and artificial. His story, then, is also the story of his network of students and collaborators, who extended and modified his work against the backdrop of the emerging Cold War. It also features a critical supporting character, an anonymous patient, K.M., whose brain would have a profound influence on the twentieth century.

Learning and the Switchboard Nervous System

Born in the fishing village of Chester Bay, Nova Scotia, in 1904, Donald Olding Hebb was the eldest child of two physicians, and he attributed much of his later character to his Scottish parents. A precocious child, he allegedly taught himself to read at a young age and was educated by his mother, who had imbibed the homeschooling methods of Maria Montessori. A young man of prodigious intellectual talent, he attained high marks in physics and mathematics but by the time he reached college in 1925 had developed aspirations to be a novelist. His literary ambitions brought him into contact with Freud, and his coursework in philosophy acquainted him with the experimental psychology of the early twentieth century. By the late 1920s, this acquaintance with psychology had transformed his ambitions. In 1928 he applied to do graduate work in psychology at McGill University with the Harvard-trained W. D. Tait, who accepted him under the condition that he familiarize himself with modern brain anatomy and the works of William James.[5]

What kind of psychology and brain anatomy would Hebb have learned in 1928? In James's 1890 *Principles of Psychology*, the state of the art in brain anatomy was that of the Austrian psychiatrist and neuroanatomist Theodore Meynert, who in the 1860s had mapped the notion of the reflex onto the anatomy of the brain and nervous system. American interpreters of Meynert gave the reflex a technological coloring; James and others came to think of the nervous system as akin to the workings of a telephone switchboard: sensory information entered via sensory nerves in the posterior of the spinal cord and thence into the brain; motor impulses extended from the anterior of the brain through the anterior of the spinal cord, initiating movement. In between, the mysteries of consciousness were thought to be produced by associations between these nerves, particularly those in the brain's frontal lobes (the so-called association cortex). In 1901, the Russian physiologist Ivan Pavlov modified the concept of the reflex through his work with dogs. Mechanical reflexes were now thought to be "conditioned" and capable of modification by pairing two stimuli together—for instance, the sound of a bell and the salivation brought about by the presence of food. By the early twentieth century, J. B. Watson and others had made the conditioned reflex the basis for the American school of psychology known as *behaviorism*. Watson, Edwin Ray Guthrie, E. C. Tolman, Clark Hull, and B. F. Skinner eventually placed the notion of conditioning at the center of this school of psychology—one that offered plenty of opportunities for social control and engineering in the Progressive-era United States.[6]

While learning about "switchboard" neurophysiology, Hebb also conducted his first psychological experiment, albeit in an informal setting. "The Scotch . . . have a name for the lad who sets out to be a preacher but does not make it: He is a *stickit minister*. I am a stickit novelist who came late to psychology," he would later reflect, "and a stickit reformer of elementary schools." In 1929, he became the principal of a Montreal elementary school and enacted an experiment to improve student performance. Could students be conditioned to enjoy learning like Pavlov's dogs could be conditioned to salivate? Students were told that "no one would be made to work, that work was a privilege, and that the reward for good behavior was to be allowed to stay in the classroom and work." Success was immediate but evanescent. Intelligence tests demonstrated that the failing students were, in fact, of "superior intelligence," and, for a time, test scores and schoolwork improved dramatically. Hebb referenced this experience for the rest of his career as the beginning of his interest in intelligence and learning and the relationship between the two.[7]

Hebb's belief in the malleability of human behavior was on full display in his master's thesis in psychology, completed at McGill University with Boris Babkin, one of Pavlov's disciples who had escaped Russia following the Bolshevik Revolution and been given asylum at McGill. Hebb noted: "[Babkin aimed] to provide me with a proper training, and I would be a proper psychological and North American representative of conditioning as it should be, Russian style."[8] Many of his later ideas were contained in embryo in his thesis, a theoretical treatise that tried to reconcile Charles Sherrington's notion of the "integrative action of the nervous system" (with its complex understanding of the reciprocal inhibition of spinal reflexes) with Pavlov's idea of the conditioned reflex. Hebb zeroed in on the role of the synapse as the site of conditioning; he speculated that, as impulses crossed the gap between nerves, the resistance to such transmission would decrease, creating "neural routes" that became stronger through repeated activity. Since the nineteenth century, many neurophysiologists (including a young Sigmund Freud) had speculated about how the nervous system might change in response to stimulation. However, Hebb went further, arguing that even supposedly inborn reflexes, like those spinal reflexes that maintain posture, might be learned.[9] Shortly after making such an extreme argument, however, he undertook research with Babkin and quickly began to develop doubts about Pavlovian ideas and methods.[10]

For a young psychologist skeptical of Pavlov and behaviorism and interested in the nervous system, there was only one choice for a PhD in the early 1930s. The University of Chicago was an eclectic space, embracing the Gestalt perceptual psychology of Wolfgang Köhler, the progressive psychobiology of C. J. Herrick, and the quantitative psychometrics of L. L. Thurstone, whose lectures on the nature of intelligence were particularly influential for Hebb. At the time, Thurstone was completing his magnum opus, *Vectors of the Mind* (1935), which encompassed his most thorough critique of Charles Spearman's concept of a single, heritable general intelligence, or *g*. Instead, he suggested the existence of multiple primary mental abilities such as verbal comprehension, word fluency, numerical and spatial ability, memory, perceptual speed, and reasoning.[11]

Thurstone's main rival in Chicago was the neuropsychologist Karl Lashley, a man who exercised an outsized influence on Hebb. Born in West Virginia in 1890, Lashley was by the 1930s perhaps the single most influential brain researcher in the United States. (We first met Lashley in chap. 2, where he served as interlocutor to Penfield's theory of memory.) He attained preeminence mainly through his extensive work with rats. He trained rats on a particular maze or problem box, then surgically destroyed portions of their brains to find the location of mental abilities or memories. After a decade of work, he announced that no such locations could be found. Instead, the brain was an "equipotential" organ that operated by a "mass action" principle—the whole brain worked to perform any behavior or mental function. Said functions could not be localized, and intelligence was ultimately a product of overall brain volume. Although he seldom discussed this publicly, Lashley, like Spearman, was ultimately convinced that intelligence consisted of a single factor (as opposed to Thurstone's multiple factors) and was the product of genetic inheritance.[12] In Chicago, Hebb absorbed the back-and-forth of these arguments while completing course papers arguing that the notion of the reflex, conditioned or otherwise, was ultimately inadequate to explain certain aspects of animal behavior.[13]

Hebb followed Lashley to Harvard in 1935, and his PhD research grew logically from Lashley's interest in the heritability of brain structure. Lashley wanted to know whether vision and visual perception were organized innately from birth or developed as an animal grew. To investigate this question, Hebb raised rats in total darkness to evaluate their ability to perceive and discriminate between simple shapes. This experiment cut right to the heart of a recurrent conflict between two key schools of psychology in the early twentieth century: behaviorism, which suggested that all animal behaviors were built up from "random movements and the accompanying sensory excitations," which were eventually organized into reflexes, and Gestalt psychology, which implied "the innateness of certain organisations of the visual field." Hebb raised rats to maturity within a "large ventilated light-tight box" within a dark room and concluded: "In the rat the figure-ground organization and the perception of . . . such geometrical patterns as the solid triangle. . . . are innately determined."[14] For the time being, he concurred with his mentor that much of mental life and behavior was the product of brain structures that were inherited, innate, and largely unchanging.

Hebb may have graduated with a Harvard PhD in 1936, but this did not instantly translate into a job. The academic market remained poor as the United States recovered from the Great Depression, and Hebb's choice of specialties—physiological psychology—had poor prospects: "Physiological psychology was in the middle of its long period of decline, between 1930 and 1950, and at the end of this postdoctoral year, job prospects were poor for a physiological psychologist." Indeed, American psychology was dominated by radical neobehaviorism, which eschewed study of the nervous system as pointless "neurologizing." Fortunately, Hebb's sister Catherine, who was then studying in Montreal, informed him that Wilder Penfield was looking for a few psychologists to "study the psychological status of his patients after brain operation" at the recently opened MNI. With his new wife in tow, Hebb returned to Montreal in 1937.[15]

An Extirpation Experiment

Hebb arrived in Montreal alongside Molly Harrower and Herbert Jasper in 1937 (fig. 4.1). While Harrower was comfortable working with patients, Hebb felt like a "babe in the woods in the clinical setting." Like most psychologists in the 1920s, he was more comfortable with rats than patients, who "usually keep on living and don't make their brains available for study."[16] Simultaneously, his impressive knowledge of brain anatomy and hard-nosed scientific attitude established his credibility with the MNI's neurosurgeons. "Hebb has done an enormous amount of work to help 'the psychologist at work' in the Institute," Harrower noted at the time. "He has systematically broken down weird ideas and prejudices."[17]

In Montreal, Hebb had a specific assignment: to study "the effects [on] intelligence [of Penfield's brain operations] as measured by tests of the standard kind."[18] Since operating on his sister in 1928 (as detailed in chap. 2), Penfield had grown increasingly interested in the psychological consequences of his brain operations. By the time Hebb arrived, however, it was still unclear what aspects of intelligence were affected by removing portions of brain tissue and

FIGURE 4.1. Donald Hebb as a Montreal Neurological Institute research fellow, ca. 1939. Reproduced by permission of the Osler Library of the History of Medicine, McGill University.

how to measure those effects. Moreover, Penfield's surgical approach, which used electric probing to map the functions of the cortex, was unhelpful for the frontal lobes. Since the late nineteenth century, physiologists had used electric probing of animal brains to map the functions of the cortex. Still, they were stymied by the frontal lobes, in which the electric probe produced no obvious or easily interpretable effects. As Penfield's close friend the surgeon Geoffrey Jefferson remarked in 1937: "[I]n the early days of the neurological renaissance . . . the frontal lobes seemed to select themselves as the seats of . . . intelligence by a process of exclusion. . . . [I]t was but natural that speculation should be busy with those large areas still remaining which were unresponsive to stimulation and often . . . curiously 'silent.'"[19]

Neurologists who studied frontal lobe injuries fared little better. Studies by the Czech psychiatrist Arnold Pick and the Italian neuropathologist Leonardo Bianchi identified a confusing list of symptoms and syndromes that seemed to follow frontal lobe injury, many wrapped in the impressionistic language of the clinician—*silliness, indifference, poor judgment, careless dressing, antisocial behavior.* Following World War I, the German Jewish neurologist Kurt Goldstein argued that the frontal lobes were responsible for the ability to think abstractly; frontal lobe patients displayed characteristic perseverative behavior and "concrete thinking." This confusing combination of symptoms came to be called *frontal lobe syndrome* or *frontal lobe signs.* This consensus was disturbed somewhat by the publication in 1932 of Richard Brickner's study of the neurosurgical patient Joe A., whose frontal lobes had been removed in 1930 because of a tumor. Not unlike Penfield's sister, Joe A. appeared both normal and abnormal—still seemingly intelligent but with odd personality changes, including grandiose behavior, childlike temper tantrums, and socially inappropriate behavior (he frequently made lewd sexual jokes in front of people he had just met). However, much of this work remained impressionistic; Joe A. was tested extensively with the Binet IQ and other psychometric tests but had not been tested before his surgery.[20] Before World War II, no study of the frontal lobes had anything like a proper control.

In 1928, a sixteen-year-old boy, later anonymized as K.M., was struck on the head while working at a sawmill in Nova Scotia. Six months after his recovery, he began to experience major epileptic attacks, usually starting with a feeling of being smothered, followed by "complicated visual and auditory hallucinations." He would then cry loudly and collapse to the floor in a tonic-clonic seizure, frequently biting his tongue and voiding his bowels. In other instances, he punched through doors and had developed "behavior problem[s]"—he had become "irresponsible, childishly stubborn, restless and forgetful."[21] His injury was sadly typical of the transition of Canada's forestry industry of the nineteenth century to one of modern industrial production.

K.M. came to the MNI in April 1938. Herbert Jasper localized his seizures to the left and right frontal lobes, and Penfield decided to remove both injured brain regions—a rare bilateral removal. K.M. also presented a remarkable research opportunity. His injury was discrete; his brain had not been compressed for years by a growing tumor, and, most importantly, he could be tested before and after the operation. "At rare intervals," Penfield and Hebb later reported, "chance presents to a neurosurgeon a lesion of the brain demanding treatment that would satisfy the exacting requirements of an extirpation experiment."[22] For Hebb, K.M. was about as close as a person could come to one of Lashley's laboratory rats.

Penfield operated on K.M. on April 9, 1938, but not before Hebb performed a series of psychological tests—in this case, the Stanford-Binet IQ test and a revision of the so-called army beta test, used extensively during World War I to test nearly two million American soldiers. Unlike the more qualitative testing that his colleague Molly Harrower had pursued, Hebb aimed for a straightforward quantification of intelligence. Numbers were what he was after, and, if he could get them, it would make calculating the cost and benefit of brain surgery infinitely easier (fig. 4.2).

The results of the operation were stunning. K.M.'s preoperative IQ score was 83. Two months after Penfield removed most of K.M.'s frontal lobes, he scored an average IQ of 95.

How to explain such a shocking result? The case of K.M. demanded careful interpretation. For several decades, the frontal lobes seemed to hold the secrets of intelligence and civilization. Yet as Hebb related to a colleague:

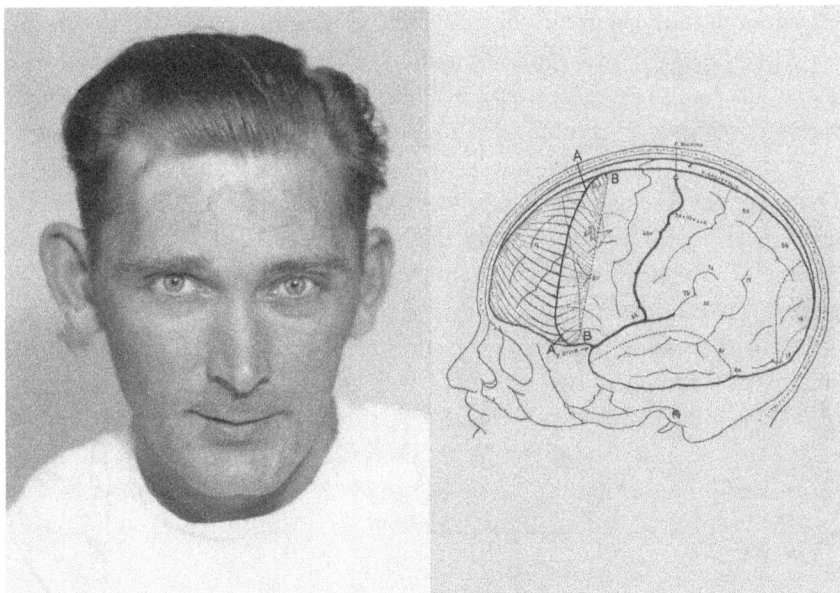

FIGURE 4.2. K.M. and the portion of his frontal lobe removed by Wilder Penfield in 1937. From D. O. Hebb and W. Penfield, "Human Behavior after Extensive Bilateral Removal from the Frontal Lobes," *Archives of Neurology and Psychiatry* 44, no. 2 (1940): 421–38. Reproduced by permission of the Osler Library of the History of Medicine, McGill University.

"[I]t is hard to exaggerate the 'normality' of the patient."[23] Moreover, by the accounts of his family, K.M. no longer suffered from bouts of anger or violent behavior; he now had "a pleasant personality": "He is no longer facetious or vulgar . . . but is considerate of the other patients and has acquired a pleasant sense of humor."[24] What had happened in K.M.'s brain?

The assembly of surgeons and psychologists at the MNI proved essential here. Hebb, Penfield, and Molly Harrower consulted on the problem of K.M. According to Harrower: "[Penfield] had placed great faith in standard I.Q. tests to show changes after removal of tumors."[25] Over time, Harrower and Hebb convinced Penfield that this was not the case. What, then, did the results from K.M. mean? Hebb and Penfield initially concluded that "human behavior and mental activity may be more greatly impaired by the positive action of an abnormal area of brain than by the negative effect of its complete absence."[26] In essence, the continued presence of damaged brain tissue—what Penfield called the *nociferous cortex*—not only deprived a patient of the psychological functions of that tissue but also interfered with the surrounding, otherwise healthy tissue. Once damaged tissue was removed, the remaining healthy brain could function normally again: "Even with the possibility that the improvement is partly due to altered outlook and fewer attacks, there could be no clearer demonstration of the detrimental effect of the presence of pathologic lesions in the brain." Hebb and Penfield were quick to add that this improvement in IQ score did not mean that there was *no* loss of intelligence with the loss of brain tissue: "It would be absurd to suppose that . . . the patient's intelligence has not been affected by the injury and subsequent operation." They clarified: "The loss of so much cerebral tissue must have had its effect." Privately, they also expressed their loss of faith in the value of existing intelligence tests themselves. Repeatedly, they noted to each other that K.M. and a handful of comparable cases demonstrated that "intelligence *as measured by current tests* is not affected by removal of tissue from the frontal poles."[27]

Notwithstanding their shared conclusions, Hebb and Penfield took slightly different lessons from K.M. For Penfield, the case proved the value of making clean excisions of tissue during surgery—in many cases, bad brain was worse than no brain. For Hebb, however, the theoretical implications were more profound and would shape the rest of his career. Intelligence tests and the very concept of intelligence needed serious revision.

The Organization of Behavior

Hebb left the MNI in 1939 and eventually rejoined his former mentor Karl Lashley, who had been appointed the director of the Yerkes Primate Laboratory in Orange Park, Florida. The puzzle of K.M.'s IQ scores continued to occupy him. "I have had a number of favorable comments on the paper on [K.M.]—I believe more firmly than ever that you made neurosurgical history when you operated there," he wrote Penfield in 1940, adding somewhat cavalierly: "The case is one that will have to be taken into account in any discussion of localization of function from now on—may you have more like it!"[28]

Hebb first tried to solve the puzzle of K.M. in a 1942 article that he later called his "peak intellectual achievement." In it, he synthesized Penfield's frontal lobe patients with studies of brain injuries in infants and children. He concluded: "[I]ntelligence itself, and not merely the ability to do well on an intelligence test, must be a product of experience." Injuries in childhood seemed to produce a "more widespread and less selective effect . . . than . . . the large adult injury." Hebb went on: "If this is so, the development and the retention of an ability may depend on the brain in different ways. An intact cerebrum is necessary for the normal development of certain test abilities, but not for their retention at a nearly normal level. In other words, *more cerebral efficiency or more intellectual power is needed for intellectual development than for later functioning at the same level.*" Intelligence was ultimately composed of two factors: "present intellectual power, of the kind essential to normal intellectual development, [and] the lasting changes of perceptual organization and behavior induced by the first factor during the period of development." These two factors were "in some sense independent of one another."[29] Intelligence—and cognition more generally—was the combined activity of the brain's various systems, interacting as they solved problems, whether on an IQ test or in everyday life. As K.M. demonstrated, the frontal lobes were vital to acquiring new skills and knowledge but were no longer as vital once those skills had been learned.

Hebb later said of his conclusions that they were a surprise: "[They] ran counter to my own assumptions at the time, as well as everyone else's." This was certainly true of his boss at the Yerkes Primate Laboratory, Karl Lashley, who had now graduated from rats to chimpanzees. As Hebb noted: "[Lashley] would study various aspects of learning and problem solving, and then I would develop measures of temperamental variables. Then he would operate on the animals' brains, we would retest them, and thus we would see what sort of brain injury affected what behavior over a broad front."[30] Hebb hoped to test his theory of intelligence formation through a small subprogram involving a chimpanzee or two, but Lashley resisted. Intelligence, for Lashley, was a single, heritable biological quantity that ultimately corresponded to overall brain mass and could not be improved through experience. In his view, the mind of man could not be bettered through education, and much of his experimental program had the implicit aim of disproving theories of learning. In the day-to-day interactions at Orange Park, his conservatism took the form of an allergic reaction to anything that smelled of Pavlovian learning or Watsonian environmentalism. Hebb's notion of intelligence as a product of interlinked, potentially separable neural systems dependent on learning was anathema to Lashley's ingrained hereditarian conception of biological intelligence. As Hebb later recalled: "When I went to Orange Park I had just published that paper of mine on early and late brain injury, and I wanted to have some controlled birth injuries [of chimpanzees] . . . to compare with the same destruction . . . made in older animals. Lashley read my paper, . . . and when I suggested the research, said 'Well, to tell you the truth, Hebb, I don't believe in that theory of yours.'"[31]

Prohibited from testing his theory of intelligence with chimpanzees, Hebb found that the animals provided a different sort of education. Five years of working with animals invigorated his criticism of behaviorism and its prohibition against discussing the mental life of animals. As he later recalled: "[I]t is unlikely that any results of brain operation [on chimpanzees] would have turned out half as interesting as the emotions and attitudes of the normal animals. . . . Mimi upset on finding a worm in her biscuit; Pan neglecting Wendy in the cage with him in favor of the young, inexperienced, but sexually receptive Soda outside the cage; Wendy properly annoyed thereat."[32] Behaviorism had banished discussion of "mentalistic" entities such as personality, emotion, or thought, but the chimpanzees at Orange Park encouraged Hebb in his conviction that behaviorism was an incomplete psychological theory, even if its methods remained useful. For Hebb, it was obvious that animals—as simple as rats and as complex as primates—were, in some fashion, thinking and that no one had yet solved the problem of thought. Intelligence was not merely the rote resuscitation of reflex behavior, conditioned or otherwise. Something more robust was needed.

In February 1944, while still at Orange Park, Hebb returned to the problem of K.M. and his frontal lobes. He realized that his initial solution—that certain portions of the brain were necessary to acquire knowledge but not to retain it—merely redescribed the problem. There was a deeper problem. How did the brain learn anything at all? To understand Hebb's answer to this question, we must first understand a key debate then occurring in American psychology over the foundations of behaviorism.

By the 1940s, behaviorism was coming under steady criticism. On one side stood radical behaviorists like Yale's Clark Hull, who insisted that conditioned reflexes could explain all behavior (including complex processes like social behavior and communication).[33] Hull ultimately sought to construct a mathematical analysis of behavior akin to Newtonian mechanics. The behavior of any organism, no matter how complex, could be described by mathematical laws; the inner life of the organism was irrelevant. On the other side stood emerging challengers like Berkeley's Edward C. Tolman. In a series of clever experiments, Tolman demonstrated the presence of "latent learning" and even "insight" in rats, which seemed to construct "cognitive maps" of mazes and transfer that knowledge without reinforcement. In different ways, he demonstrated that any psychological theory had to make space for "thinking" or "mind"—entities that could not simply be eliminated by reference to conditioning or what was coming to be called stimulus-response psychology.[34] Even rats possessed an inner life.

The other fundamental dispute was between behaviorism and Gestalt psychology. Beginning in 1910 with the work of Max Wertheimer, Wolfgang Köhler, and Kurt Koffka in Frankfurt, the Gestalt school had demonstrated that certain human perceptual processes could not be explained by atomistic notions such as the reflex. For Gestalt psychologists, the paradigmatic problem in psychology was that of perception. How could the mind work up from the atomistic pieces of perception in vision, for instance, to whole percepts of form, shape, and motion? They could not, according to the Gestaltists. The mind possessed intrinsic organizational principles that could be demonstrated through careful experimentation. Others aimed to show that the behavioral theory of learning—as a process of conditioning—was inadequate; like Tolman, the Gestaltists demonstrated that certain animals displayed a form of "insight" akin to human thought. By the early 1940s, Gestalt was the principal opponent of the then-dominant behaviorism in American psychology.[35]

The Gestalt psychologists found an ally in Hebb's mentor Lashley, whose studies seemed to show the inadequacy of reflex theories of learning. "Lashley and the Gestalters," Hebb recalled, "had abolished connectionism; the [behaviorists] no longer defended it but had withdrawn into the never-never land of the empty organism, miniature systems, and as-if theory. . . . In short, connectionism had no defenders, and the psychological world had tacitly conceded the argument that such explanations of behavior were not possible."[36] In his attempt to understand and explain K.M.'s IQ scores, Hebb realized that the true problem was the switchboard model of the nervous system that psychology had inherited from the nineteenth century. An alternative was needed.

Hebb drew on the work of the Spanish neuroanatomist Rafael Lorente de Nó, who used refined staining methods in the 1930s to examine the microanatomy of rabbit brains. De Nó made two significant discoveries. First, certain cortex areas contained neurons that could function in closed chains rather than merely as a "through-transmission, sensorimotor mechanism."

The brain was, then, "capable of purely internal activity" at the cellular level.[37] For Hebb, these reverberating loops of neurons might fire themselves for brief periods, independently of sensory input, acting as the physiological underpinnings of thought. Second, Lorente de Nó discovered that, while one neuron was incapable of firing another on its own, often the axons of multiple neurons connected to the dendrite of a single neuron. If more than one impulse arrived at a neuron *simultaneously*, those multiple presynaptic neurons might fire the subsequent neuron. In effect, this selective firing solved the problem of *set*—the preexisting organization of the nervous system into which sensory input might flow and out of which different motor responses might emerge. No longer a simple switchboard, the nervous system was, thanks to Lorente de Nó, an infinitely more complex interconnected net that could not be explained in simple stimulus and response terms. At the same time, Hebb's argument was strengthened by the work of his MNI colleague Herbert Jasper, whose electroencephalograph revealed that, far from being a simple sensory-response machine, the brain was always abuzz with internal activity. Rather than merely a series of stimulus and response reflexes, the behavior of any organism was organized by its ongoing internal neural activity.

The psychological implications of this updated neurophysiology—vastly more complex and flexible than William James's switchboard nervous system—were profound. The behaviorism of Watson and Hull, which depended on the conditioned reflex, now seemed unsustainable, even naive. The radical behaviorism of Skinner, who eschewed any attempt to explain behavior by hypothetical brain mechanisms ("physiologizing"), now also seemed passé. Indeed, for Hebb, the problem of K.M.'s postoperative IQ test had, in effect, forced the physiological issue: "In my worrying about these problems . . . I had a very definite advantage, being forced to physiologize by the need of an explanation for the brain-injury data."[38]

The capstone of Hebb's theory was a simple and elegant hypothetical model of thinking, learning, and perception—the *cell assembly*. Building on Lorente de Nó's idea of reverberatory circuits, Hebb postulated that these chains of neurons might constitute the physiological substratum of a simple perception or concept. However, he went further. If the reverberation persisted, it might induce a lasting change at the level of the synapse—the microscopic gap between the axon of one neuron and the dendrite of the next. As Hebb explained: "When an axon of cell A is near enough to excite a cell B and repeatedly or persistently takes part in firing it, some growth process or metabolic change takes place in one or both cells such that A's efficiency, as one of the cells firing B, is increased."[39] In essence, Hebb transplanted the notion of the conditioned reflex to the cellular level but allowed room for the resulting brain activity to be formed either by sensory input or by internal activity of other brain cells. A cell assembly, formed through the process of learning from sensory input, might trigger the firing of a different cell assembly, a process Hebb referred to as a *phase sequence*, which roughly corresponded to the colloquial notion of a "train of thought" (fig. 4.3).

The cell assembly, Hebb argued, might also mediate between behaviorists and Gestalt psychologists. How did atomistic bits of, for instance, visual sensation—points, lines, light, shade, color—result in perceptions of form? Moreover, how was this accomplished in a brain that functioned by conditioned neural connections? "The idea that one has to learn to see a triangle," Hebb conceded, "must sound extremely improbable."[40] Nevertheless, he explained how it could be so. A young child, for instance, might learn the form of a triangle first by fixing her eye on a single point of that triangle, causing a series of cells in the visual cortex to form a brief, reverberatory cell assembly. As the eye passed from one point to another in the triangle, the cell assemblies excited by these sensory inputs would be associated together by the multisynaptic connections to brain cells deeper within the cortex. The perception of the form of a triangle (and the ability to generalize to other types of triangles—large, small, inverted, and incomplete) rested not on the excitation of any one cell or even on any group of cells but rather on a network of learning cells diffused throughout the cortex, a network that was difficult to eradicate, even in the case of a large brain injury. Moreover, once established, the cell assemblies and phase sequences might occupy different portions of the brain from those initially necessary to establish the concept, skill, or idea.

As Hebb was quick to point out, the cell assembly, with its multiple connections and redundancies, might explain a long-observed fact in brain damage, the recovery of function: "The assembly is thought of as a system inherently involving some equipotentiality, in the presence of alternate pathways each having the same function, so that brain damage might remove some pathways without preventing the system from functioning, particularly if the system has been long established."[41] At last, Hebb had a plausible explanation for how patients like K.M. could have so much brain tissue removed yet still make such strong showings on an intelligence test. For Hebb, the learning cell assembly reconciled clinical and laboratory evidence with the phenomenological experience of thought, the existence of an independent mind, and the perplexing results of Penfield's surgeries.

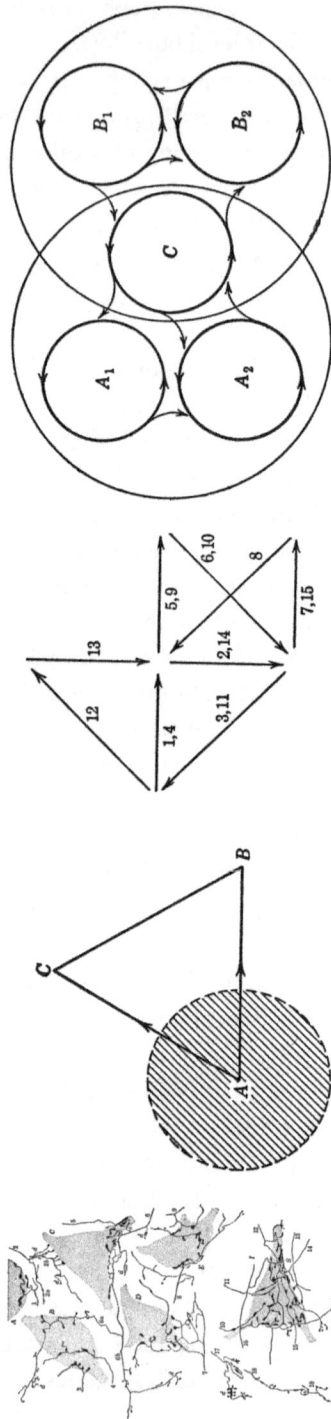

FIGURE 4.3. Donald Hebb's theory of the cell assembly. From left to right: Lorente de Nó's microanatomy of the nervous system, from which Hebb derived the notion of rever-berating neural circuitry; the hypothetical movement of eye focus along the edges of a triangle, which might lead to the creation of a "triangle" cell assembly; Hebb's theoretical cell assembly model, which might embody perceptions and ideas in neural activity, with arrows indicating reverberating circuitry; the interaction of multiple cell assemblies, correspond-ing to Hebb's notion of a "phase sequence." Donald O. Hebb, *The Organization of Behavior: A Neuropsychological Theory* (Wiley, 1949), 64, 73, 85, 130.

Behaviorism and Its Discontents

While Hebb would be the first to concede that no individual element of his theory was wholly original, its broad synthesis of a vast array of neurophysiological and psychological data to relate mind, brain, and behavior made *The Organization of Behavior* a foundational text for the cognitive revolution.[42] Hebb began writing a manuscript—initially titled *On Thought and Behavior*—in 1944. He completed a draft of the early chapters in 1946 and sent it to his Orange Park colleague Henry Nissen for comments. "Over and over again," Nissen noted on reading the manuscript, "at least once per page, I have whispered or shouted, to myself, 'Good,' 'Damgood,' 'well said,' etc. etc."[43] Hebb appreciated Nissen's positive response, which stood in stark contrast to that of Hebb's mentor Karl Lashley. "For all I knew," Hebb responded to Nissen, "[the manuscript] might stink in your nostrils as in [Lashley's]—he told me you know that the whole thing was very weak, with no value because it was so vague."[44] Hebb had shown an early draft to Lashley, proposing coauthorship. In response, Lashley returned two pages of detailed notes and then refused not only coauthorship but also to write any kind of preface or forward to the book.[45]

What was the nature of Lashley's objection? According to Hebb, the crux of his and Lashley's disagreement was over the importance—even the existence—of learning. "[Lashley] didn't really approve of me," Hebb recalled, "because he never got over the fact that when I first went to study with him I had written a paper that explained everything (including unconditioned reflexes) in terms of learning. . . . For the rest of his life, he was convinced that I was a thorough-going Pavlovian or Watsonian empiricist; I'd persuade him differently in one of our lunch-time discussions, but next week he'd have forgotten."[46] Indeed, Lashley's emphasis on the innateness and heritability of brain structure constituted the leitmotif of his entire career. In his classic article "In Search of the Engram" (1950), Lashley reviewed the results of his lifetime of research: "I sometimes feel, in reviewing the evidence on the localization of the memory trace, that the necessary conclusion is that learning just is not possible. It is difficult to conceive of a mechanism which can satisfy the conditions set for it."[47] Hebb's cell assembly provided a plausible mechanism for learning at the cellular level and further suggested that Lashley's most persuasive retort to connectionism—the holistic nature of perception—was a mirage. The whole triangle—or other percept, thought, or idea—really could be the sum of its parts if those parts were organized by an underlying learning mechanism.

Jealousy may also have been one of Lashley's motives. By the 1940s, Lashley was one of the leading opponents of behavioral psychology in the United States and was frequently invited to speak to groups of like-minded discontents. One crucial meeting occurred in 1948 at the California Institute of Technology, a few years after Lashley had read Hebb's manuscript. "Cerebral Mechanisms in Behavior," later referred to as the "Hixon Symposium," included speakers such as John von Neumann, Warren McCulloch, Ralph W. Gerard, and other major figures in neurophysiology and the emerging field of cybernetics. Lashley's address, "The Problem of Serial Order in Behavior," became a battle cry for those on the American scientific landscape who chafed against the philosophical constraints of behaviorism. In crisp and provocative language, Lashley precisely delineated several problems in psychology that seemed inexplicable by the switchboard nervous system model: language, the coordination of leg movements in insects, the songs of birds, the control of trotting and pacing in a gaited horse, the rat running a maze, the architect designing a house.[48] Indeed, many of the themes he enunciated were similar to those discussed in Hebb's manuscript—the inadequacy of reflex theory, the constantly active nature of the central nervous system, the notion of the organism as something more than a simple stimulus-response machine.[49]

Yet Lashley's proposed solution was to abandon reflex theory entirely. In its place, Lashley borrowed from nineteenth-century physics and proposed that the brain's functions were brought about not by the molar functions of neurons but by the interaction of electric wave patterns. He used a metaphor to describe this:

> I can best illustrate this conception of nervous action by picturing the brain as the surface of a lake. The prevailing breeze carries small ripples in its direction, the basic polarity of the system. Varying gusts set up crossing systems of waves, which do not destroy the first ripples, but modify their form, a second level in the system of space coordinates. A tossing log with its own period of submersion sends out periodic bursts of ripples, a temporal rhythm. The bow wave of a speeding boat momentarily sweeps over the surface, seems to obliterate the smaller waves yet leaves them unchanged by its passing, the transient effect of a strong stimulus.[50]

Lashley's wave theory may have been poetic, but ultimately it was too esoteric and vague. It also did not suggest any obvious experimental strategy. By contrast, Hebb's model allowed one to infer the nervous system's activity from an organism's behavior. The experimental strategies of behaviorism—rat mazes, for instance—could be used to probe the activity of the thinking brain. Hebb's model could function as a bridge from behaviorism to a psychology that grounded the study of mental life in neurology. Lashley had long argued for a "behavioristic analysis of consciousness."[51] Hebb had provided a way to do precisely that, yet he implied the existence of a plastic, learning nervous system, and it seems likely that Lashley resented him for it.[52]

After five years at the Yerkes laboratory, Hebb returned to Montreal in 1947 with a completed manuscript of *The Organization of Behavior*. The war had ravaged McGill's Psychology Department, and Hebb was hired to revive it, with an emphasis on experimentation. This arrangement was ideal since Hebb "expected that years of research with graduate students would be needed to get a hearing for the book once it did appear."[53] This proved unnecessary. Prepublication copies of the book circulated to key figures, including E. G. Boring, the well-known historian of psychology, who was particularly well-placed to appreciate the importance of Hebb's work. "The book has a fresh, constructive candor, which is what is needed," Boring wrote Hebb in 1950. "Lashley and Köhler with their past commitments are not able to do this thing that you, coming freshly to the field, can do. . . . I [have run] into great enthusiasm about the book."[54] This sentiment was best captured by an early review of *The Organization of Behavior* when it appeared in 1949: "There is no need to start this review with any assertion of the importance and high quality of this book. It is like the case of a movie which needs no advertising because people pass the word from one to another of 'Have you seen the picture at . . . ?'"[55]

Hebb's "bold synthesis," as a key figure in the cognitive revolution in psychology later described it,[56] not only provided a plausible neurological model for thought; it could also serve as a spur to research. Indeed, in the reverse of what Hebb expected, his book attracted students, rather than his students attracting attention to his book. His seminars attracted young students like Brenda Milner, whom we met in chapter 2, and discussions among the students "often continued late into the night."[57] In the subsequent decade, the Psychology Department at McGill functioned under Hebb as a liminal space for scientists interested in studying the mind, the brain, and behavior as a single, continuous entity. Born in his attempt to solve a clinical puzzle at the MNI, Hebb's neuropsychological theory had evolved into a general theory of brain function and intelligence that would travel through weak ties through the Cold War brain and mind sciences, where it would play a key role in the emergence of artificial intelligence, new theories of pain and pleasure, controversies over brainwashing, and debates about race and intelligence. In the remainder of this chapter, I trace those weak ties.

Sensory Deprivation

Hebb knew that his theory was speculative. Further research would certainly be necessary to substantiate it: "The [cell] assembly did not follow logically from the neurological evidence; on the contrary, its specifications put a heavy strain on the evidence, and only the known existence of delayed response, expectancies, imagery, and so forth made the argument even remotely plausible."[58] Sensation and perception were crucial links in Hebb's theory, and this suggested an experimental program: if perception was organized by the interaction of experience and the constantly active brain, what would happen if sensory input was shut off? Hebb's students studied the actions of children's eyes as they scanned letters and pictures and raised rats and dogs in isolation to determine if this more limited sensory environment affected their vision and problem-solving ability.[59]

At the same time, the emerging confrontation with the Soviet Union offered Hebb and his students the opportunity to move beyond rats and dogs. As the Iron Curtain divided Europe in 1948, the newly formed CIA canvased Western universities for weapons in the coming Cold War. Unlike the previous war, won by chemistry, physics, and engineering, the Cold War would be a battle for minds, both broadly, in the form of propaganda, and more concretely, in the form of individual spies, informants, and prisoners. The Cold War would be the psychologists' war, one that, in the late 1940s, the West already appeared to be losing. CIA leaders noted the relative ease with which the Soviet Union seemed to extract public confessions, and public events like the conversion of the Hungarian Catholic leader Jozsef Mindszenty in 1949 and the bizarre behavior of American soldiers held captive during the Korean War convinced many American intelligence officials that the Soviets and their allies had mastered a form of mind control (possibly involving hypnosis or Pavlovian conditioning). By early 1951, agents of the American and Canadian military and intelligence apparatuses were actively seeking a defense against so-called brainwashing.[60]

In June 1951, Hebb took part in a meeting at Montreal's Ritz Carlton hotel. In attendance were representatives of the Canadian Defence Research Board (DRB), along with two unnamed Americans, later identified as CIA agents. The purpose of the meeting was to discuss whether Korean War prisoners of war had been brainwashed and, if so, what defenses might be employed. Hebb proposed that one route to understanding the brainwashing procedure might be research into what he called *sensory deprivation*. By "cutting off all sensory stimulation . . . , the individual could be led into a situation whereby ideas, etc. might be implanted."[61] Evidently, Hebb's proposal was compelling.

Shortly after the meeting, he was awarded DRB contract X-38 to investigate the technique.

Historians would later debate whether Hebb was an enthusiastic Cold Warrior.[62] In fact, sensory deprivation in animals had been part of his research since the 1920s, and it was the logical next step in testing his neuropsychological theory, which suggested that continuous sensory input was necessary for normal functioning of the intact brain. The absence of sensory input would lead to disorganization of thought patterns and emotional disturbance.[63] The temptation to conduct an otherwise impossible experiment with full financial backing proved irresistible. Indeed, in the early fog of the Cold War, it is unclear who was using whom. Hebb persuaded the DRB to support his research but noted to a Rockefeller Foundation contact: "I have found it very difficult indeed to get money simply for the purpose of trying to learn a little more about how behavior is determined. To get what we have now I have had to perjure myself to some extent, by letting on to be optimistic about possible practical applications."[64] It seems entirely plausible that Hebb exaggerated the value of sensory deprivation research to the DRB to test one of the inferences of his neuropsychological theory.

Hebb and four graduate students began the sensory deprivation project in 1951 with a two-year grant from the DRB. Twenty-two male McGill students were paid $20.00 a day to lie on a bed in a cubicle in the basement of McGill's Donner Building. They wore translucent goggles that diffused incoming light into a blur, padded gloves with cardboard cuffs to prevent tactile sensation, and U-shaped foam-rubber pillows over their heads to dull sounds. They were free to leave if they chose to and permitted to eat at the bedside and use the bathroom if needed (fig. 4.4).[65]

The outcome of the experiments varied from the remarkable to the banal. Subjects typically began by napping for several hours: "Later they slept less, became bored, and appeared eager for stimulation. They would sing, whistle, talk to themselves, tap the cuffs together, or explore the cubicle with them."[66] The main impediment to collecting data was the sheer stultifying dullness of the activity itself; students often left before the twenty-four hours were up. Those who remained did report some intriguing experiences: "Nearly all . . . reported that the most striking thing about the experience was that they were unable to think clearly about anything for any length of time and that their thought processes seemed to be affected in other ways." Cold War concerns merged with Hebb's more abstract interests as the subjects were given different tests of intelligence and problem-solving before, during, and after the experiment. They were also given a task that was meant to test susceptibility to propaganda. During isolation they listened to a record of "a talk arguing for

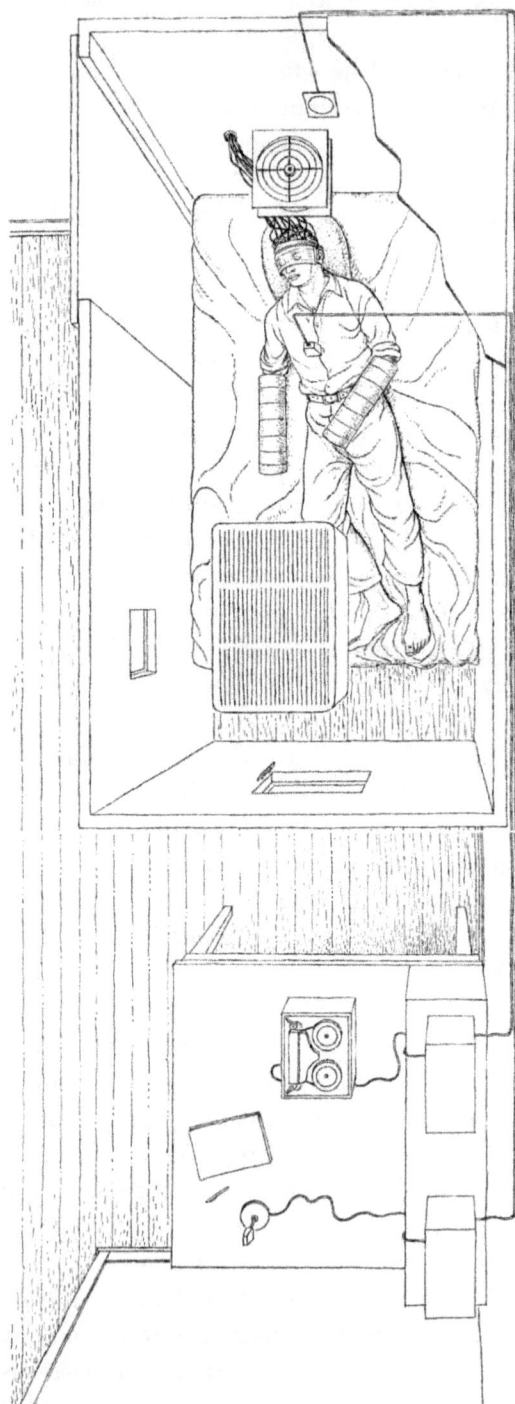

FIGURE 4.4. The experimental setup for Donald Hebb's sensory deprivation experiments. Woodburn Heron, "The Pathology of Boredom," *Scientific American* 196, no. 1 (January 1957): 52–53.

the reality of ghosts, poltergeists and other supernatural phenomena." They reported a slight increase in their belief in such phenomena, with some noting that they found themselves afraid of ghosts for days afterward.[67]

More disturbing were the hallucinations, discovered first by one of the study's overseers who also acted as a participant. Subjects reported seeing images—dots, lines, and geometric forms that soon became images of trees, babies, and "rows of little yellow men with black caps on and their mouths open." More complex scenes emerged: "a procession of squirrels with sacks over their shoulders marching 'purposefully' across the visual field, prehistoric animals walking about in a jungle, processions of eyeglasses marching down a street." Sometimes, the experience became disturbing: "Some reported feelings of 'otherness' or 'bodily strangeness'; trying to describe their sensations, they said, 'my mind seemed to be a ball of cotton wool floating above my body,' or 'something seemed to be sucking my mind out through my eyes.'" For those who made it to the end of the maximum period of ninety-six hours of isolation, the experience seemed to produce a state not quite like dreaming, not quite like wakefulness, and left many confused, disoriented, and occasionally disturbed.[68]

What was the import of this impossible experiment? Hebb presented early results to a secret meeting of the DRB in 1952 but was barred from publishing them. Until the study ended in 1955, he implored the DRB to change its decision, both because he feared misunderstandings should the study leak to the public and because he considered the results to be of great theoretical significance.[69] His concerns were prophetic. The sensory deprivation experiments were one of the worst-kept secrets of the Cold War. The Montreal press caught wind of them, and headlines such as "McGill Students Brain-Washed!" confirmed Hebb's worst fears about misinterpretation. Limited publication of the results began in 1954 with a thinly veiled cover story about research on the effects of prolonged boredom on radar operators and truck drivers.[70]

Why did Hebb consider the results so significant? In a 1954 talk to the American Psychological Association entitled "Drives and the CNS (Conceptual Nervous System)," he made it clear. In line with his interest in the neural basis of intelligence, which had preoccupied him since his studies of K.M., he was most impressed by the effects of sensory deprivation on problem-solving and concentration. When those results were combined with the simultaneous data on perceptual isolation in dogs and rats, he concluded that the main takeaway from these related studies was the importance of children having a sensory-rich environment early in life to ensure later educational success.[71] As will be seen below, these conclusions formed crucial intellectual points of departure in policies related to education, race, and intelligence.

Sensory deprivation, however, would not sit still and soon escaped the basement of McGill's Donner Building. Through different conduits, it found its way into the National Institutes of Health laboratories under the oversight of the MNI veteran Maitland Baldwin and the charismatic physician and psychoanalyst John C. Lilly. Baldwin introduced it into the heart of Cold War counterintelligence practices, conducting "terminal experiments" by locking volunteers into sensory deprivation contraptions until many became so distraught they kicked their way out. At the same time, Lilly viewed it as a technology that could aid in the quest for spiritual enlightenment. In this respect, it joined other mind-altering technologies like LSD that made the trek from the laboratory to the intelligence world to the counterculture in the 1950s. More perniciously, Hebb's results were almost immediately taken up by the CIA and incorporated into interrogation techniques that were tantamount to torture. When the CIA produced its infamous KUBARK counterintelligence manual in 1963, sensory deprivation, drawn from the Montreal studies, constituted a primary technique.[72]

Within academic psychology, sensory deprivation also proved highly productive; by 1961, over 230 articles on the subject had been published in psychology and related fields. That same year, a symposium was held at Harvard to mark the ten-year anniversary of Hebb's initial studies. The title of an earlier symposium—"Sensory Deprivation: Facts in Search of a Theory"— seemed to get the story's direction precisely backward. Sensory deprivation had emerged from Hebb's theory and produced facts that were multiple in their application. For his part, Hebb was somewhat unimpressed by the flurry of studies. "There are times when I wish I had never heard of perceptual deprivation or isolation," he informed the symposium in his opening remarks. "I had something to do with it . . . and consequently seem expected at times to explain the results. I am incapable of doing so."[73] For him, the most salient fact about sensory deprivation was that it disorganized the brain and the mind. Those who hoped to find order and meaning in that disorganization would have to look elsewhere.

Pleasure

The sensory deprivation experiments and the Cold War fears with which they intersected attracted widespread public attention to Hebb's work. Within the more cloistered world of psychology, however, the work of Hebb's students James Olds and Peter Milner scored the first significant research win for his approach. In 1953, the two men performed a series of iconic animal experiments that led to the discovery of so-called self-reinforcing brain stimulation—a finding that would transform how the concept of pleasure was understood. At the same time, the work of Olds and Milner proved iconic in two areas. Within the scientific world, it functioned as a proof of concept for Hebb's program to reconcile psychology and neurology. At the level of popular understanding, however, it was a harbinger of the promises and perils of the new brain science. While it opened new horizons for neuroscience, it was also the science of which brave new worlds were made.

By the early 1950s, Hebb's band of graduate students had added two key members. The first was Peter Milner, husband of the neuropsychologist Brenda Milner, whom we met in chapter 2. Originally trained as a physicist and electrical engineer, Peter Milner had arrived in Montreal after World War II to work on the newly opened nuclear reactor at Chalk River. Brenda introduced him to Hebb's book while it was still in manuscript form, and he was impressed enough to abandon physics to retrain in psychology with Hebb. His training in electrical engineering would eventually prove crucial (as it also would for his collaboration with IBM, discussed below) and was also quickly augmented by training at the MNI in the use of recording and stimulating brain electrodes.[74]

The second key player was the American psychologist James Olds, although to call him a *psychologist* did not do justice to the breadth of his interests. Born in Chicago in 1922, in the 1940s Olds studied psychology at Harvard, where he worked closely with Talcott Parsons, perhaps the most influential American sociologist of the twentieth century. Attracted to Hebb's circle by *The Organization of Behavior*, Olds arrived in Montreal in 1953.[75]

Olds and Milner immediately hit it off and began to work together on Milner's current research program, which involved using depth electrodes to investigate the reticular activating system (RAS) in cats. The discovery by Giuseppe Moruzzi and Horace Magoun of the RAS (which Penfield and Jasper called the *centrencephalic system*, as discussed briefly in chap. 3) presented a crucial research agenda for neurology. In 1953, the subcortical structures of the brain stem and diencephalon appeared to be critical for states of consciousness, arousal, and attention. Hebb had initially argued that neural pathways themselves might function as "a crude analogy for attention . . . like a process in the brain that opens one afferent pathway, leaving others blocked."[76] However, the discovery of the RAS simplified the issue considerably. If there was a single brain structure that mediated consciousness and attention, and if that structure in rats could be stimulated with an electric probe, it would provide a valuable experimental attack on issues of learning and intelligence. Milner began implanting stimulating electrodes in the brain stem structures of rats in the early 1950s, running them through different mazes and experimental situations to see whether he could affect their actions. What happened next, however, proved even more impressive than a remote-controlled rat.[77]

Olds and Milner's discovery of the brain's reward circuitry has been presented as a happy laboratory accident akin to the serendipity that produced penicillin or Bakelite. Indeed, there was a certain degree of luck in what happened next. Olds and Milner undertook a new set of experiments to stimulate the RAS of rats with depth electrodes. Next: "[A rat] was placed in a large box with corners labeled A, B, C, and D. Whenever the animal went to corner A, its brain was given a mild electric shock by the experimenter. When the test was performed on the animal, . . . it kept returning to corner A. After several such returns on the first day, it finally went to a different place and fell asleep. The next day, however, it seemed even more interested in corner A." Soon, Olds and Milner discovered that they could lure the rat to a different area of the box by shocking it when it approached area B: "After this, the animal could be directed to almost any spot in the box at the will of the experimenter."[78] They had not been able to produce such an effect previously, so they had the rat X-rayed and discovered, to their surprise, that the electrode had landed not in the RAS but rather in the adjacent septal area of the brain (fig. 4.5).

While the electrode landing in the septal area of the brain was an unpredictable accident, it is worth noting that two contextual factors shaped Olds and Milner's interpretation of their results. The first was their intellectual backgrounds. Milner's earliest experiments aimed to use electric stimulation of the rat's brain to test aspects of Hebb's overall theory of learning and intelligence. Olds, meanwhile, was primed to see reward and motivation because of his time studying with Talcott Parsons, who had made the issue of motivated

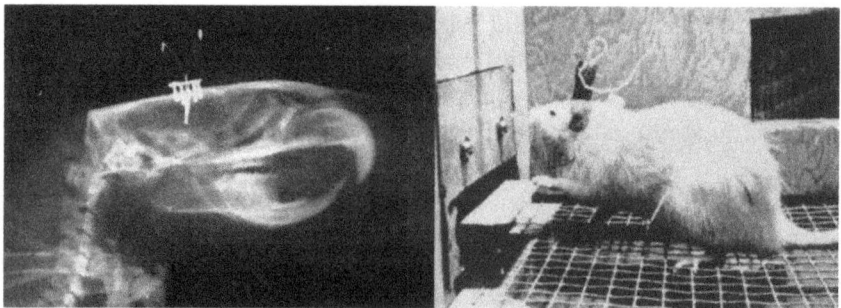

FIGURE 4.5. James Olds and Peter Milner's rat, X-rayed to show the position of the stimulating electrode in the septal region of the brain (*left*). A rat self-stimulating in a Skinner box (*right*). James Olds and Peter Milner, "Positive Reinforcement Produced by Electrical Stimulation of Septal Area and Other Regions of Rat Brain," *Journal of Comparative and Physiological Psychology* 47, no. 6 (1954): 419.

behavior a key one for sociology. As Milner later put it: "There was nothing accidental about this discovery, except the location of the electrode. It is almost certain that Olds would have found a reward site sooner or later because that is what he was primarily looking for."[79]

More important and more revealing than their intellectual backgrounds, however, was how Olds and Milner convinced the world that their results were true. Here, Olds's contribution was crucial. To the myriad other devices in the Montreal trading zone he introduced the iconic device of American behavioral psychology: the Skinner box. The eponymous device was a key tool in American psychology laboratories and carried the entangled theoretical notions of "reinforcement" and "operant conditioning." Olds, well-versed in the culture of American behaviorism, proposed that the best test of their discovery would be to place a rat in a Skinner box and give it the opportunity to stimulate its own brain. Here, the participants traded the materials of their respective laboratory cultures; Olds explained the theory behind the Skinner box, Milner translated this theory into an actual device, and the pair developed an experimental technique to gather data. The results from this hybrid practice were impressive: "The first animal in the Skinner box ended all doubts in our minds that electric stimulation applied to some parts of the brain could indeed provide reward for behavior. . . . Left to itself in the apparatus, the animal . . . stimulated its own brain regularly about once every five seconds, taking a stimulus of a second or so every time."[80] Moreover, when the current was shut off, the response rapidly vanished. Once the results had been replicated with several rats, the discovery was beyond doubt and ready for publication under the title "Positive Reinforcement Produced by Electrical Stimulation of Septal Area and Other Regions of Rat Brain."[81]

Olds and Milner relayed their results in the language of behavioral psychology, which no doubt made them more palatable to the American psychological community. Yet a certain conceptual slippage soon crept into discussions of self-reinforcing brain stimulation. News of the discovery spread quickly, unencumbered by Cold War secrecy (as the sensory deprivation work had been). The experiments were recounted in a flashy 1956 *Scientific American* article, "Pleasure Centers in the Brain," that frequently conflated the drier (but more accurate) concepts of self-reinforcement and reward with the more provocative word *pleasure*.[82] As an advertisement for the power of the emerging neurosciences, nothing could have served better than self-reinforcing brain stimulation. Indeed, Olds made the possibilities for social engineering clear in a 1958 *Science* article, "Self-Stimulation of the Brain": "[I]t is reasonable to hope that eventually it will be possible to control the reward systems pharmacologically in cases where behavior disorders

seem to result from deficits or surfeits of positive motivation."[83] In the next
decade, others looked to Olds and Milner's work as they tried to understand
the implications of the neurosciences for society. In 1955, the psychologist
Abraham Maslow speculated about the consequences for civilization if
pleasure could be achieved by simply "plugging oneself into a nearby Olds-
intermittent-stimulator socket." Elsewhere, the neurophysiologist John C.
Lilly thought that continual stimulation of the "pleasure center" could create
a society of beatific calm and nirvana.[84] More disturbingly, Olds and Milner's
pleasure centers became fodder for science-fictional visions of the brain control of deviant social behavior.[85]

Yet, beyond the science-fictional visions of Cold War America, the ultimate impact of Olds and Milner's work was on the neurosciences themselves.
Since the 1930s, psychology had forsworn the study of mentalistic entities
like *mind, emotion, motivation, reward,* and *pleasure.* Neobehaviorists like
Clark Hull and Neal Miller had even argued that pleasure itself was a kind
of illusion—merely the absence of negative sensations (such as the relief one
feels when scratching an itch). Olds and Milner's work made such hard-line
behaviorism untenable and served as a proof of concept for Hebb's agenda
and a rapprochement between the brain and the behavioral sciences. As Hebb
put it in a report to the Canadian National Research Council: "The possibilities of research opened by this work of Olds and Milner are very extensive.
It is an excellent example of the possibilities of combining physiological and
psychological methods."[86] The trading zone that had grown among surgery,
physiology, and psychology in Montreal was becoming more robust, productive, and influential.

Pain

Olds and Milner transformed the neurological conception of pleasure.
Within the next decade, another of Hebb's students would transform the neurological understanding of pain. Ironically, while Olds and Milner enshrined
the brain's so-called pleasure centers in scientific and popular discourse, in
the 1960s the work of Ronald Melzack would end the search for a localized
pain center in the brain.

Pain had been an area of triumphant progress for medicine since the
mid-nineteenth century. The discovery of ether anesthesia transformed surgery into a tolerable experience, and laboratory physiology demystified pain
mechanisms in the peripheral nervous system.[87] These developments also
made pain a topic for neurologists, who focused on the peripheral nerves;
psychologists and those interested in the central nervous system tended

to ignore the issue, which was usually discussed as a specific symptom of bodily injury. As Ronald Melzack later recalled: "For the most part . . . the general consensus [was] that the problem of pain had been largely solved. The traditional theory of pain—specificity theory—was taught in most medical schools as gospel-truth. . . . Pain was considered to be a specific sensory system with pain receptors and fibers that projected through a specific pain pathway to a pain center in the brain. . . . All that seemed to remain was to identify the pain center in the brain."[88]

Despite this sense of progress, pain remained a puzzling entity in the clinic, especially as new technologies of bloodshed left generations of veterans with curious pain syndromes. Phantom limb pain, identified in Civil War veterans by the American neurologist Silas Weir Mitchell, joined a host of other unusual pain syndromes over the twentieth century. The French surgeon René Leriche described the phenomenon of regional pain syndromes found among veterans of World War I and advocated surgical treatment. The French neurologist Jules Dejerine also proposed the possibility of surgical treatment for pain after the discovery that injuries to the thalamus frequently produced unusual hypersensitivity to pain. On the beaches of Anzio during the Allied invasion of Italy, the American anesthesiologist (and later coiner of the term *placebo effect*) Henry Beecher observed severely injured men refuse morphine, seemingly untroubled by their pain, and concluded that pain was considerably more complex than had been previously understood. By the end of World War II, some even proposed frontal lobotomy as a possible treatment for intractable pain among soldiers returning from Europe and the Pacific. By the 1950s, war had foregrounded mysterious pain that seemed to defy specificity theory, placing renewed emphasis on the role of the brain and mind in processing pain. In response, the American anesthesiologist John Bonica opened the first multidisciplinary pain clinic in Tacoma, Washington, approaching pain as a clinical problem in itself rather than as a mere symptom of illness or injury. Increasingly, pain now seemed a problem for the psychologist.[89]

Hebb had discussed pain in a chapter of *The Organization of Behavior* as part of his more general theory of learning and intelligence. Intrigued by phantom pain, neuralgia, and other complex pain syndromes, he speculated that peripheral nerve fibers did not produce pain in themselves but combined with existing activity in the cortex to produce the subjective sensation of pain.[90] In correspondence with the neurophysiologist George H. Bishop, he developed plans for a series of animal experiments that might test the proposition that the activities of the central nervous system could block, enhance, or otherwise augment the sensation of pain.[91]

Carrying out these animal experiments fell to Hebb's new graduate student Ronald Melzack. Born in 1929 in the Montreal suburb of Outremont to working-class parents, Melzack fell under Hebb's influence in the 1940s. For his PhD, he undertook experiments that paralleled Hebb's sensory deprivation work, with Scottish terriers taking the place of college students. In 1951, he took a group of Scottie puppies and raised them in complete sensory isolation for seven to ten months.[92] Afterward, the isolated dogs acted in ways that deeply surprised Melzack and his research partner, R. F. Thompson: "[W]hen these animals came out of their cages, at first they would freeze, and then they'd become excited and then they became overwhelmingly excited and would run around the room. . . . [T]hey would run around our feet and we would try to get out of the way and we'd step on a paw, a tail—no evidence of pain. I used to be a smoker in those days—I lit up a match, they'd stick their nose in the match."[93] Melzack and Thompson's conclusion was not that these dogs were impervious to pain. Instead, they concluded: "[Pain] is not a mere matter of nerve stimulation and response. Both the perception of pain and the response to it are complex processes, in which the brain plays an important part. The appropriate response to pain is acquired, at least in part, by learning."[94] Their specific findings were, in some sense, beside the point. Whatever else pain was, it was certainly not the product of a simple, telephone switchboard nervous system; the brain and central nervous system clearly played some role in regulating pain, and the autonomous activities of the brain might subject that pain to learning or other cognitive phenomena.

Much as Hebb's theory of brain function freed psychology from stimulus-response theorizing, Melzack's laboratory work freed pain from its conception as a simple injury-signaling system. Moreover, it put Melzack in an excellent position to act as an interpreter of the confusing data piling up at the interdisciplinary pain clinics then sprouting across the postwar landscape. At the University of Oregon, he spent the mid-1950s "immersed in [the] pain clinic seeing patients with chronic pain conditions, including phantom limb pain," and collecting the words that his patients used to describe their suffering. The classical medical vocabulary for describing pain (*mild, moderate, severe*) was replaced by the words that spewed forth from Melzack's patients (*burning, throbbing, flickering, quivering*, etc.). By the time Melzack left Oregon, he had collected over one hundred words to describe pain, but, as he later recollected, he "had no idea what to do with them." Subsequent fellowships at University College London and the University of Pisa provided no insight.[95]

"I got an appointment at M.I.T.," Melzack informed Hebb in a letter from Pisa in 1959. "My only regret is that M.I.T. is not in Canada." By the late 1950s, the Massachusetts Institute of Technology hoped to revitalize its Psychology

Department, and the epicenter of Cold War technoscience was "exciting and full of possibilities."[96] At MIT, Melzack met Patrick Wall, an English neurophysiologist of immaculate training who had arrived at MIT to apply Norbert Wiener's cybernetics to the study of the nervous system. Melzack and Wall began to compare notes, and what emerged from their discussions would slowly work a revolution in the midcentury world of pain. Announced in a 1965 article in *Science*, Melzack and Wall's "gate control" theory of pain represented an amalgam of Hebb and Melzack's emphasis on the role of the central nervous system in mediating pain and the emerging cybernetic conception of the nervous system then being advanced at MIT. In essence, Melzack and Wall proposed that one portion of the spinal cord—the substantia gelatinosa—"functions as a gate control system that modulates" the incoming nerve signals from the body.[97] The incoming neural signals might then act on central brain processes, which could open or close the gate, allowing for the unusual anesthesias and peculiar pain syndromes that had become so prevalent in pain clinics. The pair expressed the theory in a diagram whose similarity to an electronic engineering schematic was not coincidental. "I drew that goddamn diagram," Wall later recalled. "We first thought of it as 'gate control' . . . because we were always using triode valves [in our physiological experiments] which had a gate. That was my meaning of the word 'gate'" (fig. 4.6).[98] Indeed, a close reading of the 1965 paper reveals a subtle tension between Melzack's and Wall's areas of expertise. Melzack was clearly advocating that the brain and its central processes played some role in modulating the experience of pain, while Wall was plumping for his own cybernetically inflected understanding of the nervous system as a form of electrical engineering. Yet the take-home message was clear: "The model suggests that psychological factors

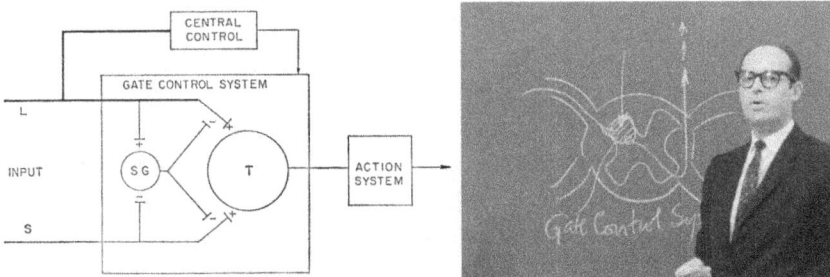

FIGURE 4.6. (*Left*) Ronald Melzack and Patrick Wall's "gate control" spinal mechanism for pain. Ronald Melzack and Patrick D. Wall, "Pain Mechanisms: A New Theory," *Science* 150, no. 3699 (November 19, 1965): 975. Melzack describing the gate control theory within the spinal cord in the 1965 Canadian National Film Board Film *The Puzzle of Pain* (*right*). Note that one resembles a wiring diagram while the other represents spinal cord physiology.

such as past experience, attention, and emotion influence pain response and perception by acting on the gate control system."[99]

By abolishing the concept of a simple signaling system for injury, gate control theory opened a neurological door for psychology, environment, and culture. This proved pivotal in the 1960s and 1970s as developed societies saw reduced infectious disease deaths and a rise in the frequency of chronic diseases and chronic pain. In the face of an onslaught of arthritis, migraines, back pain, cancer-related pain, and unspecified psychogenic pain, gate control theory seemed to provide a theoretical foundation for a multidisciplinary approach to pain science and pain management that the prior specificity theories could not. At the same time, because gate control theory gave legitimacy to psychological influences on pain, it also gave scientific legitimacy to such alternative therapies for pain as acupuncture that were then becoming popular in the United States.[100]

Gate control theory was not without its critics. Many neurologists were skeptical and noted the vagueness of the underlying neurological mechanism, a point that Melzack and Wall conceded. For Melzack, the main contribution of gate control theory was not the precise mechanism itself but the space it opened for psychology as part of a multidisciplinary understanding of pain. In this respect, Melzack was undoubtedly successful. What began with Hebb's speculations on the role of the central nervous system in mediating pain ultimately legitimated the more subjective and humanistic data that could now be used in the growing interdisciplinary field of pain science. Returning to McGill in 1963 to head the university's first pain clinic, Melzack also returned to the pain words he had collected in Oregon. He employed multiple group discriminant analysis to give order to these patient reports and transformed them into the McGill Pain Questionnaire, one of the most widely used tools in the modern study of pain.[101] When, in 1985, the American literary scholar Elaine Scarry wrote *The Body in Pain*, she credited Melzack with inspiring much of her analysis and making the voices of patients more than merely the cries of the suffering. In her view, the McGill Pain Questionnaire restored patient testimony as a diagnostic tool and turned the difficult-to-articulate aspects of pain into scientifically relevant data.[102] The same could perhaps be said for Hebb and Melzack themselves. By making the brain relevant to pain, they opened up science to the humanistic concept of suffering.

A Learning Computer

In January 1954, as his laboratory was abuzz with activity from the sensory deprivation studies and those of Olds, Milner, and Melzack, Hebb received a letter from a relatively unknown IBM engineer, Nathanial Rochester: "I was deeply impressed when I read your book. From the point of view of a computer engineer, your theory seems to present a more workable model than most." Rochester went on: "We were trying to devise a machine which would have certain properties we thought would be useful. It looked as if the sort of structure you described would be just what we needed so we simulated such a device on our large calculator." Despite positive early tests, some problems had cropped up, and Rochester suggested that he visit Hebb in Montreal to discuss whether there was some subtlety in his physiological postulate that he had overlooked. "It would be so useful to us if your idea really did work," Rochester informed Hebb.[103]

The project that Rochester was describing was the simulation of brain cells on a newly invented device—the IBM 701 digital computer—and the story of the assembly that developed between Hebb's laboratory and the world of computer engineering shows the influence of the MNI extending as far afield from the gritty reality of brain surgery as it is possible to trace. Weak ties extended Hebb's theory of intelligence from the organic brain of K.M. to the silicon and vacuum tubes of postwar computing and, ultimately, reduced Hebb's broader theory of brain function into a more compact theory of learning brain cells. While less robust than his original neuropsychological theory, the version of cell assemblies that emerged from this trading zone displayed a certain survivability that would ultimately preserve Hebb's ideas through the death of neural network research, allowing them to reemerge in their modern form in the 1980s.

Attempts to reverse engineer the human brain began in their modern form with a 1943 paper by the psychiatrist Warren McCulloch and the mathematician Walter Pitts: "A Logical Calculus of Ideas Immanent in Nervous Activity." McCulloch and Pitts argued: "Because of the 'all-or-none' character of nervous activity, neural events and the relations among them can be treated by means of propositional logic."[104] That is, the activities of individual neurons could act like logic gates, allowing neurons to calculate Boolean logical propositions: AND, OR, NOT, ELSE, and so on. Moreover, McCulloch and Pitts proposed that the combination of these activities allowed the human brain to know and manipulate the universal categories of logic and perception. The brain was, in fact, a Turing-complete computational machine—in effect, the brain *was* a computer. This notion would prove foundational for the emerging cybernetics movement—the new science of communication and feedback control that many hoped would revolutionize the biological and social sciences in the postwar world.[105]

Hebb himself, along with many other biologists and physiologists, remained aloof from this invasion of the biological sciences by engineers.[106] How, then, did his model of cell assemblies enter the world of computer engineering? The answer lies in a complex assembly of actors whose weak ties proved even more decisive than its strong ones.

Born in 1919, and educated in electrical engineering at MIT, Nathanial Rochester was one of the countless physicists, mathematicians, and engineers who contributed to the vibrant environment of MIT's Radiation Laboratory during World War II. An engineering *wunderkind*, he went to IBM after the war to spearhead the development of the IBM 701, the first mass-produced computer.[107] At IBM's laboratory for information research in Poughkeepsie, New York, Rochester hoped to develop new computation methods that might allow computers to form original solutions to mathematical problems—a kind of programmed insight. In the early 1950s, he discovered Hebb's book and tried to implement its neurological model into what he called a "Conceptor"—a machine that could learn and manipulate abstract concepts. Unlike McCulloch and Pitts's logical neurons, which would have logical relations built-in, Rochester's thinking machine would learn from its environment and recombine that learned information in original ways. Rochester hoped to build a computer that could "form and manipulate concepts, abstractions, generalizations and names"; the best way to do this would be to "find how the brain manages to do this sort of thing and copy it."[108] Hebb's theory of cell assemblies seemed tailor-made for Rochester's program.

The brain, however, could not be so easily reverse-engineered. Rochester began to simulate Hebb's cell assemblies on an IBM 701 in 1954. The program almost immediately ran into a problem. The simulated neurons did not form cell assemblies in the ways that Hebb predicted; rates of firing rose too quickly and appeared chaotic. What started as an attempt to use Hebb's theory to make a machine think had rapidly turned into a test of the theory itself.

Rochester contacted Hebb, who passed him on to a graduate student who might be conversant in Rochester's engineering language of vacuum tubes and electric circuits. Peter Milner—who was fresh from his work with Olds and had an extensive electrical engineering background—was more than capable of understanding the nature of the problem. The solution he proposed drew on his engineering training but also on his work as a nuclear physicist and his new training in neurophysiology. He realized that Hebb's model had a serious flaw: if loops of neurons reverberated in self-activating chains, and if those neurons became more efficient at self-activating, then eventually this excitation would spread across the entire cortex, with all cells firing constantly until they became metabolically exhausted. Milner realized that, in effect, Rochester had given his computer a seizure (or a nuclear meltdown, to which he drew an explicit comparison).[109]

Milner consulted with Rochester and provided him with an updated model of Hebb's cell assembly that simultaneously "made engineering sense" and was more neurologically realistic. Since the late nineteenth century, neurophysiologists knew that not all cell-to-cell connections were excitatory; the firing of one neuron could inhibit the firing of the next neuron even when other input neurons sent excitatory signals. The mechanisms behind this cellular inhibition were still unknown when Hebb formulated his theory, and he had not included it. By the 1950s, however, inhibition had been demonstrated by the Australian physiologist John Eccles. Milner was aware of this work and realized that inhibition could function to prevent the runaway firing that Rochester was encountering. Inhibition would function much like control rods in a fission reactor, preventing a meltdown.[110]

Rochester incorporated Milner's revisions in a new simulation. The results were impressive; the new model, tested on an IBM 704 whose magnetic core memory could accommodate the larger simulation, produced cell assemblies almost immediately. The next part of Hebb's theory—phase sequences, where one assembly activated the next—seemed not far off. The growing pidgin dialect between neurophysiology and computer engineering was developing quickly, with Milner acting as an informant capable of translating. As Rochester noted: "This kind of investigation cannot prove how the brain works. It can, however, show that some models are unworkable and provide clues as to how to revise the models to make them work."[111] The computer became a space where theories of intelligence and brain function could be investigated, a virtual trading zone all its own (fig. 4.7).

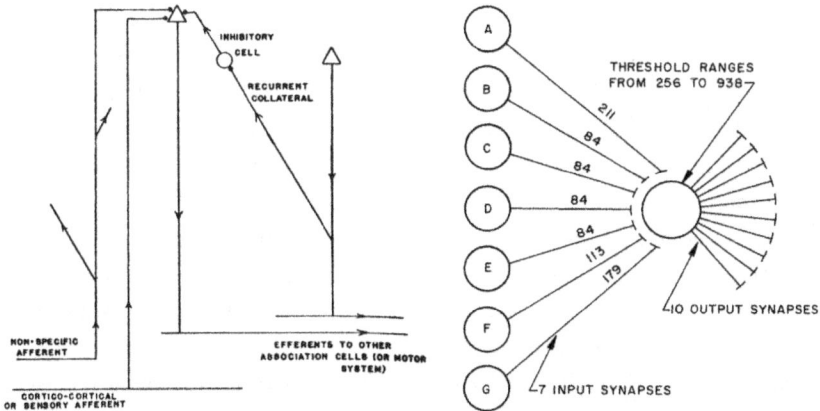

FIGURE 4.7. (Left) Peter Milner's updated cell assembly model, incorporating inhibition. Peter M. Milner, "The Cell Assembly: Mark II," *Psychological Review* 64, no. 4 (1957): 244. Nathaniel Rochester's neural model, simulated in IBM 701 and 704 computers (*right*). N. Rochester, J. Holland, L. Haibt, and W. Duda, "Tests on a Cell Assembly Theory of the Action of the Brain, Using a Large Digital Computer," *IEEE Transactions on Information Theory* 2, no. 3 (September 1956): 91.

Rochester presented his initial findings at the 1956 Dartmouth Summer Research Project on Artificial Intelligence, frequently cited as the birthplace of modern artificial intelligence research.[112] The conference is less frequently cited as the place where neural networks died (at least temporarily). Despite the modest success of Rochester's program, forces were arrayed against his approach. Since his 1954 doctoral dissertation, the mathematician Marvin Minsky had taken aim at the kind of neural models that underlay cybernetic attempts to reverse engineer the brain. Minsky argued that such models had a fatal flaw: single-layer neural networks based on the McCulloch and Pitts model could not perform several crucial logical calculations, notably "exclusive or" (XOR—the ability to categorize something as a member of one category or another but not both). Having proved this mathematically, he proceeded, over the following decades, to demolish any attempt to base artificial intelligence on neural networks. He favored an approach that did not attempt to reconstruct the brain in computing hardware; instead, he hoped to model the disembodied operations of the mind in software, an approach that came to be known as *symbolic processing*. From his powerful position at MIT, he exercised an outsized influence on the artificial intelligence community, and this influence drove neural networking to the periphery of artificial intelligence research for several decades. The vibrant trading zone that brought Hebb's theory of intelligence into the world of computer engineering vanished.[113]

Neural networks were gone but not forgotten. One actor who kept the idea alive was the Harvard-trained scientist Paul Werbos, who, in 1974, proposed a mathematical solution to the XOR problem that Minsky had identified.[114] Werbos derived his solution to the XOR problem as he tried to apply Hebb's theory. After reading *The Organization of Behavior* in 1964, he wrote Hebb: "A few days ago, while trying to apply your learning theory to a computer model, I ran into difficulties which seem to allow only one strong way out: somehow, the ingredients of a concept must inhibit each other, rather than strengthen each other."[115] Hebb responded by pointing Werbos to Milner's updated model, showing that neurons might inhibit one another and therefore that assemblies might also inhibit other assemblies.[116] Werbos spent the next decade applying this insight. In 1972, he noted to Hebb: "Many years ago, when I was a certified child mathematics prodigy, I read your book, the Organization of Behavior, and was quite impressed. It took me a few weeks to realize that your idea of 'reinforcement on connection upon positive association' was not enough to generate the complex kind of learning process your book described. . . . Ever since then, I have worked steadily in trying to find a powerful enough rule."[117] Werbos initially called his technique *dynamic feedback*. In essence, the problems in traditional neural networks could be solved by postulating an additional layer of neurons between input and output; errors were then fed back into this middle layer (the multiple layers are why this approach came to be called *deep learning*).

Werbos's ultimate goal after reading Hebb had been to "construct something like a general theory of intelligent systems," a goal that was none too pleasing to his potential MIT supervisor Marvin Minsky. Minsky dismissed Werbos's theory out of hand, noting that everyone knew that neurons followed the McCulloch and Pitts model and that no attempts to make it work had been successful. Werbos successfully defended his thesis but without Minsky's support. Minsky's dominance of the field would keep neural networks on the periphery of artificial intelligence for another decade.[118]

In the 1980s, the psychologists David Rumelhart, Geoffrey Hinton, and Ronald Williams proposed their own solution to the problem of neural networks, which they called *error backpropagation* and which was very similar to Werbos's.[119] The success of this new model, combined with the widespread availability of inexpensive desktop computers that could quickly run simulations, led to an explosion of interest in neural network research that continues to the present day. Although their work was not explicitly connected to Hebb's theory, it was clear that his model of learning as synaptic strengthening had created the intellectual terrain on which Rumelhart, Hinton, and Williams built their work.[120] What began with Hebb's attempt to understand K.M.'s natural intelligence had now been transformed, through its interaction with engineering, into the hybrid scientific field of artificial intelligence. Hebb's original theory had traveled along weak ties about as far from K.M.'s broken brain as it was possible to go.

IQ, Race, and Head Start

By the 1960s, Hebb's theory had traveled far from its origins in the postoperative IQ scores of K.M.—from natural intelligence to artificial intelligence. By the end of the 1960s, social history and racial politics would bring the matter full circle, drawing Hebb into a broader debate on the nature and nurture of intelligence.

Hebb closed *The Organization of Behavior* with a discussion of intelligence in the young and old, noting: "[C]linical research on intelligence has difficulties as a blackberry-bush has thorns."[121] The very concept of intelligence was unstable and could have two meanings: innate biological potential for learning or the fully functioning intelligence that was the product of a child's development. IQ tests measured only the second of these—a combination of innate ability and environmental stimulus that allowed for the proper formation of cell assemblies. In his book's closing passages, Hebb invoked Francis Galton's infamous formulation of the "nature/nurture" argument but flipped it on its head. IQ tests, in fact, showed not innate intelligence but rather the end result of enriched or deprived environments, themselves the product of social privilege or handicap:

> It is common to say that an intelligence test is not valid when given to a foreigner or a Negro. Not valid in what sense? as an estimate of innate potentiality, of intelligence A. It may be quite valid, on a purely empirical footing, for estimating intelligence B—the actual level of comprehension, learning, and problem-solving in this culture. Separating these two meanings of "intelligence" allows one to show where the test is valid, as well as where it is invalid. . . . Negroes living in the United States make lower average scores on intelligence tests than whites do, but we cannot conclude that the Negro has a poorer brain than the white. Why? . . . Because Negro and white do not have the opportunity to learn to speak the language with equal range and accuracy, are not usually taught in equally good schools, and do not have equally good jobs or equal exposure to cultural influences that usually require a fairly good income.

In an echo of his earlier career, he closed the book as only a former school-teacher could: "The country may be full of potential geniuses, for all we know, and it should be a pressing concern for psychology to discover the conditions that will develop whatever potentialities a child may have."[122]

Hebb's conception of intelligence as a product of innate potential and environmental stimulus might seem uncontroversial, even banal, but it illustrated a profound difference from that of his former mentor. Karl Lashley had spent most of his career arguing against any theory of learning or conception of brain plasticity. This *idée fixe* underlay his opposition to brain localization, reflex theory, and Hebb's concept of the cell assembly. Publicly, he expressed his opposition in terms of scientific positivism; his job was not to prove theories of brain function but to disprove them through careful experimentation. Privately, however, his perspective was linked to a political conservatism that, in turn, was linked to his racial politics. As he remarked toward the end of his career: "[I] [m]yself am a believer in things as they are. As I look back over the more than 40 years that I have spent in research, I am glad to see that there is not one contribution of practical value."[123] At the same time, he made liberal use of racial epithets in his private correspondence, frequently signing with the valediction "Heil Hitler and Apartheid!"[124]

Lashley's racial antagonism, vitriolic even by the standards of the 1950s, seems not to have infected Hebb, whose theory of cell assemblies did not just permit learning and brain plasticity but demanded it. Meanwhile, by the 1960s, Hebb's studies of sensory deprivation had followed a circuitous path into debates over race and education in the United States. Discovered by psychologists and education reformers, his sensory deprivation studies were enlisted as experimental support for the more contentious theory of "cultural deprivation"—the notion that children from racial minorities performed poorly in school because they came from impoverished communities, devoid of appropriate stimulation and opportunities for development. By the late 1960s, the notion of cultural deprivation formed a key theoretical justification for a new early childhood education program launched as part of the Johnson administration's War on Poverty. Head Start, as the program was called, launched in 1965 and quickly expanded from its original iteration as a summer school program into a major experiment in methods of education.[125] While Head Start drew in the loosest of ways from Hebb's sensory deprivation work, Hebb was perfectly happy to invoke the program as a practical application of his theories.

Hebb's expertise on intelligence was sought not only by proponents of Head Start but also by its opponents. In 1968, Hebb briefly corresponded with the physicist, engineer, Nobel laureate, and amateur eugenicist William Shockley.[126] The coinventor of the transistor and the "man who put the silicon in Silicon Valley," Shockley contacted Hebb about a research project he had undertaken with the backing of the right-wing Pioneer Fund entitled "Research on Methodology to Reduce the Environment-Heredity Uncertainty, Including Ethnic and Racial Aspects." "Your personal professional situation places a responsibility upon you to apply your mind to these problems and to express your views simply and directly," Shockley demanded of Hebb. Hebb did just that, unloading an avalanche of scorn: "I do not agree that this problem is unique to the Negro population; I do not agree that there is any necessary relation between skin color and native endowment; and I do not agree that there is any basis for supposing that for a Negro to be on relief, given the attitudes of this society, is evidence of a lack of native ability. I disagree fully with your attempted demonstration that this is not a function of the environment."[127]

Rebuffed by Hebb, Shockley found a different ally in Arthur Jensen, the American psychologist whose 1969 article for the *Harvard Educational Review* "How Much Can We Boost IQ and Scholastic Achievement?" took direct aim at any attempt to improve the performance of Black students. In 123 pages, Jensen laid out a case against special education in general and Head

Start in particular.[128] A touchstone in the modern debate over race and IQ, the article drew stinging rebukes from nearly every corner of American social science. Hebb's own response, published in the *American Psychologist* in 1970, was revealing. Hebb concurred with the orthodox assessment that heredity and environment could not be separated; therefore, the central premise of Jensen's argument was false. However, Hebb reserved nearly as much scorn for those who would dismiss the enterprise of IQ testing altogether: "I am appalled at the quality of the criticism from social scientists. Their reaction is dogmatic and emotional, and the hell with logic. If Jensen's argument is 'socially dangerous,' it must be more dangerous in the long run to suppress it. . . . [T]he result is to prevent criticism of the argument and, worse still convincing the racist that it is the truth that is being suppressed."[129] For Hebb, who had built much of his career around its careful interpretation, abandoning the very idea of intelligence as a scientific object posed three crucial dangers. First, it posed a danger to the validity and revisability of his theory of brain function. Second, abandoning the rigorous investigation of intelligence would ultimately place more children at risk; without the ability to identify low-achieving children for remedial education, more children would be set up for a lifetime of failure. Finally, for Hebb, the IQ test was a plank of social critique: "The IQ is not a social evil but the principal means by which to show the effect of another—economic—evil."[130] As arguments over intelligence and race continued into the sunset of the twentieth century, Hebb's more nuanced position found few adherents.

Conclusion: Cognitive Neuroscience and the Fate of K.M.

In 1974, shortly before he retired, Hebb wrote the American psychologist Ulric Neisser, who had produced the first textbook of cognitive science. *Cognitive Psychology* (1967) featured references to Hebb's work throughout, and Hebb wrote Neisser to correct a minor error of intellectual history. He had not, as Neisser had implied, argued for the complete determination of behavior by the environment: "What I did believe in was connections as the basis of behavior and cognition. . . . This is not a complaint. I am used to looking like a nativist to the learning-theory man, an empiricist to the Gestalt-minded, but I do hope that in fact I avoid those extremes and regard learning as those modifications of thought and behavior for which heredity has designed the organism. Well, more or less."[131] Neisser apologized for any misrepresentation, noting: "[Pure learning theory] . . . predominates in your book, but only because it was written so far ahead of its time. . . . *The Organization of Behavior* is a wonderful book . . . ; it turned psychology around. Perhaps this is an appropriate occasion to thank you for having written it."[132]

Hebb's legacy for neuroscience was captured, at least in spirit, in this brief exchange. Beginning with the puzzling case of K.M., his inquiry into the nature and nurture of intelligence led him to a theory of brain function that amalgamated rival positions within the human sciences—extremes of hereditarianism and environmentalism, behaviorism and cognitivism. In this respect, his scientific legacy reflected his own life of mediation and reconciliation: a novelist and a scientist, a hereditarian and an environmentalist, an antibehaviorist who trained rats in mazes, an opponent of racism who defended intelligence tests, and one of the founders of artificial intelligence who likely never touched a computer. The ultimate legacy of Hebb and his students was how they constructed their own strong assembly in Montreal and then expanded its local findings through weak ties into neuroscientific knowledge of global significance. Because their work traveled through weak ties, Hebb's students could influence a diverse array of other scientific assemblies while the origins of their work remained obscure. By extending the work of the MNI into the realm of theory, Hebb and his students made it scientifically respectable to investigate the brain and the mind as a single, continuous entity. This transformation was produced not by any individual theory or experimental discovery but rather by the summed activity of Hebb and his network of students. In many respects, the circle of collaborators that formed around Hebb resembled the cell assemblies and phase sequences of his own theory, sometimes firing together, sometimes operating in parallel, and sometimes triggering with single, weak connections vigorous activity in other assemblies. In this complex network of collaboration, coordination, and trading, Hebb and his circle transformed the experience of a neurological patient into the basis for a new scientific field—cognitive neuroscience.[133]

Hebb retired from McGill in 1976 and moved back to Nova Scotia—sailor returned to the sea. He published a final book in 1980, shortly before his death, entitled *Essay on Mind*, in which he relayed the development of his cell assembly theory: "My problem was to understand some of the effects of brain operation, and in particular how it was that a patient might have a large chunk of brain tissue removed without much effect on his IQ or his intelligence as it seemed to his family."[134] The patient who had inspired Hebb's entire career was not named. In this respect, K.M.'s fate was the opposite of his closest historical analogue, the railroad worker Phineas Gage, who nearly a century earlier had suffered a similar insult to his frontal lobes.[135] While Gage became famous in his own time, K.M. remained an anonymous brain surgery patient for the remainder of his life. What was the fate of this crucial, invisible patient?

K.M. appears sporadically in Hebb's correspondence. In 1944, the neuropsychologist Kurt Goldstein (by that time an American citizen, having fled the

Nazis in 1933) published a paper attacking Hebb and Penfield's conclusions, and Hebb felt compelled to respond.[136] Hebb asked Penfield for permission to visit K.M.: "Goldstein . . . hasn't a leg to stand on. . . . I hope to get to Nova Scotia to see my parents and while there to see [K.M.] again. It should be I think extremely important for the treatment of the brain-injured in this war to get a true picture of the psychological differences in the effect of scarring versus tumor versus clean extirpation."[137] K.M. had apparently recovered enough from his operation to enlist in the Canadian army during World War II. According to Hebb: "The medical officer examining K.M. saw the scar on his forehead, asked what caused it, was told that he had had an accident in a sawmill, and asked if it was giving any trouble. K.M. said no, it wasn't. And it wasn't either, after Wilder Penfield had taken out the remnants of both his prefrontal lobes."[138] Indeed, K.M. remained seizure free for years after his operation until a relapse while in the army led him to be discharged and returned to his family in Nova Scotia. A wartime labor shortage made it easy for him to find employment, which "he does himself, without any family guidance." K.M. evidently transitioned jobs frequently, seeking better pay. Hebb retested K.M. with several of Goldstein's psychological tests, finding none of the deficits Goldstein predicted. Satisfied that he had laid waste to Goldstein, Hebb went on to discuss the theoretical implications of the case. Useful to Hebb mainly as a source of theorizing, the person of K.M. faded from view.

K.M. survived long enough to have one final encounter with a psychologist. Brenda Milner, Hebb's student, whom we met in chapter 2, made her career by closely examining Penfield's patients using a wide assortment of psychological tests. However, unlike Hebb, she was not overly reliant on the Stanford-Binet IQ test and continued to develop new tests in conjunction with her reading of primate studies. In 1962, she made her way to Nova Scotia, where K.M. still lived. She reported: "I repeated every single one of Hebb's tests. I mean, the literal tests. . . . I used the tests that Hebb had used. Absolutely confirmed everything he got." However, when Milner administered her new card-sorting test, "[K.M.] exhibited to an unusual degree the perseverative behavior that I had come to expect from patients with lesions to the dorsolateral frontal cortex." "[I]t was," she noted, "a beautiful, visual instance of there being nothing wrong with what Penfield and Hebb reported. But they had not used the right tests essentially." For Hebb, neuropsychology was a high theoretical enterprise, meant to unravel the fundamental philosophical mysteries of the mind. For Milner, neuropsychology was a grounded practice to be undertaken in ongoing conversation with the patient. "I'm not a theoretician," Milner recalled, "but I am a good observer."[139] After her follow-up with K.M., Milner transitioned from her work on memory and inaugurated several new areas of frontal lobe research. K.M. and the frontal lobes were important after all.

Weak ties allowed Hebb's work to pass from Montreal to the broader scientific community and embed the work of the MNI (and the experience of its patients) in the fabric of cognitive science. Some of those weak ties, like sensory deprivation, brought the science of Montreal into close contact with aspects of the Cold War. In the next chapter, I examine another afterlife of the MNI that brought it into contact with the emerging Cold War, also in ways dependent on weak ties. At the same time, the story of Ewen Cameron and his brief relationship with Wilder Penfield sheds light on the context of Penfield's original creation and on the underlying tensions of interdisciplinary work among participants unwilling to share authority or understand each other's language. If Montreal emerged as a fertile contact area for practitioners from different scientific cultures, then the case study of neurosurgery's relationship with psychiatry shows what can happen when those cultures remain in isolated solitude.

5

Two Solitudes: Psychosurgery and the
Troubled Relationship between Wilder Penfield
and Ewen Cameron

In November 1963, Herbert Jasper gave an address at the Allan Memorial Institute of Psychiatry (AMI), the sister institution to the Montreal Neurological Institute (MNI) (fig. 5.1). Located on the slopes of Mount Royal, a seven-minute walk from the MNI, "the Allan" also went by another macabre moniker: Ravenscrag. Built in 1863 by the shipping magnate and financier Sir Hugh Allan, the mansion was donated to McGill University in 1940. In 1943, Ravenscrag became home to Canada's first dedicated university department of psychiatry, headed by the Scottish-born psychiatrist D. Ewen Cameron. Herbert Jasper's address, given to mark the opening of new research laboratories at the AMI, was entitled "Neurology and Psychiatry: Two Solitudes?" and began on a positive note. Those at the MNI had followed the development of the AMI with a "sense of family pride." It had been partly because of Wilder Penfield's advocacy that the AMI was established, and Penfield had helped select Cameron to direct it. Under Cameron's leadership, the Allan had been transformed into "one of the leading psychiatric research and training centers in the world." However, Jasper's tone quickly turned mournful: "The development of the Allan Memorial Institute has been largely independent of the parallel growth of the Montreal Neurological Institute. . . . At this time, over 20 years ago, there was a close association between the practice of neurology and psychiatry in Montreal. . . . Thus, the relative independence of the Allan Memorial Institute of Psychiatry is part of a larger picture of specialization in all branches of science and medicine, a trend which characterized the past quarter century. Neurology and Psychiatry have been developing, to a large degree, as 'two solitudes.'"[1] Jasper's audience would have understood his reference to Hugh MacLennan's novel *Two Solitudes* (1945). Set on the eve of World War II, the novel detailed the struggles of its protagonist to navigate his twin identities as a French speaker and an English speaker in the bilingual

FIGURE 5.1. Ravenscrag, ca. 1935, shortly before it was donated to McGill University to become the Allan Memorial Institute of Psychiatry. Courtesy McCord Stewart Museum.

but linguistically segregated city of Montreal. After the novel's publication, the phrase *two solitudes* became a shorthand expression for the dysfunction of Canadian society; Canada was a nation divided by language and incapable of communication and mutual understanding.[2] Now, Jasper chose this metaphor to describe the relations between two closely connected medico-scientific fields. Unlike the productive relationships that developed among neurosurgery, neurology, and psychology at the MNI, the divide between psychiatry and the emerging neurosciences looked as unbridgeable as that between Canada's French- and English-speaking populations.

In the previous chapters, we have seen how different scientific and medical disciplines wired together at the MNI to produce new forms of scientific knowledge in its interdisciplinary workspace. Surgeons, physiologists, chemists, and psychologists learned to coordinate and synchronize their work as they collaborated on specific projects (e.g., the functions of the temporal lobes) despite occasional disagreements on the import of their discoveries (e.g., aspects of memory formation). The interdisciplinary environment may have evolved somewhat differently than Penfield had envisioned, but, by the mid-1950s, the MNI was humming with scientific work that created a palpable sense of excitement for a new kind of brain science that could eventually unite the world's brain researchers in a shared sense of purpose. Neuroscience had grown in Montreal and diffused, in different ways, to the world through weak ties.

The absence of a relationship with Montreal's psychiatrists is all the more surprising given that psychiatry had always been a part of Penfield's plan. The MNI's funding and much of its raison d'être sprang from the Rockefeller Foundation's (RF) program in psychiatry, and, as we will see, Penfield always hoped to incorporate psychiatry and psychiatric knowledge into the MNI. Yet, as Jasper's address indicates, the city's psychiatrists, under the leadership of Cameron, did not care to wire together with researchers at the MNI. Why?

To answer this question, we need to return briefly to the story of Penfield, his institute, and the psychobiological vision of the RF in the 1930s. In doing so, we will see that two aspects of twentieth-century brain science often written of as separate—Penfield's surgeries for epilepsy and the emergence of psychosurgery (more commonly known as *lobotomy*)—sprang from the same source and overlapped in several ways. Psychosurgery and Penfield's surgical treatment for epilepsy emerged almost simultaneously, and, while the two procedures were distinct in their outcomes and rationales, they were more closely related than one might suspect. In unraveling Penfield's relationship with psychosurgery, we will begin to understand why psychiatry was unable to join the MNI's interdisciplinary community.

At the same time, we also need to understand something of the character and life of Cameron, the psychiatrist Penfield helped select for the AMI. In the 1970s, it was revealed by investigative journalists that Cameron had been a critical figure in the CIA's MK-ULTRA mind control program. His attempt to cure schizophrenia through a technique he called *psychic driving*, which employed forced confinement and liberal use of electroconvulsive therapy and psychedelic drugs like LSD, PCP, and sodium amytal, conformed to all the Cold War–era tropes of brainwashing and psychological torture. In response, his image was remade as something of an egomaniacal mad scientist. This image, however, obscures more profound lessons that can be learned from Cameron's brief collaboration on psychosurgery with Penfield and their subsequent estrangement. That aborted collaboration can help us understand the contingent nature of scientific collaboration and the obstacles that can prevent two areas of science from trading. The fallout from the failed relationship is instructive, too. Shorn of the oversight that comes from collaboration, Cameron was free to use interdisciplinarity as rhetorical cover for his own experimentation and to appropriate elements of the Montreal trading zone through weak ties.

Return to the Borderland

Psychiatry had always been a part of Penfield's vision. Even before he developed his plans for the MNI, Penfield had been influenced by the monistic psychobiology of Adolf Meyer and its attendant demands for an integrated approach to neuropsychiatry.[3] In 1916, during his medical training at Johns Hopkins, Penfield took a course in psychiatry led by Meyer, and, in 1922, having returned from graduate work with Sherrington in England, he considered joining the medical faculty in Baltimore where Meyer was establishing his Phipps Clinic: "Perhaps the most exciting event of the visit was my discussion of neurology and psychiatry with Adolf Meyer. . . . His opinion at that time is interesting. There should be, he said, 'a neuropsychiatric clinic with neurology and psychiatry on an equal footing.'"[4] The plan never came to fruition, but Penfield stayed in contact with Meyer during his time in Spain, updating the older man on his exciting discoveries with the Spanish staining methods of Cajal and Pío del Río-Hortega.[5] He later recalled: "I believed that psychiatry should be separate from neurology and neurosurgery in hospital practice for some years to come, but that all three should be joined together somehow in thought and plan, in the laboratory for scientific study, and in teaching— recalling my conversation with Adolf Meyer at Johns Hopkins."[6] By the time the MNI was founded in 1934, Penfield was clearly open to incorporating psychiatrists into his institute, provided that their approach was compatible with Meyer's vision of psychobiology: "I had been [Meyer's] student and had discovered that his own approach to psychiatry had included studies of the anatomy and pathology of the brain. . . . Would I ever meet a psychiatrist like him when I was able to do my part?"[7]

The answer to this question, at least initially, was no. Penfield collaborated briefly with the Montreal psychiatrist David Slight—from 1934 until his retirement in 1937—but afterward was unable to find a psychiatrist with whom to work, writing his neurosurgical colleague Eldridge Campbell in 1937: "[W]e are in a bad way for a psychiatrist."[8] By 1941, with the war in Europe in full swing, memos circulated among Penfield, the president of McGill, F. Cyril James, and the dean of medicine at the Royal Victoria Hospital, J. C. Meakins, about the possibility of establishing a university psychiatry department. "This war has increased the need for trained psychiatrists as well as for treatment facilities in the field," James wrote in an internal memo, adding that Ravenscrag might be an ideal site for such a facility.[9]

By 1942, discussions regarding forming a new department of psychiatry at McGill were becoming serious. Penfield followed them closely and was adamant that any new psychiatric institute should be "immediately adjacent" to his neurological institute: "In my opinion the cause of many of the psychoses will be discovered along biochemical lines. When these advances are made there will develop a gradual identity between neurology and psychiatry inasmuch as psychiatric conditions will then be recognized as diseases of the brain which may be treated directly. When that time comes, neurology and neurosurgery may well be fused with psychiatry."[10] Penfield also noted that this gradual melding of neurology, neurosurgery, and psychiatry would make securing RF funding easier. Up to that point, psychiatric cases in Montreal had primarily been the responsibility of the Verdun Protestant Hospital. Founded in 1881 as the Protestant Hospital for the Insane, the renamed Verdun Hospital (in 1965 renamed again the Douglas Hospital) had handled most of Montreal's psychiatric patients since its inception and sat then on the city's outskirts, unaffiliated with McGill. Penfield noted correctly that the RF would be unlikely to provide money for a custodial mental institution like the Verdun; any funding proposal should emphasize the new department's forward-looking research orientation. Even the location at Ravenscrag should be only temporary; Penfield hoped the Psychiatry Department would eventually be directly adjacent to his institute for easy access to laboratories and consultations on interesting cases. As the time came to decide on who would head the new department, Penfield contacted his old friend Meyer for advice:

> My own desire is to see a Department of Psychiatry developed here, quite independent from a clinical and academic point of view. However, I should like it very much if we can have a Professor of Psychiatry who will cooperate with our Department and who will recognize that we have a common interest in neuropathology and neuroanatomy and *who can talk our language as well as his own.* . . . You will remember I spoke to you about our desire to secure a good man at McGill two years ago. I wonder if you would be good enough to write to me or to Dean Meakins with any suggestions that you might have for a Professor of Psychiatry. He should be young; he must have some knowledge of basic sciences, either the biochemistry of the nervous system, or the anatomy [, or] of the neuropathology [*sic*].[11]

Meyer advised finding a psychiatrist who could speak Penfield's neurolanguage. A "native" Canadian psychiatrist would be preferable to an American import, and he ought to have good all-round training in psychiatry rather than an overspecialization in psychoanalysis. Then Meyer suggested a name: "I wonder about Ewen Cameron. . . ."[12]

Cameron

D. Ewen Cameron certainly had the pedigree to head a new, experimental psychiatric institute (fig. 5.2). Cameron had been born in Stirlingshire, Scotland, in 1901, and his prodigious medical training had taken him from the University of Glasgow to Meyer's Phipps Clinic at Johns Hopkins to Switzerland to work with Eugen Bleuler, who coined the word *schizophrenia*. In 1929, he moved to Canada to serve as the primary psychiatrist for the entire province of Manitoba, where he established ten clinics. He then moved to Massachusetts in 1936 and New York in 1938, where he became psychiatrist-in-chief at Albany Hospital.[13] His intellectual orientation also fit the bill nicely. Neither a strict organicist nor a psychoanalyst, he advocated a kind of relentless experimental empiricism, outlined in his 1935 *Objective and Experimental Psychiatry*, in which he noted: "[T]ime has brought us so great a development of instrumentation and so much larger an understanding of experimental methods that we are now in a fair way to realize our dream of analyzing human behavior objectively, dispassionately and, above all, of being able to predict and control."[14] For Penfield and the RF, this sort of scientific orientation was highly appealing. A formal offer was extended to Cameron in 1943 to come to Montreal to open Canada's first university department of psychiatry.

Cameron arrived in Montreal in September 1943 to begin building the Allan into a world-class research clinic, but he did not have his first formal meeting with Penfield until March 1944. On the sixth floor of the MNI building, Penfield and Cameron finally compared notes. Evidently, Penfield found the meeting so disappointing that he followed up the next day with a letter to Cameron that is remarkable in its candor and worth quoting at length:

> I regret your conclusion that you can develop your department best quite independent of ours, and I feel that your simile of the twins "in utero," one of which must be weak because the other uses up the nourishment, does not apply. On the contrary, if a satisfactory relationship is established each of the twins should be able to help the other secure nourishment in the form of money and men and add strength in other ways.
>
> It had always been my hope to see McGill develop a strong Department of Psychiatry, and I have refused the suggestion of a combined department with one head. I have always known that our own work would be better if we had . . . cooperation. . . . Our laboratories on the 7th floor of the M.N.I. are nevertheless open to you while you are developing your own.

FIGURE 5.2. D. Ewen Cameron (1901–67), 1945. Courtesy McCord Stewart Museum.

You know psychiatry; you are our elected authority here, and we accept your decision to go it largely alone, although I am personally disappointed. I can assure you, however, that any time in the future we will gladly establish the cooperation you consider unwise now.

Psychiatrists in Montreal, without being organized in a university department, have attended our Wednesday meetings, and contributed a good deal to

them, as well as in occasional contact. I hope the current of the new organization will not lessen this. Our neurologists in public service and out-patient
department will consider psychoses to lie in the field of your department, in
spite of the fact of their previous interest and training in the field. Neuroses
they will consider a common problem, and I should hope, as time passes,
that the majority of the neurotics will be referred to those men, in either department, who get the best results. For the sake of record now I would like
to express my belief that any Department of Psychiatry which loses contact
with those who work primarily with brain lesions is weaker for that lack, and,
conversely, any Department of Neurology (in which I include Neurosurgery)
which has not close contact with those who work in the field of psychiatry is
the weaker for that lack.

The time will come when neurologists will find in the brain the cause for
many psychiatric disturbances, and psychiatrists will likewise find in that organ some of the things they are looking for. Thus, looking into the future,
their fields must eventually merge to some extent, and far-seeing academic
organization should provide for it.

Good luck to you. These things may work themselves out gradually, perhaps after you and I have handed on the professorships to our successors.[15]

Penfield sent a carbon copy of his letter to J. R. Fraser, the dean of the
McGill Medical Faculty, noting: "I give up, of course, my cherished hope of a
Psychiatric Institute built close to the Neurological Institute and the combination of associated work. This may come at McGill, but it will not come in
my lifetime."[16]

In the previous chapters, we saw how and why scientists formed productive
trading relationships at the MNI. Molly Harrower, Brenda Milner, Herbert
Jasper, and Donald Hebb all found ways of contributing to and extracting
valuable experiences from their work at the MNI. Yet here, raw professional
ambition seemed to block the creation of any kind of trading between neurosurgery and psychiatry. In this respect, Cameron's turf guarding was of a piece
with the transformation of psychiatry that had occurred in the first half of
the twentieth century. Largely through the efforts of Adolf Meyer, psychiatry
had emerged from the world of the asylum and the state or provincial mental hospital and "found itself" as an expansive medical specialty, returned to
medicine with somatic therapies like malaria fever treatment, insulin coma
treatment, and electroconvulsive therapy. It had also "found itself" as an institution increasingly influential on social issues such as crime, juvenile de-
linquency, alcoholism, and other areas of social dysfunction.[17] A member of
a newly respectable and expanding specialty, a psychiatrist of Cameron's ambitions was reluctant to surrender any authority to a rival medical discipline,

one that was also experiencing impressive growth. As Cameron himself put it: "[W]hen I came here in 1943 McGill was universally regarded as a very difficult place for a psychiatrist because of strong organic biases represented by Pathology, by the Neurological Institute and indeed by the whole curriculum, yet . . . there were powerful forces in favor of a strong psychiatric development [including] the general forward movement of psychiatry everywhere. . . . [T]he psychiatric tide was running strongly and hence the . . . wrong thing to do would have been to centralize it as neuropsychiatry was centralized in the MNI."[18] For Cameron, collaboration meant subordination, and he was unwilling to surrender the toehold he felt psychiatry had obtained by entering into any relationship with Penfield and the MNI. His hesitancy here was not entirely unreasonable; at the time of his arrival, Penfield and his institute had become a considerable force in McGill politics, and it was not impossible that his overtures were disingenuous and his ultimate aim was to swallow psychiatry whole.

Despite Penfield's assurances that psychiatry and neurology would dance together at arm's length rather than in a tight embrace, Cameron's decision to go it alone seemed intractable. Initially seen as a promising collaborator, Cameron was instead intent on making Ravenscrag, a short walk from the MNI, a fortified psychiatric castle to ward off a possible siege from neurosurgery or neurology. Yet, within a year of their disastrous meeting, Penfield and Cameron would indeed collaborate on a series of experimental psychosurgical operations. To understand how psychosurgery could constitute a potential bridge between the two solitudes of psychiatry and neurology, we must first understand the complex relationship among Penfield, the research emanating from the MNI, and the emerging practice of lobotomy.

Penfield and Psychosurgery

Penfield's radical treatment for epilepsy was not the only form of radical brain surgery that had grown from the RF's efforts in psychiatry; lobotomy, the first form of brain surgery meant to treat mental illness, grew from the same institutional and intellectual soil as Penfield's most important contribution to medicine and science. Moreover, Penfield's career and that of lobotomy overlapped in several more direct ways. New information about the frontal lobes from the MNI proved crucial for the creation of lobotomy and psychosurgery despite Penfield's best efforts to ensure that such data were interpreted carefully.

The origins of lobotomy are typically attributed to events at the Second International Neurological Congress, held in London in 1935. At this

congress, the American psychologist Carlyle Jacobsen presented his research on chimpanzees. Jacobsen had been working with the neurophysiologist John F. Fulton, whose laboratories at Yale were, after the MNI, the
greatest recipient of the RF's funding in psychiatry. Jacobsen had severed
the frontal lobes of a chimpanzee from the rest of the cortex to determine
their role in learning and memory. In presenting his results, he noted, as an
aside, that the operations seemed to have dramatically reduced the emotional outbursts of one particularly troublesome chimpanzee, Becky. In his
infamous words, it seemed to have "joined a happiness cult . . . and . . .
placed its burden on the Lord!"[19] Sitting in the audience of Jacobsen's talk,
the Portuguese neurologist Egas Moniz asked whether the reduction in
anxiety and agitation observed in the chimpanzee could be replicated in
human patients. While no one at the conference dared suggest such a radical procedure, Moniz would attempt just such an experiment later that year
with a series of patients suffering from "agitated depression." Psychosurgery
was born.[20]

While the proximal origin of lobotomy was Moniz's appropriation of
Jacobsen's monkey work, the expansion of lobotomy in the 1930s and 1940s
drew on Penfield's research in important ways. Penfield's 1935 paper on his sister and the frontal lobes (see chap. 2) constituted, in the words of John Fulton,
"the most crucial case of frontal area extirpation that will probably ever appear in neurological literature"[21] and formed a crucial touchstone in discussions of the frontal lobes in the 1930s and 1940s. One character who observed
this discussion carefully was the American neurologist Walter Freeman, who
would later adapt Moniz's leucotomy procedure into the so-called standard
lobotomy with the neurosurgeon James Watts. Freeman had, in fact, been
a close follower of the saga of Penfield's sister and wrote Penfield in 1936:
"I remember listening to [your] presentation which was the highlight to me
of the Association [for Research on Nervous and Mental Diseases] meeting
[in 1932], and the idea that such extensive removal of the frontal lobe could be
accomplished without serious intellectual deficit must have sunk in because
when the report of [Egas] Moniz's work [on leucotomy] came to me it seemed
that there must be something to it."[22] Of course, Penfield had encouraged
caution in interpreting the results of his sister's operation; the psychological
deficits produced by frontal lobe operations were real but not easily found
by existing psychological tests. For Penfield, this meant that there was something *wrong with the tests*. Freeman, by contrast, read the paper as suggesting
that deficits produced by frontal lobe operations were minimal and acceptable—so acceptable that, when combined with Moniz's reports, the study of
Penfield's sister served as a spur to psychosurgery itself.

By 1936, news of the Moniz psychosurgery operations had spread around the world, and Freeman had developed the standard lobotomy with the neurosurgeon James Watts. Penfield greeted the news with varying degrees of disbelief and hostility. To Alan Gregg, he remarked in 1936 that he "was much disturbed when he learned of Moniz' operation on frontal lobes," although he also commented that "frontal lobe removal for tumours frequently produce[s] placidity and lack of initiative which is often pleasing to the family."[23]

By 1937, however, Penfield was on the offensive. In September of that year, he drafted an editorial with Stanley Cobb to be published anonymously in the flagship journal *Archives of Neurology and Psychiatry*. Entitled "Experimentation in Clinical Medicine," the editorial lambasted psychosurgery along with other somatic therapies for psychiatric disorders: "Human experimentation may result in tragedy and can only be justified if it is conducted in a crucial, scientific manner and if it is preceded by thoughtful elaboration of a reasonable hypothesis. Two recent examples of clinical experimentation do not seem to satisfy these conditions fully. The first is 'psychosurgery' which after short trial has been described in the public press. This bastard term was used by Watts and Freeman."[24] It went on to express concern about the mission creep that seemed to characterize lobotomy's application, applied first to patients with severe agitated depression but later to patients with psychoses and "that vague group called schizophrenia": "The bounds of science and wisdom seem to have been overstepped." Moreover, lobotomy patients had been insufficiently studied to determine potential long-term effects, which might include epilepsy from the lacerations on their brains created by the procedure. Considerably more long-term study was needed before the operation received widespread acceptance, especially by a public "too eager to accept untried remedies indiscriminately."[25]

For unknown reasons, the editorial was never published. Privately, Penfield remained ambivalent about lobotomy but also seemed impressed by its possible therapeutic utility. In May 1940, three years after the aborted editorial, with reports filtering in about lobotomy's success, he remarked to a colleague:

> [I]n regard to the Moniz and Freeman operation, I may say that the procedure has filled me with astonishment. I have been unwilling to undertake it myself. . . . So far as I have been able to tell, they have applied it to certain cases which were not hopeless. However, I have no right to speak with any authority; I have not investigated it, and of course I do know that a large frontal operation sometimes makes people more placid. Perhaps many of us would be better off if we were converted into nit-wits by some such procedure.[26]

As lobotomy was being developed, Penfield had continued his frontal lobe operations, now assisted by the psychologists Molly Harrower and Donald Hebb. In 1937, he conducted a major operation on a young epileptic patient, K.M, during which he removed much of the young man's frontal lobes. With Donald Hebb, he studied the effects of this operation and made the bizarre discovery that K.M.'s IQ scores actually *increased* following the operation (see chap. 4). The pivotal frontal lobe operation on K.M. supplanted the case of Penfield's sister as the most important frontal lobe extirpation operation, and the paper he published with Hebb[27] was immediately noticed by the lobotomist Walter Freeman. In 1940, while Penfield and Hebb's paper was in galley proofs, they were contacted by Freeman, who had heard about the findings and wanted to use the research to "liven up" his forthcoming *Psychosurgery* (1942).[28] Penfield wrote Hebb: "I am not very keen about Freeman claiming too much about our paper and I should hate to see him use any of our illustrations. On the other hand, we do not want to insult him, but if you could find it possible not to have any spare copies of the illustrations . . . I should do so [*sic*]." Two days later, Hebb dispatched a letter to Freeman containing the agreed-on excuse, adding in a note to Penfield: "I think it sounds enough like innocent ignorance that he can't take offense."[29]

Penfield did not trust Freeman to make responsible use of his studies of K.M., but his confidence in his own abilities was growing. That same year, his close friend the Harvard neurologist Stanley Cobb featured Penfield and Hebb's study in his annual "Review of Neuropsychiatry" for the *Archives of Internal Medicine*. Cobb summarized the results, noting: "[T]he authors conclude from their psychologic examinations that removing a third of both frontal lobes need not cause gross deterioration."[30] This sentence prompted a candid and revealing response from Penfield:

> I wept when I saw your quotation of what we said about [K.M.] . . . I immediately looked up the manuscript to see if we had really said that and found that we had. The implication, of course, is that we conclude that such frontal removals had no effect upon intelligence or personality. Certainly, taking the sentence out from its context gives that impression. You are quite right to criticize it. . . . What I meant to point out was that the methods which we used for psychological study showed no change. What I believe is that these methods are inadequate. . . . I also believe that the observations that I was able to make in my sister's case are more accurate and more to the point than the accepted psychological tests would have been in her case. I believe that in that case, as in the cases of large single frontal removal that Evans and I studied, there was "impairment of those mental processes which are prerequisite to planned initiative."

The danger of overinterpreting the evidence from K.M. led Penfield to con-
ceal the information from Freeman and rebuke his close friend. Yet, on the
very next page of his letter to Cobb, he wrote: "I liked very much your discussion of the lobotomies. I have had a number of doubts recently, thinking
perhaps that I was too conservative, and wondering if there was not some way
of making selective frontal removals or frontal incisions that would not give
rise to important defects but that would benefit the psychotic patient. I have
had in mind the possibility of going down to see Freeman and humble myself
before him so as to see his results."[31]

Penfield's objections to lobotomy were ultimately about precision and scientific rigor; the Freeman and Watts operation seemed reckless. However,
with his great sense of precision and scientific prowess, could Penfield improve
on the procedure, asking the patient to sacrifice less for the same treatment?
If he could, he would add another surgical procedure to his armamentarium
and further legitimate his radical approach to brain surgery. Unlike his innovation with epilepsy, however, such work would demand collaboration with
a psychiatrist, if for no other reason than to provide appropriate patients.
Penfield had mentioned as much to J. R. Fraser, the dean of the McGill Medical
Faculty, as plans were made for the new department of psychiatry: "[I]n many
institutions certain operations are carried out upon the brain to combat . . .
mental changes and this has been called psychosurgery. I, for one, am not
thoroughly convinced that this has a permanent place, but there is definitely
an occasional case who should have an operation upon his frontal lobe. Such
operations could only be carried out in the Neurological Institute."[32] After his
first overture to Cameron had been rebuffed, it was the lure of collaboration
itself that finally overcame his resistance to psychosurgery. If a project undertaken with Cameron could yield a better lobotomy—one more in tune with
Penfield's cautious approach and scientifically grounded method—so much
the better. And, if this better lobotomy could unite psychiatry and neurology
under the banner of neurosurgery, it would cohere the psychobiological vision that had given birth to the MNI in the first place.

Gyrectomy

Within a year of their disastrous sit-down meeting, Penfield approached
Cameron to collaborate on a small series of experimental psychosurgeries,
dubbed *gyrectomies* (removal of specific brain gyri). Rather than making
blind disconnection through trephined holes (as in a standard lobotomy),
this procedure would leave more brain tissue intact, theoretically reducing
postoperative intelligence and personality deficits (fig. 5.3). "I think I would

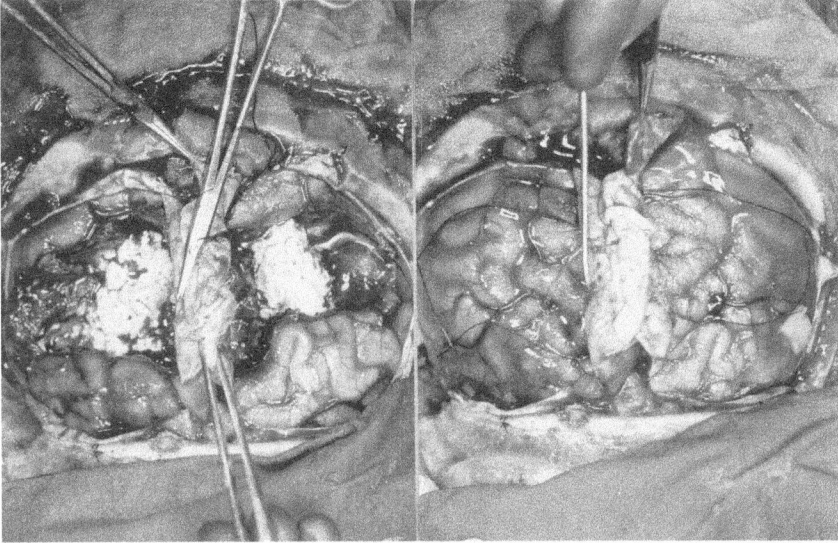

FIGURE 5.3. An example of Wilder Penfield's gyrectomy procedure. On the right, the cortex to be removed is dimly outlined in black thread. On the left, the ablation has been carried out. Reproduced by permission of the Osler Library of the History of Medicine, McGill University.

be glad to do a series of these patients," Penfield confided to a colleague at the nearby Verdun Protestant Hospital. "It might be only a trickle, but this is a type of operation that I am very anxious to tackle, and after the war we could perhaps swell the trickle into a larger stream."[33] Cameron selected seven candidate patients and handed them over to Penfield for operation. In the spring of 1944, Penfield carried out his first gyrectomy on a twenty-three-year-old man named H.M. who had suffered for years from an "inability to concentrate and to work."[34] He had previously received fifteen treatments of electroconvulsive therapy without benefit and had attempted suicide at least once. The diagnosis was "chronic anxiety neurosis with feelings of unreality." Penfield first operated in May and in July extended his removal of frontal lobe tissue. After H.M. had recovered from the operation, "his depression . . . disappeared and he [became] president of the local radical political club in which he takes a great interest."[35]

Other patients did not fare as well. A.L., a woman of thirty-three, complained for six years of "listlessness, fear of crowds and of heights." After several rounds of electroconvulsive and insulin coma treatment, her "chronic anxiety neurosis" remained unchanged. The gyrectomy, performed by Penfield in October 1944, relieved her of fear temporarily, but, following a major epileptic seizure, she was readmitted to the AMI and given a standard

lobotomy two years later. I.C., a twenty-seven-year-old woman, seemed initially relieved of her "obsessive compulsive neurosis" but was later readmitted
and lobotomized. All told, Penfield performed the new gyrectomy procedure
on seven patients.[36]

How would the success or failure of the new procedure be evaluated? For
most lobotomy practitioners, success or failure was judged primarily in terms
of whether patients returned to their social role. Lobotomy was developed
within the context of Meyerian psychobiology, which emphasized pragmatism and eclecticism in therapy, and any procedure that held out hope of re-
lieving mental distress and returning patients to productive citizenship was
worth pursuing. In an era before randomized controlled trials, whether the
patient had recovered was left mainly for the psychiatrist to judge.[37]

Yet, for Penfield, this simply would not do. Certainly, Penfield had included an element of this framework in his own thinking. Reports on his patients were replete with comments about whether they had returned to work
or reconciled with their families and social circles. However, Penfield also
included more objective measures of success. Were seizures controlled? Were
they reduced in frequency? If so, by how much? Could the patients be weaned
from anticonvulsive medication? Moreover, since his earliest operations on
epileptic patients, Penfield had displayed a thirst for information beyond
what could be seen with the naked eye; this had motivated his histological
work in Spain, his adoption of Foerster's operating techniques, and his eventual collaborations with psychologists like Donald Hebb and Molly Harrower
and physiologists like Herbert Jasper. In the case of his collaborations with
psychologists, the assembled actors could eventually synchronize their work
on specifics even if they disagreed on broader theory or interpretation. Did
the psychological test show a gain or a loss of IQ in the patient? Did the test
indicate the presence of a tumor in the frontal lobe? If so, was one found in
exploratory surgery? Was the patient's memory affected by the operation? If
so, in what way? These synchronizing activities between surgery and psychology grew from necessity, and collaboration on these concrete problems
slowly created a shared language among the surgeons and psychologists and
gradually wired the groups together. By contrast, if Penfield and Cameron
could not agree on whether their gyrectomy procedure worked or on any
concrete way to know, then they could not properly synchronize their work.
Mere clinical impressions would not be enough to recommend the operation.

Both Hebb and Harrower had left the MNI by 1944, but Penfield consulted with Hebb from a distance and pestered Cameron to add a psychologist to the team.[38] Penfield and Cameron eventually recruited Robert Malmo,
then conducting war work at the National Institutes of Health in Bethesda.

Malmo was selected because of his impressive experimental work with primates, and he arrived to work on the project after four operations had already been performed. Penfield later informed Hebb: "Malmo is making a study of the gyrectomies and the whole future of operations of this sort now hangs in the balance as we are reviewing the cases already done."[39] "I naturally assumed," Malmo later recalled, "that they would not have proceeded without thorough preoperative psychological testing. You can imagine my astonishment and deep disappointment in finding on arrival for work in July 1945 that 4 of 6 gyrectomies already operated had had no preoperative testing, and that the preoperative testing of the other two . . . was meagre (only Wechsler Intelligence Scales, and incomplete ones at that)."[40] Nevertheless, he persisted, conducting an extensive battery of psychological tests on the remaining patients and also comparing the cases to a series of separate lobotomy patients.[41]

Beyond exposing an overall lack of coordination between the AMI and MNI, the gyrectomy project also displayed a deeper aspect of Penfield's vision of medicoscientific collaboration, one that made it unique among RF-funded efforts. Virtually all RF-funded psychiatric initiatives displayed a deep commitment to crossing disciplinary lines; however, this interdisciplinarity was mainly about talk rather than action. Ideas and data crossed disciplinary boundaries easily, but successful collaborative practice was rare.[42] This was in keeping with Adolf Meyer's ideas about psychiatry; psychiatry was supposed to be pragmatic and eclectic and not too dogmatic in its theory. Any data or ideas that seemed useful should be studied or tried.

On the other hand, Penfield's interdisciplinarity was about practice rather than theory; colleagues worked together on shared problems, bringing their independent expertise to bear. In this regard, Cameron could not have been a worse collaborator. Penfield outlined the problem in a letter to a colleague just as the gyrectomy experiments concluded: "When [psychiatry and neurology] are completely isolated we, for our part, miss very much all contact with psychiatry and psychology. At present the Allan Memorial Institute seems to be so far away that it takes a long time to get a consultation and it is impossible to have spontaneous discussions on mutual interesting subjects. . . . I believe that the rough and tumble of active discussion between those interested in these two fields might well lead to some real advance."[43] Cameron had only the most limited interest in collaboration, lest the growth of psychiatry as an independent specialty be compromised. The two solitudes to which Jasper would later allude had begun to grow and frustrate the search for a better lobotomy.

Meanwhile, word of Penfield's gyrectomy procedure had gotten out. Fred Mettler, a professor of neuroanatomy at Columbia University and later in

charge of the Columbia-Greystone project in New York—the most extensive research initiative yet launched on the effectiveness of psychosurgery—
contacted Penfield in 1947. Mettler was employing a nearly identical
procedure called a *topectomy* that was derived from early reports of Penfield's
own operation.[44] He explained that, in his study, surgeons were prevented
from knowing the results of the operation and the psychological examiners
were prevented from knowing the areas removed or whether brain tissue
had been removed at all (an early example of a double-blind study). Penfield
mentioned this "strange sort" of evaluative procedure to Cameron, adding:
"[F]rom [this] choir of ignorance may emerge a chorus of true tones not influenced by any wishful thinking." Then Penfield asked Cameron point blank:
"When shall we get together to put our cards on the table about gyrectomy?"[45]

Penfield had reason to press Cameron for an answer. The men had agreed
to present the gyrectomy cases at the 1947 meeting of the Association for
Research in Nervous and Mental Disease (ARNMD). This meeting was ex-
ceptionally important; the years after World War II had seen an explosive
growth in the number of lobotomies performed. At the same time, Walter
Freeman's invention of the transorbital lobotomy (the infamous "ice pick"
procedure) threatened to take control of psychosurgery out of the hands of
trained neurosurgeons since it required no formal surgical training to perform. The psychiatrist could render the patient unconscious with electroconvulsive therapy, insert the ice pick–like instrument through the soft bone
above the eyeball, and twist the device to sever neural tissue in the frontal
lobes, all without the assistance of a trained brain surgeon. John Fulton, initially so pleased to promote lobotomy, now organized the ARNMD meeting to regain some control over the enterprise by refocusing discussion on
the physiological underpinnings of psychosurgery in the hope of developing a better, more precise operation.[46] The gyrectomy seemed like the ideal
candidate.

When Penfield, Cameron, and Malmo presented their findings at the
ARNMD meeting, the form of their presentation was revealing: unable to
come to a consensus, all three men presented separate papers. Penfield concluded: "[G]yrectomy is a difficult, long, and somewhat dangerous procedure.
The therapeutic results have been variable. This operation is not proposed
at the present time as an acceptable substitute for . . . 'frontal lobotomy.'"[47]
Malmo, for his part, presented convincing evidence that gyrectomy did lead
to detectable postoperative intelligence loss.[48] Cameron, by contrast, interpreted the results within the Meyerian adjustment framework; success was
judged by whether the patients could reacclimatize to their home and work.
He closed on a more hopeful (or perhaps ominous) note: "This operation

clearly is one which has no greater value than the lobotomy. . . . However, we may just as clearly state that it is reasonable to explore this new field of surgery and psychiatry through further modifications of this and other operations."[49] For Penfield, the failure of the gyrectomy cases was the nail in the coffin for psychosurgery. For Cameron, it was just the beginning.

"And they hate psychiatrists"

Cameron's enthusiasm was not dampened. In April 1947, Cameron wrote Penfield about possibly continuing traditional lobotomies (as opposed to gyrectomies) on a limited basis for psychoneurotics, asking: "Could you let me know when you wish to start?"[50] He received no reply. Three months later, Penfield responded to a letter from Freeman and Watts asking if he would support a conference on psychosurgery in Lisbon. Penfield said that he would not and sent a private note to Henry Alsop Riley, the president of the American Neurological Society, asking him not to support Freeman's proposal. For his part, Herbert Jasper concurred, noting to Penfield: "I am in wholehearted agreement with your reaction to Freeman's proposal": "How easy it is to lose one's perspective!"[51] Penfield carried on cordial written communication with Cameron, but his private asides were more candid and reflected his disappointment in their relationship. In 1949, in response to a question from the Harvard Medical School about Cameron, whom they hoped to recruit, Penfield noted: "Cameron is a good teacher, an industrious worker, and a man of considerable force. He is anxious to cooperate with other departments within the university *but also always does so in his own way*. He has had very little in common with our department."[52]

By the early 1950s, Cameron began to propose new methods for replacing lobotomy, methods that became more experimental and fanciful. In 1952, he wrote Penfield about the possibility of using ultrasonic waves to disrupt brain tissue, a technique then being developed by the cybernetic psychiatrist Warren McCulloch. Penfield responded with a touch of sarcasm:

[I]t seems to me that ultrasonic technique might be quite useful if the cortex of the brain were laid out smoothly on a curving surface instead of following innumerable cracks which extend into the substance of the brain. I do not believe that any technique, however well worked out, . . . could ever give anything like the localization of destruction that actual gyrectomy produces, and inasmuch as none of the gyrectomy patterns that I was able to work out seemed to produce a result that you felt was useful, I cannot imagine that any diffuse process of that sort would fare better. . . . [T]he research approach does not seem to me to be a very good one.[53]

Cameron responded to Penfield's dismissal with additional suggestions, adding: "I remain convinced that there is gold in this particular hill, but how to come at it is quite a question."[54]

Penfield, for his part, washed his hands of psychosurgery. To a colleague in 1954 Penfield noted: "I have carried out no further gyrectomies, and, aside from a very few lobotomies carried out for a year or two following that, I have given up such procedures altogether. Dr. Ewen Cameron . . . and I have decided not to do any lobotomies or gyrectomies for the time being. Other members of my staff carry out the procedure of lobotomy when requested to do so, on chronic patients in psychiatric hospitals, but the number of such operations is comparatively small."[55] The collapse of the gyrectomy project and Penfield's further refusal to investigate the subject destroyed the fragile bridge between the AMI and the MNI. No common ground could be found on whether the procedure worked or what its potential might be.

The two solitudes that now enveloped psychiatry and neurology in Montreal became evident only a few years later, in a 1956 profile of Penfield in *Maclean's* magazine. "Penfield and his staff are a close-knit bunch," related one unnamed Montreal psychiatrist, "and they hate psychiatrists." According to the *Maclean's* reporter: "Penfield smiled when he heard that, and admitted it was half true. 'We certainly don't hate psychiatrists,' he said. 'It's true that we are concerned chiefly with treating ills of the brain and nervous system by surgery and medicine; but psychiatry and neurology are closely related—and someone may suddenly make a discovery that will bring us all together.'"[56] While Penfield yearned for a transcendent discovery that might reunite neurology and psychiatry, had he been aware of Cameron's next steps, he likely would have welcomed their continued estrangement. At almost the same time that Cameron gave up his search for a better lobotomy, he began a new research program that displayed an entirely different style of interdisciplinarity, one that would inaugurate a medical nightmare.

The Sleep Room

Penfield and Cameron were certainly men of strong egos and opinions. Both enjoyed being called "the Chief" by their staff and exercised firm control over the activities of their institute. Yet, at the MNI, subordinates could voice their disagreement without fear of reprisal, and Penfield was willing to admit error. As Molly Harrower put it when recalling an exchange with Penfield: "[T]he most remarkable characteristic of these discussions . . . was his ability to admit that his original point of view was wrong."[57] Penfield's institute may have been hierarchical, but it was not a dictatorship. Cameron, too, styled himself "the Chief" of the AMI, but his style of leadership was perhaps best captured in an anecdote. According to a former patient, a frequently heard crack among nurses and other AMI figures when Cameron passed through the wards was, "There but for the grace of God, goes god."[58] According to one former colleague, Cameron resented any colleague who came off as a strong father figure and avoided the MNI "as if he feared the godlike figure of Penfield." As a result: "[H]e was inclined to be stubborn, oppositional, and competitive with figures of authority, and was not given to emphasize collaboration."[59] Cameron's view of psychiatry was monotheistic: there was only one God.

Cameron was now free to pursue his own forms of medical innovation, shorn of the implicit oversight that comes with collaboration. He began a new form of therapy in 1953 just as his relationship with the MNI completely deteriorated. This experimental treatment displayed his own ideas about interdisciplinarity, ideas that were the mirror image of those at the MNI. At the MNI, interdisciplinarity was the product of extended and deep collaborations. Cameron, who hoped to build a psychiatric empire, approached other disciplines as one might approach a buffet—take from any adjacent discipline as desired.

Virtually all Cameron's innovations shared a quality of impatience. As early as 1946, he attempted to reform the clinical practice of psychiatry in Montreal by transforming the AMI into a "day hospital" that could return patients to their homes more quickly and avoid the expense of long-term hospitalization.[60] Beyond the format of the mental hospital itself, Cameron had for years nursed a desire to automate psychotherapy. New postwar technologies like the magnetic tape recorder suggested how he could transform the long, drawn-out process of psychotherapy into an automatic procedure akin to dialysis or a blood transfusion. Simultaneously, the growing postwar array of psychoactive drugs—sodium amytal, sodium Pentothal, phencyclidine, and, most dramatically, LSD—provided a broad pharmacopoeia that could augment any psychiatric treatment. Somatic therapies like insulin coma treatment and electroconvulsive therapy too appealed to Cameron's technophilia.[61]

In 1953, these different impulses combined into a new therapy meant to secure Cameron the Nobel Prize for curing schizophrenia. Called either *depatterning* or *psychic driving*, this therapy had as its goal essentially to brainwash people suffering from schizophrenia out of their delusions and disruptive behavior patterns, typically through some kind of automated means. In practice, a patient would be given a brief psychotherapy session that was recorded on a primitive magnetic tape recorder. The patient was then forced to listen to this tape for several hours to several weeks. The goal was "the penetration of defenses, the elicitation of hitherto inaccessible material and the setting up of a dynamic implant."[62] Needless to say, some patients resisted the "treatment," and Cameron employed several techniques to overcome this reaction. Patients were often confined to a so-called sleep room where they were strapped to a bed and placed on a steady diet of psychoactive drugs, including Desoxyn, sodium amytal, PCP, barbiturates, Nembutal, Seconal, and LSD, along with regular and severe electroconvulsive therapy—at least two to four sessions daily and sometimes more—leading to "an organic brain syndrome with acute confusion, disorientation and interference with learned habits of eating and bladder and bowel control."[63] Finally, a therapeutic "disorganization" took place in which the patient might experience partial or total amnesia and incontinence and would become helplessly dependent on the AMI nursing staff.[64] Between 1953, when Cameron developed the procedure, and 1964, when he left the AMI, over one hundred patients endured the psychic driving treatment, which often left them with severe amnesia. One patient was kept in a coma for eighty-six days. Another woman was unable to identify her children after she was released.[65]

Shortly after Cameron published his initial reports on psychic driving, the CIA identified his work as of possible value to its emerging research on psychological warfare. His early reports suggested that, in the context of de-patterning and psychic driving, the therapist ought to play a role similar to that of an interrogator. Indeed, psychic driving seemed to induce the kind of breakdown and rebuilding that many in the agency's MK-Ultra mind control program thought was necessary for interrogations and the creation of sleeper agents. Through a front organization—the Society for the Investigation of Human Ecology—the CIA began funneling money to support Cameron's research in 1957.[66]

In the 1970s, as part of his investigation into the CIA's mind control research, the American journalist John Marks first surfaced the details of Cameron's CIA connections. The ensuing scandal involved investigations by the Canadian government, including a royal commission, alongside personal lawsuits. Cameron's image was remade from a psychiatric pioneer into a caricature—a mad scientist unleashed by the forces of Cold War hubris.[67]

Yet the macabre spectacle has obscured more profound questions about Cameron. One seems especially pertinent here: given that both the MNI and the AMI grew from similar intellectual and institutional sources (psychobiology, the RF, and the emergence of neuropsychiatry), how did the approaches of the two institutes differ? In the context of the mid-twentieth-century brain and mind sciences—the same context that had produced Penfield's MNI—what shaped Cameron's radically different approach?

Two details about Cameron's work can help answer these questions. The first stems from an aspect of his psychic driving treatment. Along with the drugs, electroconvulsive shocks, and forced sleep, Cameron also employed a variation of Donald Hebb's sensory deprivation procedure (see chap. 4). Early reports of Hebb's experiments were published at about the same time that Cameron began to develop his psychic driving procedure. Cameron employed "an adaptation of Hebb's psychological isolation" to "reduce the defensiveness of the individual while applying driving." A patient would be "isolated . . . from incoming stimuli by putting him in a dark room, covering his eyes with goggles, reducing auditory intake, and preventing him from touching his body—thus interfering with his self image." Cameron cited Hebb's 1954 paper on the subject but made some additions to the procedure. Notably, he added straps and restraints to "cut down on expressive outflow."[68] Put another way, unlike Hebb's well-paid college students, patients at the Allan could not stop the experiment.

Even though both Hebb's and Cameron's experiments received funding from the intelligence community, and despite their proximity, perhaps the most revealing fact is that they appear to have exchanged absolutely no communication about the sensory deprivation studies. Despite their proximity, Hebb and Cameron corresponded almost not at all, and it appears that Hebb was largely unaware of the extent to which Cameron was using his techniques until all was revealed in the 1970s. Asked by John Marks about Cameron's experiments in 1976, Hebb remarked: "That was an awful set of ideas Cameron was working with. It called for no intellectual respect. If you actually look at what he was doing and what he wrote, it would make you laugh. If I had a graduate student who talked like that, I'd throw him out. . . . Look, Cameron was no good as a researcher. . . . He was eminent because of politics."[69] Indeed, Hebb complained about Cameron at the time. Since his arrival in Montreal, Cameron had ambitions to build a psychiatric empire with the Allan at its center. When a system of federal-provincial mental health grants was inaugurated in 1948, he pounced on the opportunity to fund his vision. Hebb also hoped for some of the federal-provincial money to fund his research, but he found that Cameron had actively subverted his attempts to get grants. In a 1955 letter of complaint to the principal of McGill, Hebb laid out at great length his view of the relationship between his department and the AMI. Parochial department infighting soon gave way to a disquisition on the nature of collaboration itself. "Psychology should be to psychiatry as physiology is to medicine or physics to engineering," he proposed. "If [there] is a collaborative undertaking, between psychology and psychiatry, the principle [should] also be maintained that its guidance is collaborative."[70] Amid snide comments, Hebb had articulated the fundamental difference between his and Cameron's vision of interdisciplinarity. For Hebb (and his former MNI colleagues), collaboration had to be between those whose expertise was acknowledged; authority may have rested in a leader, but subordinates could contribute and disagree. For Cameron, collaborators were technicians at best, there to provide techniques that he could excise from their experimental or theoretical milieu, repackage, and apply at will. Cut off from one interdisciplinary scientific community, Cameron was isolated but also free to appropriate tools and techniques as he saw fit, with no oversight from their originators.

A second detail of Cameron's work also elucidates his interdisciplinary style, one not often discussed in light of the more scandalous revelations about psychic driving. In 1956, while the psychic driving experiments were in full swing, Cameron embarked on a new line of research for a different disorder—senility. The memory loss associated with aging had been a preoccupation of his since the 1940s, and, with his now-consolidated position at the Allan, he felt that the time was ripe for a new attack on the problem. He had recently read of the research of Holger Hydén, the Swedish scientist who attempted to demonstrate that RNA was the physical substrate of memory. He combined this with his reading of the curious research of J. V. McConnell, the American psychologist who had seemingly demonstrated hereditary transmission of memories in the planarian flatworm. McConnell did this by training the worms in several maze-based and Pavlovian conditioning tasks, then grinding up the trained planarian and feeding them to a new generation of worms to see whether they would inherit the memories of the previous generation. If the memories had been inherited, the worms would presumably complete the tasks more efficiently; this is precisely what appeared to happen.[71] McConnell's widely publicized results were later disputed,[72] but they appeared convincing enough to Cameron.

As we saw in chapter 3, the work of Hydén and McConnell partially inspired the RNA memory theory of Francis Schmitt at MIT, which provided much of the impetus for his Neuroscience Research Program. Schmitt would pursue the theory of molecular memory to transcend the differences between branches of the brain and mind sciences—a transdisciplinary approach, in contrast to the patient interdisciplinary work at the MNI. With their growing expertise in the neurology of memory, experts at the MNI like Penfield, Milner, and Hebb found it difficult to take Schmitt's RNA memory work seriously. Given his isolation from this group, it should perhaps come as no surprise that Cameron not only took the RNA theory seriously but also attempted to transform it into a form of therapy. Beginning in 1956, he conducted a series of studies at the AMI to investigate a possible treatment for memory loss in the elderly. If RNA were the substrate of memory, could RNA supplementation be used as a therapy for memory loss?

Cameron and his colleagues at the AMI tested this hypothesis with a mechanism that was elegant in its simplicity, if not its sophistication; they simply gave a group of twenty senile, presenile, and arteriosclerotic patients RNA supplementation, both orally and intravenously. While earlier experiments with DNA supplementation produced no results, Cameron was encouraged by some limited success when he turned to RNA, although "the earlier intravenous solutions were so apt to produce severe shock-like reactions . . . they were stopped."[73] Adverse side effects appeared frequently in the RNA memory study, including drops in blood pressure, hyperventilation, and stomach upset and cramping. Given his evident tolerance for the side effects of his psychic driving treatment, one wonders how severe these side effects must have been to give Cameron pause. By 1961, a new laboratory technician had evidently improved the intravenous RNA solution, ending the shock-like side effects (the nausea remained, and blood pressure changes were compensated for with aramine), and the intravenous therapy resumed. Cameron attempted to demonstrate success in his RNA-based treatment through Wechsler intelligence tests and claimed that electroencephalogram records of the patients showed an ill-defined "reduction in pathology."[74] All told, approximately thirty patients underwent some variation of Cameron's RNA memory therapy.

Beyond the horror they often elicit from modern readers, Cameron's psychic driving and memory studies also illustrate a more profound point about interdisciplinarity in midcentury science. Despite their veneers of theoretical legitimacy, both Cameron's studies were absurd hashes that were produced not from thoughtful innovation or respectful collaboration but rather from appropriation and outright theft. In this respect, Cameron's ambition to build a psychiatric empire produced a relationship with neighboring disciplines more akin to imperial plunder than constructive trade.

In the years after the abortive collaboration on psychosurgery, Cameron and the AMI present a sort of dark mirror image of Penfield and the MNI. Divorced from the kind of tacit supervision that comes from collaboration, Cameron was free to pick and choose the bits of technique and theory that most fit his agenda and combine them in ways that would have been unrecognizable to their originators. Of course, participants did not always agree on aspects of their research within the scientific assemblies that grew at the MNI. As we have seen, neurosurgeons, neurologists, and psychologists often came to different understandings of the theoretical import of their work. Yet, at the MNI's peak, the collaborators had synchronized enough of their work to develop a mutually intelligible language of practice. Cameron, by contrast, seemed interested in no other language than his own. When he first arrived in Montreal, and before their relationship had deteriorated, Penfield advised Cameron that he ought to learn French. In the bilingual city of Montreal, many potential patients would speak French as a first language, and Penfield had gone to great lengths to ensure that he could conduct neurological examinations in the patient's native tongue. Yet Cameron remained a monoglot during his tenure in Montreal.[75] This unwillingness to learn the local language mirrored his unwillingness to learn the scientific language of his collaborators.

Cameron's contemporaries noticed the analogy between multilingualism and interdisciplinarity. Returning to Herbert Jasper's address at the AMI in 1964, his words seem both prescient and also strangely naive:

> Neurology and Psychiatry have been developing, to a large degree, as "two solitudes." However, during the past few years and, I would predict, in the next quarter century, this trend is to be reversed as there is a healthy scientific revolution taking place in our studies of the nervous system in relation to the mind and behaviour, just as there is a local revolution occurring in the relationships between French and English cultures in our Province. . . . I would not like to give the impression that I believe that solitude has been, or is, necessarily something to be avoided. In science, as in medicine, specialization has become and will continue to be a practical necessity, and solitude a rare and highly cherished privilege. There is no magic formula in "interdisciplinary research." Cooperation among scientists, no more than among peoples, can be prearranged or organized. It must come out of natural desires and needs. Pre-arranged marriages are seldom successful. However, specialization and isolation in science carries the seeds of its own destruction as it does so often in society.[76]

This prediction of a healthier relationship between neurology and psychiatry failed to materialize, much as the optimism about improved French and English relations in Quebec gave way to the increased linguistic isolation and separatism of the 1960s. The two solitudes became entrenched. Meanwhile, Cameron's solitude led to precisely the destruction that Jasper predicted. A colleague of Cameron's came close to spelling this out in a memoir published shortly after the revelations of Cameron's various abuses: "[Cameron] could not tap the scientifically restraining milieu provided by a congenial group of colleagues in evaluating ideas and experiments."[77]

Growing skepticism of MK-Ultra's value at the CIA led to the termination of much of Cameron's funding in 1964. Cameron left Montreal shortly thereafter, if not in disgrace, then to a certain deafening silence, and took a position at the Albany Medical School in New York, where he continued his research on memory. He died of a heart attack while hiking in the mountains of New York State in 1967.[78] Perhaps the definitive statement on Cameron was made by his former second-in-command at the AMI, Robert Cleghorn, who eventually put an end to the psychic driving experiments. In an address to the "interdisciplinary group" at the AMI (revealingly titled "Pitfalls of Thinking Big—Megalomania"), Cleghorn noted:

> The writer's attention was drawn to this topic originally by observing the sad phenomenon of scientists whose outlook seemed to have become poisoned by discovery, pseudodiscovery or hope for discovery. As individuals, they seemed to have become impervious to contrary evidence, resentful of criticism and to have exaggerated expectations of rewards and recognition. Their findings, when they had any, they often elaborated in a grandiose way; their data became woven into a system of almost delusional proportions. Such behavior is not infrequently a besetting sin of late middle life, at a time when men begin to write definitive books and, if never before, to show that they think big.[79]

It is hard to imagine that Cleghorn was describing anyone other than his former boss. If Jasper's address displayed the possibility and potential of interdisciplinarity embodied at the MNI, then Cleghorn's revealed Cameron and the AMI as interdisciplinarity's dark mirror.

Coda: Two Solitudes and the Psychopharmacological Revolution

There is one additional coda to the story of Montreal's two scientific solitudes. While Cameron was attempting to cure schizophrenia through his depatterning experiments, one of the most important developments in the treatment of the disease was taking place only a few miles away. At the Verdun Protestant Hospital—the same psychiatric hospital meant to house the refuse patients from the newly created Allan Memorial Institute—the German-born psychiatrist Heinz Lehmann was conducting studies on the drug Largactil (chlorpromazine), recently marketed in France by the pharmaceutical giant Rhône Poulenc for the possible treatment of schizophrenia. Lehmann could read the French-language sales material that described chlorpromazine primarily because he had learned French in Montreal following his marriage to a French Canadian woman.[80] He is widely credited with introducing chlorpromazine to the North American market, one of the most consequential developments in the birth of the modern field of psychopharmacology. The drug was not without its drawbacks, and the scientific foundations of psychopharmacology remained shaky for much of the rest of the century; nevertheless, the discovery of chlorpromazine represented the first major therapeutic breakthrough in psychiatry in decades.[81]

Despite Penfield's professed belief that significant breakthroughs in psychiatry would come from chemistry and pharmacology, Penfield and Lehmann carried on no correspondence or collaboration; Cameron's consolidation of power at the AMI had so thoroughly poisoned the well between psychiatrists and neurologists in Montreal that the introduction of chlorpromazine took place almost entirely without Penfield's knowledge. Counterfactual histories must remain, at best, interesting thought experiments for the historian and can never constitute evidence in themselves. Nevertheless, it is hard not to wonder how the history of mental health care might have evolved differently in the twentieth century had Penfield found a more humble collaborator.

The previous chapters have shown how participants at the MNI wired together neuroscience from elements of their pragmatic collaborations and how the products of that interdisciplinary trading shaped the scientific landscape of the twentieth century. In the case of Cameron, the inability and unwillingness of participants to trade also had ripple effects that shaped not only activities in Montreal but broader developments in the history of neuroscience, psychiatry, and the Cold War. The weak ties connecting Montreal to the rest of the scientific world could be as crucial to its long-term impact as its close collaborations.

In the next chapter, I examine one final story of the strength of the MNI's weak ties. The career of David Hubel, who began as an MNI technician in the 1950s, constitutes one of the institute's most important afterlives. Hubel and his microelectrode would travel along weak ties from Montreal to the halls of Harvard University, transformed by each new setting, yet retaining an essential family resemblance to the work he began a decade earlier in Montreal. The end result of his movement along the weak ties of the postwar brain sciences would be what could be considered neuroscience's first Nobel Prize.

Eye, Brain, Vision: David Hubel, Microelectrodes, and the Science of Seeing

In this brief chapter, I examine one final legacy of the Montreal Neurological Institute (MNI), its brand of interdisciplinary brain science, and one final character, the physician and neurophysiologist David Hubel. In Hubel's life and work, we will see how the strong assembly of actors at the MNI formed and shaped scientific practice. Hubel's career also shows how weak ties between assemblies could transfer technologies from one scientific culture to another and how those technologies were transformed when they moved into new laboratory environments. Herbert Jasper's contribution to neuroscience occurred through institutions, while Hebb's intervention had been at the level of theory. Here, the scientific instrument takes center stage. Hubel's scientific triumph in the 1960s of mapping and explaining the mechanisms of vision in the brain is historically explicable only if we first understand his experiences at the MNI and how they led to his construction and use of a key new neuroscientific instrument—the tungsten microelectrode. This instrument, which could record the electric activity of single neurons in the cortex, emerged partially from an existing laboratory tradition of single-cell recording in animals. However, Hubel's use of the equipment also grew from clinical practices and experimental traditions at the MNI. This hybrid of clinical and laboratory research, brought together by one of the MNI's most promising alumni, inaugurated a new era in the study of human vision and, for the first time since the establishment of the neuron doctrine in the late nineteenth century, showed that an aspect of the human mind (perception) could be understood as a function of underlying neural mechanisms. It also culminated in Hubel and his research partner Torsten Wiesel winning the 1981 Nobel Prize in Physiology and Medicine.

From Physics to Medicine

Born in Windsor, Ontario, in 1926, David Hubel was a precocious child. The son of an American chemical engineer, he possessed dual citizenship from birth. This would aid his movement across the border during his career but also make him subject to compulsory military service in both countries. Transplanted to the mainly French-speaking Montreal suburb of Outremont a few years later, he grew up "speaking a half–French Canadian half-English polyglot which no one else could understand."[1] Interested in science from an early age, he terrified neighbors in Outremont with homemade explosives cooked up in his basement with a Lott chemistry set. He also developed a passion for electronics, constructing his own makeshift radio equipment. This interest in electronics eventually led him to enroll in physics and mathematics at McGill in 1943. (Ironically, he was interviewed for admission at the Massachusetts Institute of Technology [MIT] by F. O. Schmitt but could not attend because of World War II.) After graduating from McGill with a degree in physics in 1947, he applied for medical school entirely on a whim and was shocked to be accepted.[2]

More interested in medical research than clinical practice, Hubel was informed that the MNI, a short walk from the McGill campus, provided an opportunity to do both. As he later noted:

> The MNI was perched high on the hill to the southeast of Mount Royal, a sort of ivory tower that medical students seldom climbed. I decided to grab the bull by the horns and made an appointment to see Penfield himself. Finally, the day arrived. I borrowed the family car, parked it on University Street, and in a state of some terror climbed up to the fourth floor of the institute. Penfield was at his most charming, and when I told him of my physics background, he immediately took me up to see Herbert Jasper, who in turn, immediately offered me a summer job doing electronics in his physiology group. (When I got back to the car I found it running, with the keys locked inside. I took the streetcar home to get a spare key, and 90 minutes later was back. It was a stressful afternoon.)

The positive association with Jasper continued after Hubel's medical training and would ultimately orient the entirety of his later career. He completed a year's neurology residence and an additional year as Jasper's assistant in clinical electroencephalography (fig. 6.1). Jasper was a perspicuous mentor. "His scientific outlook was wonderfully broad," Hubel recalled, "and he had a clarity of mind and skepticism that made him stand out among brain scientists": "The first time we spoke, the day of the locked car, he asked me what I had read in the field. I told him I had just read *Cybernetics*, by Norbert Wiener. He gave me an odd look, and said, 'Did you understand it?' I thought I had, even if through a glass, darkly, and when I said so, he grinned. It was clear that he thought that Wiener's brain science was off the wall, but he was nice enough not to want to put me down."[3]

Hubel's training at the MNI remained grounded in clinical realities, far from the bloodless speculation of the American cybernetics movement. In 1953, Hubel took over much of the clinical electroencephalograph (EEG) work at the institute, examining hundreds of patients, and sitting in on countless temporal lobe surgeries. At the same time, the MNI still aimed to integrate clinical and basic research, and interns were expected to stay abreast of developments in basic neurophysiology. A prophetic assignment to survey and summarize the existing research on the visual system would profoundly shift Hubel's life course.[4]

How does the brain see the world? This question had animated philosophers and artists, neurologists and psychologists for decades, if not centuries. The discovery of language localization by Pierre Paul Broca and motor localization by Gustav Fritsch and Eduard Hitzig in the nineteenth century kicked off an era of brain localization research, and, by the 1880s, the English physiologist David Ferrier had localized much of the brain's visual activities to the occipital lobe in the posterior of the brain. Yet the actual neurophysiology of vision languished as a research topic for much of the early twentieth century. Beyond the fact that nerve impulses activated ganglion cells in the eye's retina and that these signals then passed down the optic nerve bundle, through the lateral geniculate body in the thalamus, and on to the primary visual cortex in the back of the brain, little was known.[5]

One critical advance came in the late 1940s with the work of the American neurophysiologist Haldan Keffer Hartline, who used minute electrodes to record the electric impulses from the optic nerve of the *Limulus* horseshoe crab (whose optic nerves were large and easily accessible). Hartline discovered that retinal ganglion cells, activated by the retina's rod and cone cells when struck by light, were interconnected so that cells responded only when light fell on specific parts of the retina. In contrast, those same ganglia were inhibited when light fell on surrounding patches of the retina. Hartline termed these the *excitatory field* and the *inhibitory surround* of the ganglion cell. In effect, the retinal ganglion cells acted to sharpen contrasts and borders in a visual scene.[6] The Hungarian émigré neurophysiologist Stephen Kuffler (who is examined in more detail below) refined Hartline's initial discoveries, using glass microelectrodes to extend these discoveries from the retinal ganglion cells to the lateral geniculate

FIGURE 6.1. David Hubel as a Montreal Neurological Institute fellow, ca. 1953. Reproduced by permission of the Osler Library of the History of Medicine, McGill University.

bodies, showing that these way stations for stimulus from the eyes transmitted excitatory and inhibitory data in the same fashion in mammals and other more complex animals.[7] It was Hartline's and Kuffler's 1952 papers describing their work[8] that Hubel had discovered while researching vision at the MNI, and it was these papers that would prepare him intellectually for his later work.

In 1952, Herbert Jasper still sat at the center of the global network of electroencephalographers (see chap. 3), and it was through this network that Hubel eventually found himself in the United States. That same year, a young neurologist, Charles Luttrel, visited the MNI to learn electroencephalography from Jasper, who passed him off to his talented young technician. Luttrell subsequently arranged for Hubel to take a neurology residence at Johns Hopkins. However, before he could take this residency, Hubel would have to discharge a responsibility that came with his dual Canadian and American citizenship: the army doctor draft, which mandated that newly minted physicians could be drafted into the military. In 1955, Hubel arrived at the Walter Reed Army Institute of Research, a posting that, while brief, would greatly alter the direction of his research.[9]

Glass and Tungsten

Hubel's time at Walter Reed proved decisive. Opened in 1950, the Walter Reed Army Institute for Neuropsychiatry and Neurophysiology was a legacy of World War II and the countless cases of battle neurosis that the American army observed during that conflict. The institute's first director, the neuro-psychiatrist David Rioch, had been a protégé of Adolf Meyer's and had an eclectic medical background to match. Trained initially in surgery and neu-roanatomy, Rioch trained in psychoanalysis with Harry Stack Sullivan at the Chestnut Hill Lodge Psychiatric Hospital in Maryland, becoming its research director in 1943. Rioch observed firsthand the devastating psychiatric effects of combat during World War II and convinced General George Marshall to create a division of the Walter Reed Army Medical Center to engage in basic research on the problem of battle stress from a neurological point of view.[10] Rioch's open, multidisciplinary approach was itself a legacy of the same Meyerian psychobiology that contributed to the birth of the MNI, a fact noted by Hubel when he arrived: "In the neuropsychiatry division [of Walter Reed] David Rioch had assembled a broad and lively group . . . notably M. G. F. Fuortes and Robert Galambos in neurophysiology, Walle Nauta in neuroanatomy, Joseph Brady and Murray Sidman in experimental psychology and John Mason in chemistry. . . . As in Montreal, the focus was on the entire nervous system, not on a subdivision of biological subject matter based on methods."[11]

At Walter Reed, the focus of research was on investigating the underlying neurophysiology of stress and fatigue, and Rioch oriented early research in this area toward the brain stem structures that were the main preoccupation of neurophysiologists following the discovery of the centrencephalic and reticular activating systems only a few years earlier.[12] At the same time, research funding was plentiful, and Rioch encouraged animal experimentation. Hubel, brought on at Walter Reed because of his expertise in EEG, was instead schooled in the foreign discipline of animal experimentation by his immediate supervisor, Mike Fuortes. Fuortes's research was of a piece with much of the work done in laboratory neurophysiology in the 1950s. Following the overall reductionist path pioneered by E. D. Adrian in England, the axonologist group at Washington University, and Ralph Gerard in Chicago, those investigating the functions of the nervous system used increasingly fine micropipette electrodes to record electric impulses from nerve fibers and cells first within the peripheral nerves and, eventually, within the spinal cord. These glass microelectrodes could record electric signals from individual nerve cells, providing for the first time a picture of what those cells were doing.[13]

The sensitivity of these remarkable instruments came at the cost of fragility. Glass electrodes—first developed for use in botany—were seldom strong enough to record from any portion of the nervous system other than carefully prepared dissections of the spinal cord.[14] At the same time, the prospect of recording from individual cells in the cortex was enticing to anyone interested in neurophysiology. Hubel's mentor at the MNI, Herbert Jasper, had been suitably tempted and began a research program in single-unit (i.e., single-neuron) recording when Hubel was training at the MNI. Jasper's purpose was to reconcile the slow wave recordings EEG he made of whole brain regions with the deeper activity of the centrencephalic/reticular system and, ultimately, with the electric activity of individual cells.[15] At Walter Reed, Rioch and his team hoped to use a similar technique to understand the role of the reticular system in sleep and wakefulness as part of their broader research project on stress and fatigue. Both Jasper and Hubel, however, were initially unsuccessful at recording from any brain structures; glass electrodes, hand pulled down to only one μ in width, could not withstand even the normal pulsing of the brain or the breathing of an animal and would break before any readings could be obtained. Aware of these problems from his time in Montreal, Hubel knew he would need a more robust material if he were to extend recording from the spinal cord into the cortex.

Here, Hubel's presence at a military hospital was decisive. Hubel knew that steel—the obvious choice for a tough, flexible electrode—would not work; even steel became too brittle at the sizes demanded for single-unit recording. Leon Levin, the physical chemist and instrument maker at Walter Reed, suggested that Hubel try tungsten, which he could electrochemically "sharpen . . . with alternating current in a bath of concentrated sodium nitrite," a task that Hubel was able to complete owing to his background in physics and electronics: "The results were spectacular; within days I was able to make a pointed wire that looked ideal and was strong enough to pierce, with a little care, my thumbnail" (fig. 6.2).[16]

FIGURE 6.2. A microphotograph of one of David Hubel's early tungsten microelectrodes. Note the insulation that runs nearly to the tip, providing the ability to do precise recordings from single neurons. From David H. Hubel, "Tungsten Microelectrode for Recording from Single Units," *Science* 125, no. 3247 (1957): 550. Courtesy US Army.

While he had succeeded in producing a remarkable new instrument, other challenges remained, and here we can see the importance of weak ties in connecting the MNI's approach to other centers. Herbert Jasper, then in the midst of his work on single-unit recording from the cortex, heard of his former student's electrode and traveled to Walter Reed to learn how to make it. (It was Hubel's electrode that eventually enabled Jasper to make single-unit studies of conditioning in the monkey, and it was these studies that formed a bridge to the Soviet physiologists that brought forth the formation of the International Brain Research Organization, as we saw in chap. 3.)[17] However, recording from the cortex of even an anesthetized animal demanded more than a single instrument, and Jasper and his colleagues had mastered this part of the problem by drawing on their familiarity with neurosurgical methods. Using an implanted hollow screw in the skull, the electrode and an advancer could be attached to an animal and negate the effects of movement.[18] Hubel's assemblage of instruments was, in a very real sense, also an assembly of wired-together actors. Trading between the Montreal assembly and the Walter Reed assembly, by way of the weak tie of David Hubel, had created an entirely new form of investigation—a fusion of clinical electroencephalography, spinal cord physiology, and neurosurgical technique.

The results of Hubel's ingenuity and his training at the MNI and Walter Reed were spectacular. By 1958, Hubel was able to record from single brain cells. Crucially, his earliest recordings displayed an important transition from the overall focus at Walter Reed. His early experiments recorded from different areas of the cortex, including the deeper brain stem structures that had formed the core of research on sleep, stress, and fatigue.[19] Eventually, Hubel began recording from the striate cortex—the most anterior part of the occipital lobe, which received fiber projections from the lateral geniculate nucleus and, ultimately, from the retina—in sleeping and alert cats: "There were clear differences, but nothing that shed light on the nature of sleep."[20] Influenced by his earlier reading of the Kuffler and Hartline papers on the retina and lateral geniculate, he then began to wonder whether more complex cell activity related to vision might be occurring in the cortex itself. Certain cells in the cortex responded to changes in light, just as was predicted by Kuffler and Hartline's work, but others were totally unresponsive:

> I slowly became convinced that cortical cells required for their activation fancier stimuli than simply turning on or off the room lights. I started casting about for ways to make them react. My first successes came one day when out of desperation I waved my hand back and forth in front of a cat. My electrode

was lodged between two cortical cells that gave unequal amplitude spikes that I could easily tell apart, neither of which reacted to turning on and off the room lights. But to my amazement they responded vigorously to the hand-waving, and my amazement increased when I saw that one of the cells was responding to left-to-right movement and the other to right-to-left. Clearly the cortex must be doing something interesting![21]

Hubel's discovery that certain cortical cells responded selectively to move-ment in particular directions was profound. If it was possible to correlate an external stimulus, such as a particular visual phenomenon, to the activities of individual brain cells, it might be possible to unravel the functional physiol-ogy of vision within the brain itself and understand how the brain came to build up the visual world from the atomistic impulses of sensation. Moreover, the functional localization within the cortex, revived in the 1930s by Penfield and Jasper, could now be transformed into a more sophisticated investigation of the functions of systems within and between those regions. Hubel's micro-electrode could finally allow precise physiological investigation of the most complex functions of the brain.

What the Cat's Eye Tells the Cat's Brain

Meanwhile, word of Hubel's electrode had gotten out.[22] His publication of the method in *Science* had attracted the attention of a newly arrived postdoctoral fellow at Kuffler's laboratory at Johns Hopkins, Torsten Wiesel, a Swedish physician-turned-physiologist who had been hired to continue his work on the visual system. "My first task," Wiesel recalled, "was to learn how to make different types of microelectrodes, which was a relatively new technique for recording impulse activity from single nerve cells."[23] In the late 1950s, Wiesel traveled to Walter Reed to learn the electrode technique directly from Hubel. Here, Hubel's conversance with the laboratory culture of Kuffler, which fo-cused on vision (and was housed within the Johns Hopkins Wilmer Institute of Ophthalmology), allowed him to trade technique and material culture with Wiesel. This trading was effective enough that word spread back to Kuffler and his Hopkins colleague Vernon Mountcastle, who had also been attempt-ing electrode recordings from the cortex. Mountcastle extended an offer to Hubel to join Wiesel at Hopkins, and Hubel accepted, arriving in 1958.[24]

What followed was one of the most productive working relationships in modern science (fig. 6.3). Wiesel, having mastered the experimental appa-ratus of Kuffler's laboratory, could provide experimental setups that allowed for careful, controlled stimulation of the retina. With his new recording technique, Hubel could provide the instruments needed to investigate the complexities of the visual cortex. Once the pair had set up their equipment in 1958, the primary challenge was to discover the visual stimuli that would activate the recorded cell. Fishing for the right stimuli to activate a particular cell proved challenging. As Hubel recalled:

> The major breakthrough (to use that hackneyed term) came in our third or fourth experiment. We had isolated a big stable cell which for some hours was unresponsive to anything we did. . . . The ophthalmoscope had been designed for retinal stimulation and recording and was wonderful at generating spots of light of calibrated intensity or dark spots against a light background—but for cortical work it was a horror. . . . Spots of light were produced by a set of thin wafers the size of microscope slides, made either of brass with holes of various sizes to pass the light or, for black spots, glass slides to which thin metal circles of various sizes had been glued. These wafers, glass or brass, were inserted into a slot in the ophthalmoscope. . . . We struggled, and seemed to be getting nowhere, when suddenly we started to evoke brisk discharges. We finally realized that the discharges had nothing to do with the dark or light spots but were evoked by the action of inserting the glass slide into the slot. The cell was responding to the faint shadow of the edge of the glass moving across the retina, and it soon became clear that the responses occurred only over a limited range of orientations of the edge, with a sharply determined optimum and no response to orientations more than 30 degrees or so from the optimum.[25]

Even more remarkable than Hubel's informal discovery of a cell that responded to the selective motion of his hand was Hubel and Wiesel's discovery that individual cells in the cat cortex responded selectively to lines of specific orientation. Cells in the cortex, in effect, "saw" or responded to different elements of the visual world—line, shade, form, motion.

In subsequent experiments, the pair confirmed that cells that were selective for a particular orientation were organized into discrete columns within the cortex while cells that varied in their preference for orientation were arranged perpendicularly to the previous column. Over months of experimenting, Hubel and Wiesel sketched a picture of a visual cortex that organized perception hierarchically and rationally, from simpler sensations to increasingly complex perceptions. Up to this point, ideas about how the visual cortex constructed a picture of the world were vague at best.[26] The most robust ideas came from Hubel's former McGill colleague, Donald Hebb (see chap. 4), who argued that visual perceptions were built up from experience. Cells corresponding to points of visual sensation became associated into assemblies that represented lines, shapes, forms, and, ultimately, more complex perceptions

FIGURE 6.3. David Hubel and Torsten Wiesel during their early cat experiments, ca. late 1950s. Courtesy Countway Library for the History of Medicine, Harvard University.

and ideas.[27] Simultaneously, the cybernetic psychiatrists Warren McCulloch and Walter Pitts advanced the idea that visual perception and auditory perception were the product of the averaging of neural inputs through complex calculations computed by the neurons themselves. In each case, however, both theories of perception relied on the idea that the cortex was, for the most part, randomly organized and perception was a function of learning. These hypotheses, however, remained just that—theoretical constructions rather than experimental observations. Hubel and Wiesel, by contrast, had demonstrated the functional organization of neurons in a flesh-and-blood animal. Moreover, they had demonstrated that visual perception was dependent not on learning but rather on the innate structure of the visual cortex, which they could now explain in increasingly fine-grained detail.

In this respect, the studies of Hubel and Wiesel contrasted tellingly with those occurring simultaneously at MIT's Research Laboratory of Electronics (RLE). Founded in 1946 to preserve the wartime research of MIT's Radiation Laboratory, the RLE had become, by the 1950s, the vehicle through which Norbert Wiener hoped to translate the theoretical insights of cybernetics into practical research achievements. As Patrick Wall, an early recruit to the neurophysiology section of the RLE, recalled: "We were going to explain the nervous system in the ways that Wiener knew it worked."[28] Here, a small circle of researchers was attempting to investigate and model the nervous system in accordance with cybernetic theory. Two members of this group, Jerome Lettvin and Humberto Maturana, undertook a new series of experiments in the mid-1950s that resembled Hartline's and Kuffler's early work on the retina. Lettvin and Maturana used electrodes to record the activity of frog optic nerves and their response to different stimuli.[29] While they came to conclusions similar to Hartline's and Kuffler's, they also made more grand claims— that the retina itself, rather than the cortex, contained feature detectors for line, shade, curvature, and movement and even a potential "bug detector." They mined similar territory as Hubel and Wiesel, yet the fate of their paper "What the Frog's Eye Tells the Frog's Brain" reflected the different laboratory culture and assembly of actors at MIT. At MIT, their results were interpreted within the framework of cybernetics even if those results functioned as a significant challenge to that body of theory. They served as a major obstacle to the neural networking theories of Lettvin's RLE colleagues Warren McCulloch and Walter Pitts, who had argued that the visual cortex analyzed the visual world through the actions of logical neurons. As Lettvin himself put it, the frog experiments suggested that "[t]he eye speaks to the brain in a language already highly organized and interpreted" and that much of visual perception was analog rather than a function of the digital logic of the cortex.[30] By contrast, Hubel and Wiesel, who began from the empiricist and less theoretically driven laboratory culture of Kuffler, interpreted their work in more positive and productive terms and, as we will see below, almost immediately began to examine its possible clinical implications. Additionally, because of their more robust training in neuroanatomy, they were able in subsequent publications to point to serious problems and limitations of Lettvin and Maturana's work. As Hubel recalled: "We obviously had much in common with Lettvin and Maturana. . . . [They] carried their unusual approach to extremes, perhaps because they lacked an advisor like Steve [Kuffler], who insisted that we make at least a few measurements for the sake of scientific respectability."[31]

The prestige and excitement resulting from Hubel and Wiesel's discoveries would follow them to Harvard in 1960, when that university recruited their mentor, Steven Kuffler, to open a new department of neurobiology, the first department bearing such a name in the country. While Kuffler and others continued to focus on aspects of the peripheral nervous system and cell membranes, Hubel and Wiesel emerged as the "brain boys."[32]

Sensory Deprivation at Harvard

Hubel and Wiesel arrived at Harvard in 1960 as part of the new Department of Neurobiology. Their early work on visual perception succeeded because of the complex trading between the MNI's hybrid clinical/physiological environment and Kuffler's laboratory-based approach. At Harvard, they re-imported clinical concerns back into the laboratory. In June 1961, Hubel and Wiesel embarked on a new experimental program designed to determine which aspects of visual perception were innate and which were acquired during growth. This work was inspired both by theoretical concerns, such as those that motivated Hebb in his earliest studies on rats raised in darkness (see chapter 4), and by more immediate medical concerns:

> Our motives for studying kittens were clearly tied to our clinical backgrounds. We wanted to know whether the response properties of the cells we had been studying in adults were innate or were developed in early life by some process analogous to learning. We wanted to learn whether we could alter the responses by modifying the animals' early experience. . . . Our experiments were undoubtedly influenced by the observations, described in detail by [Marius] von Senden, . . . that children with congenital cataracts have substantial and often permanent visual deficits after removal of the cataracts. In animals, behavioral studies . . . had shown that dark rearing, or being raised in an environment devoid of contours, can similarly lead to severe visual impairments. It seemed important to establish whether these impairments were occurring in the eye or further downstream. Behavioral tests of vision are just one means of assessing the system, and we felt that our knowledge of cortical physiology might supply another useful index.[33]

Drawing on the same studies that had informed Donald Hebb's theoretical speculations about perception and cell assemblies, Hubel and Wiesel began their own variation on sensory deprivation. First, the men sewed shut one eyelid of a newborn kitten and removed the sutures after three months. Comparing the recordings of the experimental kitten with normal controls, Hubel and Wiesel determined that "after monocular deprivation the cortex was far from normal." Some cells in the cortex showed normal orientation selectivity, as though nothing had occurred. However: "There was no avoiding the conclusion that at least some cells had acquired their connections by mechanisms that must be innate."[34]

Still, it was also clear that the deprivation profoundly affected the kitten's vision; other cells showed "sluggish responses and poor orientation selectivity." When an opaque contact lens was placed on the good eye, the kitten frequently tumbled off a table onto a cushion, indicating that it was, for all intents and purposes, functionally blind. Hubel and Wiesel examined the kitten's brain under the microscope and discovered that the lateral geniculate layers supplied by the closed eye were thin and pale and had clearly not developed in the same way as those supplied by the opposite eye. In experiments on additional kittens, they discovered that cells in the striate cortex that would normally respond to visual information from both eyes now responded only to visual information from the eye that had not been sutured shut; in effect, the striate cortex had redistributed during development to favor the unsutured eye. The visual cortex reorganized itself during development on the basis of input from the environment. At the same time, some cortical cells would not respond to stimulus to either eye, indicating that, if deprived of stimulation during development, other areas of the visual cortex would never recover.[35] In effect, Hubel and Wiesel had turned their blinded kittens into a laboratory model for normal childhood visual development.

Subsequent experiments confirmed and extended Hubel and Wiesel's results and answered additional questions about the types of deprivation that might lead to permanent vision problems. If cortical cells were exposed only to diffuse light during development, they would not respond normally to form and shape later. If the kitten's eye was sewn shut at a later stage, vision problems would be less severe, and adult cats whose eyes were sutured and then later opened showed no loss of vision at all.[36]

While Hubel and Wiesel's research on visual deprivation had all the trappings of basic laboratory physiology—animal experimentation, careful controls, and highly artificial experimental setups—it ultimately had important clinical implications. Before their work, treatments for childhood vision ailments like congenital cataracts, strabismus (misalignment of the eyes), and amblyopia (lazy eye) lacked an underlying understanding of the mechanisms of visual development. The cause of these disorders was thought to be in the eye or the optic nerve, and surgery for congenital cataracts was not typically done on children; this form of surgery never restored complete vision, a fact that had no clear explanation.[37] By showing the crucial importance of visual experience to proper visual development, Hubel and Wiesel placed much of childhood ophthalmology on a rational basis, and their research had practical clinical implications; early surgery for congenital cataracts, for instance, went from a rare to a recommended intervention. While it may not have unlocked the secrets of memory, as Francis Schmitt's Neurosciences Research Program (NRP) hoped to do with its transdisciplinary discussion groups, Hubel and Wiesel's melding of clinical and laboratory research not only unlocked many of the secrets of vision but also unlocked vision for many children.

Indeed, there is an irony to Hubel's presence at Harvard in the 1960s. Simultaneously, down the road at MIT, Francis Schmitt was developing the NRP, with its emphasis on uncovering a molecular mechanism for memory formation (see chap. 3). He believed that discovering a transcendent memory molecule would unite physiology and psychology, mind and matter. Barely a mile away, Hubel and Wiesel were achieving that goal—not through a transcendent, totalizing discovery but through methodical, piecewise work. At Harvard's new Department of Neurobiology, a very different form of interdisciplinarity took root, one grounded in experimental practice and the material culture of the laboratory. The electrical investigation of the cortex that began for Hubel in the MNI's EEG laboratory had taken him from Montreal to Washington, DC, to Harvard along a series of weak ties that transformed his experimental practice and also transplanted a modified version of the MNI's interdisciplinary style to a new location. In its first decade of existence, the NRP never invited Hubel to attend its meetings, and it seems likely that Schmitt was not even aware of his work; the lack of any weak ties between them meant that the Harvard and the MIT assemblies—and their neurosciences—were worlds apart.

Neuroscience's First Nobel

The collaboration between Hubel and Wiesel continued for nearly two decades at Harvard, and, in the 1960s and 1970s, the implications of their work began to influence areas far afield from neurobiology and medicine. Beyond the immediate applications in ophthalmology and pediatrics, Hubel and Wiesel's studies of the neurophysiology of visual perception played a subtle but profound role in ushering forth the new field of cognitive science. In his 1967 textbook that surveyed the emerging field of cognitive psychology, Ulric Neisser noted that, for cognitive psychologists, "neuroanatomy and physiology" were "outside the limits" of the field; discussions of the mind could proceed independently of knowledge of the brain's physiology. However, Neisser made an exception for the work of Hubel and Wiesel, whose studies implied an innate structuring of brain function that was copacetic with the cognitivist's conception of the mind as an organ for information processing.[38] A decade later, the linguist Noam Chomsky, ordinarily unwilling to make any pronouncements on the neurological apparatuses that might underly his notion of the "deep structure" of language, also made an exception for Hubel and Wiesel. He noted that their work had demonstrated a "grammar" of vision innate to the human brain and that a comparable mechanism might soon be discovered for language.[39] For many in the field of cognitive science, Hubel and Wiesel's work suggested that the human brain was not just *like* a computer or an information processor but *really* worked in similar terms.

Conversely, when the study of artificial neural networks was finally revived in the 1980s, Hubel and Wiesel's work provided much of the needed physiological knowledge to enable convolutional neural networks to

recognize and perceive objects. In 1980, the Japanese computer scientist
Kunihiko Fukushima drew on Hubel and Wiesel's work to construct a new
neural network architecture that could learn and recognize form and shape;
today, Fukushima's work sits at the center of the neural networks that enable
all forms of visual pattern recognition, from mobile phones to satellites.[40]
While many cognitive scientists invoked Hubel and Wiesel to argue that
brains worked like computers, in the field of neural networking research their
discoveries made computers work more like human brains.

This outcome likely came as a surprise to Hubel himself. His career, which
had taken him from Montreal to Washington to Baltimore to Cambridge,
remained immersed in the flesh, blood, and nerves of actual humans and animals. Theoretical concerns were largely absent from his work, which was empirical to its core. Reflecting on the life of his friend and collaborator, Torsten
Wiesel noted: "David and I approached the visual cortex as explorers of a new
world. Neither of us had any preconceived ideas about what we would find
on our journey; instead, we let our discoveries dictate what questions to ask
next. At times we felt more like naturalists of a bygone era."[41] Of course, to
say that instruments and empiricism drove Hubel's career is not to say that it
was without shaping influences. In this respect, no influence could have been
more profound than that of the MNI, whose interdisciplinary brain science
shaped so much of his approach.

Hubel and Wiesel's collaboration continued until 1980, when increasing
administrative and teaching responsibilities forced an end to their partner-
ship.[42] The following year, the men were informed that they would share the
1981 Nobel Prize in Physiology or Medicine. While there had been other
Nobels awarded for work on the nervous system—Charles Sherrington and
Edgar Adrian in 1932, Joseph Erlanger and Herbert Spencer Gasser in 1944,
Walter Rudolf Hess in 1949, Ragnar Granit, Haldan Keffer Hartline, and
George Wald in 1967, and Bernard Katz, Ulf von Euler, and Julius Axelrod in
1970—these prizes were awarded for the kind of discipline-bound laboratory
physiology that had characterized the field for over a century. By contrast,
Hubel's discoveries shared a clear family resemblance to the interdisciplinary
neuroscience that was born at the MNI nearly a half century earlier. It would
seem fair to conclude, then, that Hubel's 1981 award was neuroscience's first
Nobel Prize.

Conclusion: Reassembling the Origins
of Neuroscience

Normal Neuroscience

In March 1982, David Hubel returned to his hometown. The occasion was the annual Hughlings Jackson Memorial Lecture, sponsored by the Montreal Neurological Institute (MNI), which Hubel had been asked to give in light of his recent receipt of the Nobel Prize. Hubel's MNI mentor Herbert Jasper wrote a note of congratulations to his former student: "Your remarkable studies with Torsten [Wiesel] on the organization of the visual system in its development have made a very great impact upon all of neuroscience, it seems to me. The principles that you have demonstrated are now being found applicable to the entire cerebral cortex in a remarkable manner."[1] Hubel replied in a handwritten note: "I've always been grateful for the start I got with you at the MNI. . . . More than anything else I owe to you a debt for an apprenticeship in clear thinking at a time and in an era when that must have been rare. How lucky I was to have started training, through nothing but accident, in Montreal."[2]

Hubel's training in Montreal certainly began at a fortuitous time. His apprenticeship at the MNI may have been brief—only a few years—but it occurred during the period in which the institute was at the height of its scientific powers, as Penfield, Jasper, Milner, and their colleagues unraveled the mysteries of temporal lobe epilepsy and placed much of their Montreal procedure on a sound scientific footing. Moreover, while Hubel participated in the strong assemblies of the MNI during its most active era, he also formed critical weak ties between Montreal and other crucial research sites. The microelectrode that he developed (what he later described as "the single most important tool in the modern era of neurophysiology")[3] was a product of those weak ties. And, much like Hubel's microelectrode, interdisciplinary neuroscience was a product of Montreal's strong assemblies and weak ties.

Hubel left Montreal during the MNI's heyday in the 1950s, the same decade when Penfield and his institute entered popular culture. Lavish coverage in magazines like *Time* and the *Saturday Evening Post* announced that Penfield was leading the world into an age of the explorable brain. This new age promised a revolution in brain science and brain control, one that was disconcerting to many onlookers. The science fiction author Philip K. Dick alluded to this new brain science in his 1962 novella *Do Androids Dream of Electric Sheep?* in which characters could use a so-called Penfield Mood Organ to stimulate their brains to produce different desired emotional states artificially. The revolutionary brain science assembled in Montreal had entered popular consciousness, and the future of that science seemed full of remarkable and frightening possibilities.[4]

Ironically, just as the contributions of the MNI were becoming well-known, its heyday was coming to a close. By the 1960s, its most exciting and revolutionary period was ending as many of its most crucial characters retired or left. Penfield, rapidly approaching seventy, retired from the MNI in 1960. His later years saw him enter the realm of revered elder statesman of neurosurgery and neuroscience. In his post-MNI "second career," he continued writing for the public and became a visible cultural commentator asked to weigh in on issues ranging from the perceived breakup of the nuclear family to the language policies of decolonizing nations.[5] This second career prefigured the growing cultural importance that the brain sciences would enjoy in the coming decades.

Soon after Penfield's retirement, Herbert Jasper decamped from the MNI for the Université de Montreal in 1964. The French-language university had inaugurated its own neuroscience program and tapped Jasper to head it. However, the reasons for his departure had more to do with local developments than with career ambitions. Changes to health care financing in Quebec forced the MNI to speed up its throughput of patients, meaning that less time could be spent on each one. Simultaneously, the Ottawa-based Medical Research Council began supporting the MNI through a block-grant system, which meant that sums of money from the federal government were used primarily to pay the salaries of neurosurgeons, with the leftovers devoted to laboratory research.[6] Consequently, laboratory-based research at the MNI that dealt with more basic science evaporated.

Reflecting on the situation shortly before his death, Herbert Jasper noted of the MNI: "I'm a little disappointed in some parts of its development. I miss the close teamwork we had in the early days, where we had Penfield, who was interested in all the clinical and research laboratories, and he brought us all together. And our ward rounds included research labs as well as the neurologists and neurosurgeons and the psychologists. That spirit of teamwork, I don't see today at the Neurological Institute. And that disappoints me."[7] Jasper's departure for the Université de Montreal indicated a broader shift in Montreal's science and technology landscape in the 1960s. The "quiet revolution"—the sociopolitical transformation of Quebec that followed the end of the Catholic political machine of Maurice Duplessis—had brought a greater emphasis on the funding of science in the province but often with the caveat that French-language science was to be of paramount importance. Science became a priority of the provincial government with the establishment of additional French-language universities, such as the University of Quebec at Montreal. Jasper's move to the French-speaking Université de Montreal thus fit with a broader pattern. As Montreal became a modern "city of knowledge," it also became more inward looking.[8] Although Jasper continued to do creative scientific work at the Université de Montreal, the most productive working relationship of his life was gone.

It would be wrong, however, to conclude that, because its most revolutionary period was behind it, the MNI fell into decline. Instead, it would be more accurate to say that it entered a period of "normal science," with its major paradigms and practices established and plenty of intriguing puzzles to solve.[9] Perhaps the most poignant example of this is Brenda Milner, who spent the remainder of her career at the MNI, retiring only in her early nineties. Milner later referred to the tenure of the institute's second director, Theodore Rasmussen, as a particularly productive period in her career as her neuropsychological testing became increasingly interwoven into the institute's operations. Milner's career stands as a living testament to the transformative power of the MNI to forge new scientific fields and careers.

The new scientific identities forged at the MNI also settled into a period of normal science. Perhaps the best example of this ties directly to the fate of the amnesic patient H.M. In 1960, Brenda Milner took on one of her first graduate students, a young woman from Connecticut, Suzanne Corkin. In 1962, Corkin examined H.M. during one of his rare visits to the MNI and discovered through a bizarre twist of fate that she had grown up across the street from William Beecher Scoville, the surgeon who had performed H.M.'s operation in 1953. After completing her training with Milner, Corkin was hired by the Massachusetts Institute of Technology (MIT) Psychology Department, which was in the process of aggressive expansion in the 1960s under the direction of the German immigrant psychologist Hans-Lukas Teuber. Teuber

encouraged Corkin to use her family connections, and, in the 1970s, she became the lead researcher on patient H.M., a position she would hold until her
death in 2016. In the intervening period, Corkin's MIT-based Laboratory of
Behavioral Neuroscience produced dozens of scientific papers about H.M.
and comparable patients along with a popular book on the topic. While the
output of Corkin's laboratory was prodigious, it was also predictable; all the
work done there fit within a paradigm developed by two other women, Molly
Harrower and Brenda Milner, at the MNI. However, rather than forging a
new science from a helping profession, Corkin could now take center stage as
a scientific celebrity in her own right.[10]

If the MNI transitioned to a period of normal science in the 1970s, this
could also be said for neuroscience more generally. The founding of the
Society for Neuroscience and the proliferation of journals and other forms
of scientific communication presaged the establishment of neuroscience as
a stable field. There was a certain irony in this; Francis Schmitt's brand of
neuroscience lived or died on the existence of the memory molecule that he
hoped would transcend the different neurological and psychological sciences.
The weakness of basing an entire field on a single hypothesis became obvious
when no such molecule could be found, and Schmitt and the Neurosciences
Research Program (NRP) had to pivot the aims of their project drastically.
At the same time, however, the 1970s saw a dramatic surge in research on the
functions of brain neurotransmitters. Driven mainly by research emerging
from the psychopharmacological revolution of the 1950s, neurochemistry
came to form a major plank of neuroscience in the 1970s as dozens of newly
discovered neurotransmitters were found to play crucial roles in mediating
everything from memory formation to mood.[11] The explosion of neurotrans-
mitter research was largely unanticipated by those working at the MNI, and
its scientists played only a small role in this field of research (as we saw in
chap. 3 with the work of Herbert Jasper and K. A. C. Elliott on the neurotrans-
mitter gamma-aminobutyric acid). However, while Schmitt's fragile neuroscience had to fold when new discoveries rendered its perspective moot, the
more robust interdisciplinary neuroscience birthed at the MNI could absorb
these new developments without toppling. By the 1970s, neuroscience was
here to stay.

Wired Together—History and Neuroscience

This book has attempted to reassemble the origins of modern neuroscience by
tracing the lives of the men and women who built the MNI and gave it life over
the middle decades of the twentieth century. Reassembling the complicated

genealogy of neuroscience reveals five things. First, dynamic new interdisciplinary fields of science do not simply spring into existence from one man's insight or force of vision. This was certainly true in the case of F. O. Schmitt, whose attempt to create a reductionist, transdisciplinary neuroscience at MIT was not only not the beginning of modern neuroscience but also not even successful in its more modest goal of discovering a molecular memory mechanism. However, a similar point could be made about Penfield and the MNI. To be sure, Penfield's vision of an institute for exploration of the brain was crucial for creating an environment in which a new interdisciplinary field could grow. However, returning to the image with which this book began—of Wilder Penfield in the operating room, stimulating a patient's brain with an electric probe—we can now see that the nameless subordinates around him were every bit as important to his scientific enterprise as the surgeon himself was. This is, of course, not to diminish Penfield's accomplishments but rather to reframe them. The achievement of Penfield and the MNI was not merely to make one man's vision a reality but rather to create a space where others could participate, bringing their unique abilities to bear on scientific and medical problems. Bridging disciplinary divides necessitated creating new assemblies of individuals, providing them with concrete problems to solve and places to synchronize their work. It also involved leadership, compassion, and the humility required to admit error and ask for help. Penfield's willingness to acknowledge that he was out of his depth, that his interventions were imperfect, and that he needed the help of outsiders was one of the reasons his collaborations with psychologists and physiologists succeeded and his collaborations with psychiatrists failed. Interdisciplinarity demands not only leadership and vision but also humility and a broad knowledge of another person's scientific language.

Second, the assemblies out of which neuroscience emerged in Montreal included not only surgeons and scientists but patients as well. Historians of medicine have, of course, long observed that patients played a crucial role in the production of medical knowledge, from the cadavers of revolutionary Paris hospitals to ethically scandalous experiments in the twentieth century. Indeed, it was not rats or squid that sought treatment in Penfield's institute but human beings, often vulnerable, scared, and seeking comfort. Patients like Penfield's sister Ruth and those anonymized by their initials—F.C., P.B., K.M.—proved crucial in spurring scientific discovery and often participated in the process directly. This is not to suggest that Penfield took advantage of his patients or exposed them to unnecessary risk. By all accounts, he treated their suffering with great humanity and seriousness and took every precaution he could to avoid exposing them to unnecessary risk. It is worth noting, however, that much of the neuroscientific knowledge generated at the MNI came about because of efforts to lessen the adverse effects of Penfield's surgeries. Indeed, the assembly of actors at the MNI often did their best work while trying to mitigate the risks of Penfield's radical operations. Including these patients as part of the assemblies that made neuroscience possible not only might challenge us to think more deeply about the ethical issues involved in brain research but also might encourage us to see safety and compassion as spurs to scientific discovery rather than as hindrances.[12]

Third, understanding the origins of neuroscience in Montreal can highlight some residual tensions that emerge in any interdisciplinary science. One is the difficulty in scaling up interdisciplinary collaborations. In her recent study of the history of scientific cooperation, Lorraine Daston closes by noting that, despite all attempts to forge international scientific cooperation at a global level, the possibility of creating true collaboration still depends on scientists "being in the room together."[13] Certainly, this was true of the MNI, where surgeons and scientists could form strong assemblies together and inaugurate a tradition of brain science that culminated in discoveries about memory, new theories of mind, and even a Nobel Prize. The example of this institute also created a belief, palpable and hopeful, that a new science of the brain was possible if interdisciplinary work could be scaled up to the global level, uniting a new field of science—neuroscience—in a common project to understand the brains of humans and animals. Yet, as we have seen, innovations from the MNI frequently transformed as they spread along weak ties to the global brain research community, often in ways that made them unrecognizable from their originating forms. Hebb's theory of intelligence, Hubel's microelectrode, and, perhaps most consequentially, Jasper's vision of an interdisciplinary brain research organization all changed in important ways as they made their way from Montreal to the rest of the world. Tracing the family tree of the MNI can show us how its strong assemblies could travel along weak ties to shift even the broadest landscapes of science; yet it also reveals the challenges faced by any effort to scale up or transplant a local scientific collaboration to a global level.

Fourth, the history of the MNI sheds light on one of the key tensions within neuroscience itself—the conflict between its concrete scientific achievements and its grander transcendent ambitions. Neuroscience grew from the psychobiology of the 1930s and the close collaborations funded by philanthropic organizations like the Rockefeller Foundation (often with idiosyncratic agendas of their own). Interdisciplinary sites like the MNI were rare and fragile but also innovative and exciting and inaugurated traditions of research that can be traced like the descendants on a family tree. The crowning achievement of Herbert Jasper, the formation of the International Brain Research Organization (IBRO), attempted to make the world's brain scientists into members of this family, with the shared sense of kinship and purpose this implies. Yet Jasper and IBRO discovered very quickly that this was impossible—the brain researchers of the world were not a family but a diverse community, with each member or group possessing different methods, priorities, and traditions that could not be easily coordinated with one another. This is not to belittle the MNI's or IBRO's achievements. To scale up the MNI's interdisciplinary approach, IBRO first had to create the social world of neuroscience through the laborious task of surveys and correspondence. This act transformed neuroscience from a family tree into a map, creating a shared sense of community. This community created the stability that neuroscience now enjoys as an established feature on the modern scientific landscape.

Yet the cost of that stability may have been creativity. Today, the Society for Neuroscience has thousands of members at hundreds of sites around the globe who struggle (sometimes successfully) to integrate their work. Today's neuroscience is impressively—even bewilderingly—*multidisciplinary*, with a dazzling assortment of approaches, techniques, and levels of analysis. Computational models of the brain sit next to molecular and genetic research, and neuropsychologists still run tests on surgical patients and stroke victims. Yet grouping these approaches under the label *neuroscience* may mislead us into thinking that integrating these approaches is simple or even desirable. The history of the MNI suggests that, if modern neuroscience wants to stay wired together, it may need to recognize that true interdisciplinarity comes through careful local collaborations rather than grand and transcendent pronouncements of interdisciplinarity.

Finally, understanding the real history of neuroscience as the product of these challenging local collaborations could benefit neuroscience itself. This point is perhaps best illustrated by a final anecdote. In many ways, the role of the MNI in creating modern neuroscience was obscured and minimized by a certain historical mythmaking initiated by F. O. Schmitt and perpetuated by his student and colleague George Adelman. In 1991, Adelman was invited to attend a conference at the MNI and felt compelled to write Herbert Jasper a letter afterward:

> [T]he talks [at the MNI conference] gave me a new perspective on Dr. Penfield's and your accomplishments in developing the Montreal Neurological Institute. I am beginning to realize much more clearly that you and he, in a very real sense, were primary founders of the field of neuroscience. Given my "basic" and NRP bias, my mind set/impression was that the problem of

"how the brain works" was attacked originally by basic scientists from a primarily philosophical-psychological base. It should have been more obvious to
me that clinicians who, closely involved with the faulty brain and wanting to
know how to fix it, would have been among the first to want to know how it
worked. . . . [You] further reminded me that I must broaden my views and pay
more than lip service to the importance of the feedback from clinical neuroscience to basic neuroscience for a true understanding of the mechanisms of
brain and mind.[14]

Adelman's confession might serve as a reminder to today's neuroscientists
not only of the value of close collaboration but also of the value of history.
Knowing the real history of this remarkable and endlessly fascinating area of
science can help us understand better its genuine achievements and the best
ways to overcome its limitations. Perhaps we can help neuroscience become
a truly interdisciplinary science of mind and brain if we first understand how
it was wired together.

Acknowledgments

Wilder Penfield's autobiography was entitled *No Man Alone*. When I first read this book at the beginning of my studies in the history of science, I dismissed the title as an exercise in false modesty. The more I learned about Penfield and his institute, however, the more I came to appreciate the truth of the title. I also became keenly aware that it could apply to my own experience of scholarship. In that spirit, I have a relatively long list of people to thank.

This book would have been impossible without the steady guidance, encouragement, and knowledge of Anne Harrington. Anne nurtured this book from impossibly vague beginnings and was a continual source of inspiration, probing questions, and thoughtful guidance. Along with her contributions, she served as an inestimable mentor in my becoming a historian and graciously agreed to read and comment on an early draft of this book. It is an immeasurably better product because of her support and wisdom, and I am endlessly grateful for her mentorship and friendship.

I also owe an enormous debt of gratitude to Rebecca Lemov and David Jones, who provided invaluable support and inspiration during my time at Harvard and beyond. Rebecca's endless curiosity and insight and David's exacting commentary and knowledge improved my scholarship in ways too numerous to count, and I am deeply in their debt. My time at Harvard was also significantly enriched by mentorship from Eram Alam, Allan Brandt, Janet Browne, Alex Csiszar, Peter Galison, Evelynn Hammonds, Matthew Hersch, Elizabeth Lunbeck, Hannah Marcus, Naomi Oreskes, Sarah Richardson, Steven Shapin, Victor Seow, Gabriela Soto Laveaga, Nadine Weidman, and Benjamin Wilson, along with Andrew Jewett, Ahmed Ragab, Sophia Roosth,

Melinda Baldwin, Bruce Moran, and Hannah Rose Shell. Allan, in particular (and unsurprisingly), served as a remarkable mentor during my postgraduate time at Harvard, and I value his contributions immensely. I was also lucky enough to spend time with the extraordinary figure of Everett Mendelsohn in his final years at Harvard and remain in awe of his knowledge, experience, and generosity.

A remarkable community enriched my years at Harvard, and I owe thanks to the many friends I made there. Dani and Andrew Inkpen proved to be remarkable friends, and Lisa Haushofer offered endless support and friendship on both sides of the Atlantic, as did my close friend Julia Reed. In moments of self-doubt and confusion, Brittany Luby and Erika Dyck kept me on the path and provided encouragement and friendship.

In my final years in Massachusetts, I somewhat belatedly discovered two impressive scholars, Danielle Carr and Andreas Killen, who shared my fascination with the midcentury brain sciences and provided crucial encouragement and thoughtful commentary. Jon Swinton graciously shared an early draft of his biography of Molly Harrower with me, and I greatly appreciate his insight and encouragement. Luke Dittrich shared his extensive expertise on the history of neurosurgery and the famous patient H.M. with me. Delia Gavrus shared her considerable knowledge and kindly read some of the early scholarship that grew from this research. I was also lucky enough to meet the towering figure of Michael Bliss in the Osler Library while he perused the records I had just finished with (Penfield's recently released letters and diaries). A single question from him ("Did Penfield ever do any lobotomies?") not only shifted the direction of my research but also concisely demonstrated his masterful historical instincts. I only regret that I never got the chance to answer his question in person.

While working on this book, I found myself transplanted to Zurich to join Flurin Condrau's bustling team at the Institut für Biomedizinische Ethik und Medizingeschichte. Flurin provided an ideal environment for the completion of this book and served as a valuable and wise counselor. His commitment to the history of medicine offers an exemplary model of scholarly mentorship, and I am deeply grateful for his support. My Swiss colleagues Leander Diener and Manuel Merkofer kindly commented on early drafts of this book and helped me manage the challenging transition to life in Switzerland.

Over six years as a historian in training, three years of teaching, and the writing and completion of this book, several friends outside the academy kept me sane. I owe a deep debt of gratitude to Nancy Theberge, Colin and Crystal Phillips, Jennifer and Jason Park, and Ryan and Hillary Davidson.

Liana Rosenkrantz Woskie was present for the writing of nearly every word in this book as our Zoom-based work sessions blossomed into a deep friendship. This friendship has meant the world to me, and this book would not have been completed without her support and companionship.

The research for this book was supported financially by generous grants from the Social Sciences and Humanities Research Council of Canada, the Harvard Weatherhead Center for International Affairs, and the Mary Louise Nickerson Fellowship in Neuro History. Numerous archivists and librarians in the United States and Canada managed my endless requests for material, but I want to single out the staff at the Osler Library of the History of Medicine in Montreal, whose knowledge and assistance was invaluable.

Portions of this book were previously published as "'The Sleeping Beauty of the Brain': Memory, MIT, Montreal, and the Origins of Neuroscience," *Isis* 112, no. 1 (2021): 22–44; and "Two Solitudes: Wilder Penfield, Ewen Cameron, and the Search for a Better Lobotomy," *Canadian Bulletin of Medical History* 38, no. 2 (2021): 253–84. I would like to thank the editors of those journals for their permission to republish portions of this earlier work. Karen Darling and the good folks at the University of Chicago Press lived up to their reputations for professionalism, enthusiasm, and expert editorship. Joseph Brown displayed surgical precision in his copyediting and corrected countless errors. I would also like to thank the anonymous reviewers of this book, whose thoughtful commentary improved the final result tremendously.

A number of surgeons, scientists, and their families shared resources and expertise with me and contributed to this book in countless ways. I was privileged to speak with the remarkable figures of Brenda Milner and the late William Feindel about their experiences at the MNI, and I am deeply grateful for their contributions. Richard Leblanc, whose scholarship on Penfield and the MNI has been invaluable, provided critical insights and saved me from making several embarrassing errors. Thanks also to Mary Ellen Hebb for her assistance and permission to use some of her late father's book.

Finally, my family has contributed to this book in more ways than I can name. My sister, Eva, provided years of steadfast support, encouragement, and good humor during my time in graduate school and beyond, and this book would never have been written without her. And most of all my parents, Ken and Glenda, shaped this book in ways that I cannot begin to list. My mother, one of the earliest female neuroscientists in Canada, and my father, a psychologist with a deep understanding of the past, nurtured my interest in the history of the brain and mind sciences since I was a child, and their influence can be felt on every page. This book is for them.

Notes

Introduction

1. *Heritage Minutes: Wilder Penfield, Historica Canada*, 2016, https://www.youtube.com/watch?v=pUOG2g4hj8s.

2. See https://doodles.google/doodle/wilder-penfields-127th-birthday.

3. Wilder Penfield, "The Epilepsies: With a Note on Radical Therapy," *New England Journal of Medicine* 221, no. 6 (1939): 209–18.

4. Lawrence S. Kubie, "Some Implications for Psychoanalysis of Modern Concepts of the Organization of the Brain," *Psychoanalytic Quarterly* 22, no. 1 (1953): 46.

5. Some excellent examples of this scholarship include Anne Harrington, *Medicine, Mind, and the Double Brain: A Study in Nineteenth-Century Thought* (Princeton University Press, 1989), and *Reenchanted Science: Holism in German Culture from Wilhelm II to Hitler* (Princeton University Press, 1996); R. Smith, *Inhibition: History and Meaning in the Sciences of Mind and Brain* (University of California Press, 1992); Katja Guenther, *Localization and Its Discontents: A Genealogy of Psychoanalysis and the Neuro Disciplines* (University of Chicago Press, 2015); and E. Clarke and L. S. Jacyna, *Nineteenth-Century Origins of Neuroscientific Concepts* (University of California Press, 1992). Some examples of newer histories examining the emergence of modern neuroscience include F. W. Stahnisch, *A New Field in Mind: A History of Interdisciplinarity in the Early Brain Sciences* (McGill-Queen's University Press, 2020); Fabio De Sio, "One, No-One and a Hundred Thousand Brains: J. C. Eccles, J. Z. Young and the Establishment of the Neurosciences (1930s–1960s)," in *Progress in Brain Research*, vol. 243, *Imagining the Brain: Episodes in the History of Brain Research*, ed. Chaira Ambrosio and William MacLehose (Elsevier, 2018), 257–98; and Joelle M. Abi-Rached, "From Brain to Neuro: The Brain Research Association and the Making of British Neuroscience, 1965–1996," *Journal of the History of the Neurosciences* 21, no. 2 (April 2012): 189–213.

6. Judith P. Swazey, "Forging a Neuroscience Community: A Brief History of the Neurosciences Research Program," in *The Neurosciences, Paths of Discovery*, ed. Frederic G. Worden, Judith P. Swazey, and George Adelman (MIT Press, 1975), 529–46; George Adelman, "The Neurosciences Research Program at MIT and the Beginning of the Modern Field of Neuroscience," *Journal of the History of the Neurosciences* 19, no. 1 (January 15, 2010): 15–23.

214 NOTES TO PAGES 5–17

7. Swazey, "Forging a Neuroscience Community," 529.

8. The twentieth century saw a proliferation of discipline-crossing scientific initiatives alongside attempts by philosophers to restore a sense of unity to science. In understanding the rise of interdisciplinarity in science, I have found invaluable Julie Thompson Klein, *Interdisciplinarity: History, Theory, and Practice* (Wayne State University Press, 1990); Roberta Frank, "'Interdisciplinary': The First Half Century," *Issues in Interdisciplinary Studies* 6 (1988): 139–51; Peter Galison, "The Americanization of Unity," *Daedalus* 127, no. 1 (1998): 45–71, and "Meanings of Scientific Unity: The Law, the Orchestra, the Pyramid, the Quilt and the Ring," in *Pursuing the Unity of Science* (Routledge, 2016), 12–29; and Harvey J. Graff, *Undisciplining Knowledge: Interdisciplinarity in the Twentieth Century* (Johns Hopkins University Press, 2015).

9. On trading zones, see Peter Galison, *Image and Logic: A Material Culture of Microphysics* (University of Chicago Press, 1997). On the genesis of trading zones and some criticisms, see Peter Galison, "Trading with the Enemy," in *Trading Zones and International Expertise: Creating New Kinds of Collaboration*, ed. Michael E. Gorman (MIT Press, 2010), 25–52.

10. I owe this distinction to the excellent scholarship of Julie Thompson Klein. See Klein, *Interdisciplinarity*.

11. Mark S. Granovetter, "The Strength of Weak Ties," *American Journal of Sociology* 78, no. 6 (1973): 1360–80.

12. Granovetter, "The Strength of Weak Ties," 1360.

13. I examine Hebb's theory in detail in chap. 4. The "fire together, wire together" popularization of Hebb's theory is often attributed to Carla J. Shatz, "The Developing Brain," *Scientific American* 267, no. 3 (September 1992): 60–67. However, it may have first appeared in Siegrid Löwel and Wolf Singer, "Selection of Intrinsic Horizontal Connections in the Visual Cortex by Correlated Neuronal Activity," *Science* 255, no. 5041 (1992): 209–12.

14. Harrington, *Reenchanted Science*, xxiv. In the history of science, biography has a complex history of advocates and detractors. For a useful overview of its pros and cons, see Joan L. Richards, "Introduction: Fragmented Lives," *Isis* 97, no. 2 (June 1, 2006): 302–5; Mary Terrall, "Biography as Cultural History of Science," *Isis* 97, no. 2 (June 1, 2006): 306–13; Theodore M. Porter, "Is the Life of the Scientist a Scientific Unit?," *Isis* 97, no. 2 (June 1, 2006): 314–21; and Mary Jo Nye, "Scientific Biography: History of Science by Another Means?," *Isis* 97, no. 2 (June 1, 2006): 322–29. Mott Greene notes that scientific biographies tend to focus on the lives of outsized figures (e.g., Newton, Darwin, Einstein) and do not reflect the collaborative or cumulative nature of science. See Mott T. Greene, "Writing Scientific Biography," *Journal of the History of Biology* 40, no. 4 (December 3, 2007): 727–59. In this book, I use biographies of outsized figures like Penfield and lesser-known figures like Herbert Jasper, Donald Hebb, Molly Harrower, and others to get at this collaborative and cumulative process. It is impossible to do justice to all the figures who contributed to the MNI. Readers wishing to learn more should consult the authoritative William Feindel and Richard Leblanc, *The Wounded Brain Healed: The Golden Age of the Montreal Neurological Institute, 1934–1993* (McGill-Queen's University Press, 2016). On the outsized figure of Penfield, see the invaluable Richard Leblanc, *Radical Treatment: Wilder Penfield's Life in Neuroscience* (McGill-Queen's University Press, 2020).

Chapter 1

1. Wilder Penfield (WP) to Madeline Ottmann, April 18, 1930, C/G 4, box 43, WPP.

2. Wilder Penfield, *No Man Alone: A Neurosurgeon's Life* (Little, Brown, 1977), 247–58.

3. Wilder Penfield, "The Significance of the Montreal Neurological Institute," in *Montreal Neurological Institute Annual Reports, 1934–1960* (Montreal Neurological Institute and Montreal Neurological Hospital, 1934), 11.

4. Penfield, *No Man Alone*, 3–25; J. Lewis, *Something Hidden: A Biography of Wilder Penfield* (Formac, 1983), 1–28.

5. On the laboratory revolution in medicine, see A. Cunningham and P. Williams, *The Laboratory Revolution in Medicine* (Cambridge University Press, 2002); and Martha Gardner and Allan M. Brandt, "The Golden Age of Medicine?," in *Companion to Medicine in the Twentieth Century*, ed. Roger Cooter and John V. Pickstone (Routledge, 2000), 21–37. On the RF's goals in medicine, see William H. Schneider, "The Men Who Followed Flexner: Richard Pearce, Alan Gregg, and the Rockefeller Foundation Medical Divisions, 1919–1951," in *Rockefeller Philanthropy and Modern Biomedicine: International Initiatives from World War I to the Cold War*, ed. William H. Schneider (Indiana University Press, 2002), 7–60.

6. Conklin belonged to a generation of American biologists—including T. H. Morgan and Ross Harrison—who turned away from the fieldwork of naturalists and toward the laboratory. R. E. Kohler, *Landscapes and Labscapes: Exploring the Lab-Field Border in Biology* (University of Chicago Press, 2010); Penfield, *No Man Alone*, 18–20.

7. WP to Jean Jefferson Penfield (JJP), April 30, 1911, "1909–1913 Princeton Personal Corresp WGP to JJP Typescript Extracts," D C/D 33-2/1 1, box 41a, WPP.

8. On the growth of modern surgery, I have found the following enormously helpful: T. Schlich, *The Origins of Organ Transplantation: Surgery and Laboratory Science, 1880–1930* (University of Rochester Press, 2010), "Asepsis and Bacteriology: A Realignment of Surgery and Laboratory Science," *Medical History* 56, no. 3 (July 2012): 308–34, and "Negotiating Technologies in Surgery: The Controversy about Surgical Gloves in the 1890s," *Bulletin of the History of Medicine* 87, no. 2 (2013): 170–97; Gert H. Brieger, "From Conservative to Radical Surgery in Late Nineteenth-Century America," in *Medical Theory, Surgical Practice: Studies in the History of Surgery*, ed. Christopher Lawrence (Routledge, 1992), 226–27; and Peter Kernahan, "Franklin Martin and the Standardization of American Surgery, 1890–1940" (PhD diss., University of Minnesota, 2010). On neurosurgery, see M. Bliss, *Harvey Cushing: A Life in Surgery* (Oxford University Press, 2007); and Samuel H. Greenblatt, T. Forcht Dagi, and Mel H. Epstein, eds., *A History of Neurosurgery: In Its Scientific and Professional Contexts* (American Association of Neurological Surgeons, 1997).

9. As one surgeon explained in 1888: "Abdominal surgery is now the field where the most brilliant successes are to be attained. . . . It is from the work we are now doing & hope to do in abdominal surgery (& . . . cerebral surgery as well) that the Hospital must gain its position among the hospitals of the world at the end of the next ten years." M. H. Richardson quoted in Charles E. Rosenberg, *The Care of Strangers: The Rise of America's Hospital System* (Basic, 1987), 148–49.

10. Penfield, *No Man Alone*, 30–31.

11. WP to JJP, January 15, 1915, "Typescript Extracts WGP to JJP 1913 to 1916," D-C/D 33-2/2, box 41b, WPP.

12. Bliss, *Harvey Cushing*, 86, 110; William Feindel, "Osler and the 'Medico-Chirurgical Neurologists': Horsley, Cushing, and Penfield," *Journal of Neurosurgery* 99, no. 1 (2003): 188–99.

13. Gerald L. Geison, *Michael Foster and the Cambridge School of Physiology: The Scientific Enterprise in Late Victorian Society* (Princeton University Press, 1978).

14. John F. Fulton, "Sir Charles Scott Sherrington, O.M. (1857–1952)," *Journal of Neurophysiology* 15, no. 3 (May 1, 1952): 167. See also Judith P. Swazey, "Sherrington's Concept of Integrative Action," *Journal of the History of Biology* 1, no. 1 (1968): 57–89.

15. John F. Fulton, "The Historical Contribution of Physiology to Neurology," in *Science, Medicine, and History: Essays on the Evolution of Scientific Thought and Medical Practice Written in Honour of Charles Singer*, ed. E. Ashworth Underwood (Oxford University Press, 1953), 543. See also Swazey, "Sherrington's Concept of Integrative Action"; Gordon M. Shepherd, *Foundations of the Neuron Doctrine* (Oxford University Press, 1991); and Brian Bracegirdle, "The Microscopical Tradition," in *Companion Encyclopedia of the History of Medicine*, ed. Roy Porter and W. F. [William F.] Bynum, 2 vols. (Routledge, 1993), 1:102–19.

16. Penfield's observations on British medicine are in keeping with a trend identified by John Pickstone: "[P]hysiologists were the vanguard of academic medical science in Britain; from about 1860 they had promoted themselves as liberal educators, training medical students to be analytical observers of nature. By the end of the century most medical schools had physiology laboratories for teaching and for research." J. V. Pickstone, *Medical Innovations in Historical Perspective* (Palgrave Macmillan, 1992), 6.

17. WP to JJP, January 25, 1915, "Typescript Extracts WGP to JJP 1913 to 1916," D-C/D 33-2/2, box 41b, WPP.

18. WP to JJP, January 31, 1915, "Typescript Extracts WGP to JJP 1913 to 1916," D-C/D 33-2/2, box 41b, WPP.

19. C. S. Sherrington, *Mammalian Physiology: A Course of Practical Exercises* (Clarendon, 1919), 44, 49.

20. W. Feindel, "The Physiologist and the Neurosurgeon: The Enduring Influence of Charles Sherrington on the Career of Wilder Penfield," *Brain* 130, no. 11 (April 5, 2007): 2758–65.

21. WP to JJP, January 31, 1915, "Typescript Extracts WGP to JJP 1913 to 1916," D-C/D 33-2/2, box 41b, WPP.

22. Penfield, *No Man Alone*, 26–52.

23. Harvey Cushing, "The Special Field of Neurological Surgery," *Bulletin of the Johns Hopkins Hospital* 16, no. 168 (March 1905): 77–78.

24. Samuel H. Greenblatt, "The Emergence of Cushing's Leadership: 1901 to 1920," in Greenblatt, Dagi, and Epstein, eds., *A History of Neurosurgery*, 167–90; Bliss, *Harvey Cushing*.

25. Cushing, "The Special Field of Neurological Surgery," 78.

26. Bliss, *Harvey Cushing*, 359–60.

27. Wilder Penfield, "Sir Charles Sherrington, OM, GBE, FRS," *Nature* 169, no. 4304 (April 26, 1952): 690. By contrast, Cushing had a poor opinion of Sherrington, commenting in a private letter: "He operates well for a 'physiolog' but it seems to me much too much. I do not see how he can carry with any accuracy the great amount of experimental material he has under way. . . . As far as I can see, the reason why he is so much quoted is not that he has done especially big things but that his predecessors have done them all so poorly before." Quoted in Bliss, *Harvey Cushing*, 149–50.

28. Schlich, *The Origins of Organ Transplantation*, 149–51.

29. While Cushing embraced several aspects of a physiological approach (e.g., he was instrumental in introducing blood pressure monitoring into the operating room for the relief of intracranial pressure), the consensus opinion among historians is that his approach was still conservative. The range of diseases for which he prescribed surgery remained mainly tumors and cysts, and he aimed to retain as much brain tissue as possible. See Bliss, *Harvey Cushing*; Delia Gavrus, "Opening the Skull: Neurosurgery as a Case Study of Surgical Specialisation," in *The Palgrave Handbook of the History of Surgery*, ed. Thomas Schlich (Palgrave Macmillan, 2018), 435–55; and Greenblatt, "The Emergence of Cushing's Leadership."

30. Wilder Penfield, "Alterations of the Golgi Apparatus in Nerve Cells," *Brain* 43, no. 3 (November 1, 1920): 290–305. Penfield also completed a study within the classical Sherringtonian paradigm on spinal reflexes in decerebrated cats that produced an eighty-one-page research report. See H. Cuthbert Bazett and Wilder Penfield, "A Study of the Sherrington Decerebrate Animal in the Chronic as Well as the Acute Condition," *Brain* 45, no. 2 (1922): 185–265.

31. Penfield, *No Man Alone*, 43–52; Schlich, *The Origins of Organ Transplantation*, 31–52, 146–64; Arthur E. Lyons, "The Crucible Years, 1880 to 1900: Macewen to Cushing," in Greenblatt, Dagi, and Epstein, eds., *A History of Neurosurgery*, 153–66.

32. As Penfield told his mother: "[The surgeon's] easy going carelessness with regard to asepsis must be responsible for occasional failure. Br[itain] in surgery at present saves so few patients. Most of the cases are beyond hope anyway and the only possible thing is to relieve them from headache and to save their eyesight for a little while, and the number of immediate deaths are so enormous in comparison with other branches of surgery that many physicians advise against turning their cases over to the surgeon even when they do realize that they have a brain tumor." WP to JJP, January 22, 1921, D-C/D 33-3 1917–1928, box 41a, WPP.

33. WP to JJP, January 22, 1921, D-C/D 33-3 1917–1928, box 41a, WPP.

34. R. D. McClure, the chief surgeon, flatly denied the possibility of doing animal research at the otherwise "splendid modern hospital." Privately, Penfield commented that the problem was not so much McClure as the superintendent, Ernest G. Liebold, Henry Ford's friend and private secretary. According to Penfield: "[Liebold] is hopeless. I told him I wanted to continue research. He was ridiculous. He said, well we have a good laboratory and if there is any problem that needs solving, we will just turn it over to specialists there and we will get things solved. He talked about medicine as any conceited ignoramus might." WP to JJP, June 8, 19, 1921, D-C/D 33-3 1917–1928, box 41a, WPP. Elsewhere, Penfield speculated that animal research was not carried on because the hospital was "Ford Driven" and because "no animal experimentation is allowed as Henry [Ford] is prejudiced against it." Penfield, *No Man Alone*, 55. Liebold's judgment was open to question; a rabid anti-Semite, he had been crucial in Ford's republication of the forged Protocols of the Elders of Zion and was later an open Nazi sympathizer.

35. Notably, the chief of surgery, Allen O. Whipple, was enthusiastic about Penfield's training in physiology and pathology and willing to teach him the ropes of general surgery. Whipple would later develop a surgical procedure for pancreatic cancer very much in line with the physiological conception of surgery. Penfield, *No Man Alone*, 53–73.

36. WP to JJP, July 3, 1921, D-C/D 33-3 1917–1928, box 41a, WPP.

37. WP to JJP, October 2, 1921, D-C/D 33-3 1917–1928, box 41a, WPP.

38. James Lawrence Pool and Charles Albert Elsberg, *The Neurological Institute of New York, 1909–1974: With Personal Anecdotes* (Pocket Knife, 1975).

39. WP to JJP, December 4, 1921, D-C/D 33-3 1917–1928, box 41a, WPP.

40. WP to JJP, November 20, 1921, D-C/D 33-3 1917–1928, box 41a, WPP.

41. John Hughlings Jackson, "On the Anatomical, Physiological and Pathological Investigation of Epilepsies," *West Riding Lunatic Asylum Medical Reports* 3 (1873): 315–39. See also David A. Steinberg and George K. York, *An Introduction to the Life and Work of John Hughlings Jackson with a Catalogue Raisonné of His Writings* (Wellcome Trust Centre for the History of Medicine, University College London, 2006).

42. Owsei Temkin, *The Falling Sickness: A History of Epilepsy from the Greeks to the Beginnings of Modern Neurology* (Johns Hopkins University Press, 1994), 303–82. A succinct discussion of the history of brain localization theory and Hughlings Jackson's role in it can be

found in Anne Harrington, "Beyond Phrenology: Localization Theory in the Modern Era," in *The Enchanted Loom: Chapters in the History of Neuroscience* (History of Neuroscience 4), ed. Pietro Corsi (Oxford University Press, 1991), 207–34. See also Harrington, *Medicine, Mind, and the Double Brain*, 206–34; Susan Leigh Star, *Regions of the Mind: Brain Research and the Quest for Scientific Certainty* (Stanford University Press, 1989); Steinberg and York, *The Life and Work of John Hughlings Jackson*; and Michael Hagner, "The Electrical Excitability of the Brain: Toward the Emergence of an Experiment," *Journal of the History of the Neurosciences* 21, no. 3 (July 2012): 237–49.

43. These included psychiatric asylums as well as such specialized institutions as New York's Craig Colony, founded in 1896. Ellen Dwyer, "Stories of Epilepsy, 1880–1930," in *Framing Disease: Studies in Cultural History*, ed. C. E. Rosenberg and J. L. Golden (Rutgers University Press, 1992), 248–72.

44. A. E. Osborne, "Responsibility of Epileptics," *Medico-Legal Journal* 11, no. 2 (1893): 210–11.

45. Dwyer, "Stories of Epilepsy," 256–57.

46. Dwyer, "Stories of Epilepsy," 249–53. Epilepsy skirted the boundary between neurology and psychiatry, and early surgical therapies were of a piece with other surgical approaches to mental illness in the early twentieth century. For instance, under the belief that mental illness resulted from focal infections of bodily organs, the psychiatrist Henry Cotton directed surgeons to removed ovaries, testicles, tonsils, spleens, portions of colon, and teeth from patients at the New Jersey State Lunatic Asylum from 1907 to 1930. See A. Scull, *Madhouse: A Tragic Tale of Megalomania and Modern Medicine* (Yale University Press, 2007).

47. This *Stückchen-Pathologie* (bit pathology) never caught on in Europe but became increasingly popular in the United States. Raffaele Lattes, "Surgical Pathology at the College of Physicians and Surgeons of Columbia University," in *Guiding the Surgeon's Hand: The History of American Surgical Pathology*, ed. Juan Rosai (American Registry of Pathology, 1997), 41–60.

48. Penfield, *No Man Alone*, 90–95; Percival Bailey, *Intracranial Tumors* (Charles C. Thomas, 1933); Sherise Ferguson and Maciej S. Lesniak, "Percival Bailey and the Classification of Brain Tumors," *Neurosurgical Focus* 18, no. 4 (April 2005): 1–6.

49. Penfield, *No Man Alone*, 90–95.

50. WP to JJP, March 1924, D-C/D 33-3 1917–1928, box 41a, WPP.

51. The purported backwardness of Spanish science even gave rise to a debate, the *polémica de la ciencia española*, in which Cajal was a participant. In 1897, one year before the disastrous Spanish-American War cost Spain its American colonies, Cajal addressed the Academy of Science of Madrid, calling for more state sponsorship of science and more connection with the scientific nations of Europe. Agusti Nieto-Galan, "The History of Science in Spain: A Critical Overview," *Nuncius* 23, no. 2 (2008): 211–36; Wilder Penfield, "The Career of Ramón y Cajal," *Archives of Neurology and Psychiatry* 16, no. 2 (1926): 213–20.

52. Santiago Ramón y Cajal, *Recollections of My Life* (MIT Press, 1989), 57.

53. Michael D. Gordin, *Scientific Babel: How Science Was Done Before and After Global English* (University of Chicago Press, 2015), 105–86.

54. As Thomas Schlich has observed, before World War I, internationalism was a key practice among surgeons (who traveled to different clinics to learn surgical technique) and a virtue of the surgical community, whose members viewed their profession as peaceable, scientific, and forward-looking. This trend came to an abrupt halt during World War I, but Penfield's trips were an early attempt to revive this international exchange. See Thomas Schlich, "'One and the Same

the World Over': The International Culture of Surgical Exchange in an Age of Globalization, 1870–1914," *Journal of the History of Medicine and Allied Sciences* 71, no. 3 (July 2016): 247–70. On internationalism in science before World War II, see Catharina Landström, "Internationalism between Two Wars," in *Internationalism and Science*, ed. Aant Elzinga and Catharina Landström (Taylor Graham, 1996), 46–77; and Elisabeth A. Crawford, "The Universe of International Science, 1880–1939," in *Solomon's House Revisited: The Organization and Institutionalization of Science*, ed. Tore Frängsmyr (Science History Publications, 1990), 251–69.

55. Wilder Penfield, "Pío del Río-Hortega, MD, 1882–1945," *Archives of Neurology and Psychiatry* 54, no. 5 (1945): 413–16; Miguel Prados and William C. Gibson, "Pío del Río-Hortega, 1882–1945," *Journal of Neurosurgery* 3, no. 4 (1946): 275–84; Amandip S. Gill and Devin K. Binder, "Wilder Penfield, Pío del Río-Hortega, and the Discovery of Oligodendroglia," *Neurosurgery* 60, no. 5 (May 2007): 940–48.

56. Penfield, *No Man Alone*, 94.

57. For the first confirmation of the new cell type, see Wilder Penfield, "Oligodendroglia and Its Relation to Classical Neuroglia," *Brain* 47, no. 4 (1924): 430–52. See also Gill and Binder, "Wilder Penfield, Pío del Río-Hortega, and the Discovery of Oligodendroglia." The identification of oligodendroglial tumors was largely reliant on Penfield's improved staining methods. See Percival Bailey and Paul C. Bucy, "Oligodendrogliomas of the Brain," *Journal of Pathology and Bacteriology* 32, no. 4 (1929): 735–51; and Leblanc, *Radical Treatment*, 37.

58. Pío del Río-Hortega and Wilder Penfield, "Cerebral Cicatrix: The Reaction of Neuroglia and Microglia to Brain Wounds," *Bulletin of the Johns Hopkins Hospital* 41, no. 5 (November 1927): 278–303; Wilder Penfield, "The Mechanism of Cicatricial Contraction in the Brain," *Brain* 50, nos. 3–4 (1927): 499–517; Wilder Penfield and Richard C. Buckley, "Punctures of the Brain: The Factors Concerned in Gliosis and in Cicatricial Contraction," *Archives of Neurology and Psychiatry* 20, no. 1 (1928): 1–13.

59. On scientific virtues, see Lorraine Daston and Peter Galison, *Objectivity* (Zone, 2007). Daston and Galison use *epistemic virtue* to mean valuable but distinct scientific goals—e.g., objectivity. I use *scientific virtue* in a related sense, meaning a social or moral ethic thought necessary for the effective conduct of science. Here, Penfield's internationalism was a route to the techniques and perspectives that might be locked away in other countries. Penfield likely absorbed this virtue from Osler, who frequently recommended "brain dusting" trips to Europe for medical students. Michael Bliss, *William Osler: A Life in Medicine* (University of Toronto Press, 2002), 320.

60. Wilder Penfield, "Proposed Laboratory of Neurocytology," C/G 3-1/1, box 43, WPP.

61. Penfield, *No Man Alone*, 118.

62. Wilder Penfield, ed., *Cytology and Cellular Pathology of the Nervous System, 3 vols.* (Paul B. Hoeber, 1932), 1:vii.

63. WP to Isabella Rockefeller, May 21, 1927, C/G 2-1, box 43, WPP. Penfield scratched out *religion* in the draft and replaced it with *philosophy*.

64. M. A. Entin, *Edward Archibald: Surgeon of the Royal Vic* (McGill-Queen's University Press, 2004); Wilder Penfield, "Edward Archibald 1872–1945," *Canadian Journal of Surgery* 1, no. 2 (1958): 167–74.

65. Edward Archibald to WP, June 8, 1927, A/M 11/1-2, box 4, WPP.

66. WP to JJP, March 1924, D-C/D 33-3 1917–1928, box 41a, WPP.

67. Bliss, *William Osler*, 80–121; J. Hanaway and R. L. Cruess, *McGill Medicine: The Second Half Century, 1885–1936* (McGill-Queen's University Press, 2006), 110–200.

68. WP to Edward Archibald, July 17, 1927, A/M 11/1-2, box 4, WPP.

69. WP to JJP, July 17, 1927, D-C/D 33-3 1917–1928, box 41a, WPP.

70. Archibald to WP, August 9, 1927, A/M 11/1-1, box 4, WPP. Archibald went on: "There is nobody in Boston who can fill Cushing's shoes, and I shall expect at the very least that for the eastern half of the continent you will dispute the honours of the game with Dandy, with Grant following as an indifferent third. Not one of them all, that I know of, is doing any fundamental work in histology, and scarcely anybody in physiology or experimental surgery. With your training along those lines you will repeat Cushing's career, and do better than he did along the experimental line." Ibid.

71. WP to JJP, January 29, 1928, D- C/D 33-3 1917–1928, box 41a, WPP.

72. Katja Guenther has also examined the relationship between Penfield and Foerster, in particular their differing use of the reflex concept and the ambitions of neurology and neurosurgery. Guenther, *Localization and Its Discontents*, 96–125, 153–184. Foerster's technique was not wholly original, nor was Penfield's adoption of it. The neurosurgeons Fedor Krause and Harvey Cushing both experimented with electric stimulation and removal of epileptogenic scar tissue early in the century, although neither pursued it extensively. In a 1932 letter to Penfield, Cushing noted: "I, too, just 30 years ago was extirpating a cortex [*sic*] for epilepsy. If I had had the industry and ability that you and Foerster combine, I might have gone ahead with it and made something out of it. But I soon dropped it for things I thought I could do better." Quoted in William Feindel, Richard Leblanc, and Jean-Guy Villemure, "History of the Surgical Treatment of Epilepsy," in Greenblatt, Dagi, and Epstein, eds., *A History of Neurosurgery*, 474.

73. WP to JJP, February 26, 1928, D-C/D 33-3 1917–1928, box 41a, WPP.

74. Wilder Penfield, "Impression of Neurology, Neurosurgery, and Neurohistology in Central Europe," W/U 17, box 155a, WPP.

75. Wilder Penfield, "Impression of Neurology, Neurosurgery, and Neurohistology in Central Europe." In a private letter, Penfield put it more bluntly: "I have little respect for Parisian science right now." WP to JJP, August 26, 1928, D-C/D 33-3 1917–1928, box 41a, WPP.

76. Penfield, "Impression of Neurology, Neurosurgery, and Neurohistology in Central Europe." The Kaiser Wilhelm Institute für Hirnforschung deserves a brief digression. Founded in 1914 by Oscar and Cecil Vogt, the institute was perhaps the leading brain research center in Europe. The Vogts employed a dozen assistants to perform careful histology of human and animal brains, launching the specialized field of cytoarchitectonics, in which thin slices of brain were examined microscopically and cell types described in an attempt to correlate areas of the brain with psychological functions—a kind of microlocalization. Penfield was impressed by the Vogts' enterprise: "To prepare one brain for study requires the exclusive work of one technician for one year." Yet he also felt the institute suffered from having no clinical component and viewed the rococo localization of cytoarchitectonics with skepticism. Ibid.

77. WP to JJP, February 26, 1928, D-C/D 33-3 1917–1928, box 41a, WPP.

78. Penfield, "Impression of Neurology, Neurosurgery, and Neurohistology in Central Europe."

79. William H. Schneider, "The Model American Foundation Officer: Alan Gregg and the Rockefeller Foundation Medical Divisions," *Minerva* 41, no. 2 (2003): 155–66, and "The Men Who Followed Flexner"; S. C. Wheatley, *The Politics of Philanthropy: Abraham Flexner and Medical Education* (University of Wisconsin Press, 1988), 167–96.

80. "Public-Health Statesman," *Time*, November 26, 1956.

81. Theodore M. Brown, "Alan Gregg and the Rockefeller Foundation's Support of Franz Alexander's Psychosomatic Research," *Bulletin of the History of Medicine* 61, no. 2 (1987): 155–82; Jack D. Pressman, "Human Understanding: Psychosomatic Medicine and the Mission of

the Rockefeller Foundation," in *Greater Than the Parts: Holism in Biomedicine, 1920–1950*, ed. Christopher Lawrence and G. Weisz (Oxford University Press, 1998), 189–208; Schneider, "The Men Who Followed Flexner," 38–39.

82. Alexander Forbes and Alan Gregg, "Electrical Studies in Mammalian Reflexes: 1, The Flexion Reflex," *American Journal of Physiology* 37, no. 1 (1915): 118–76.

83. Vincent quoted in Schneider, "The Men Who Followed Flexner," 160.

84. Pressman, "Human Understanding"; Brown, "Alan Gregg and the Rockefeller Foundation's Support of Franz Alexander's Psychosomatic Research"; Schneider, "The Men Who Followed Flexner," 31–43.

85. On Meyer's transformative role in American psychiatry, see S. D. Lamb, *Pathologist of the Mind: Adolf Meyer and the Origins of American Psychiatry* (Johns Hopkins University Press, 2014). See also Jack David Pressman, *Last Resort: Psychosurgery and the Limits of Medicine*, Cambridge History of Medicine (Cambridge University Press, 1998), 18–46.

86. Stanley Cobb, *Borderlands of Psychiatry* (Harvard University Press, 1943), 10. The concept of disciplinary borderlands frequently prefigured the concept of interdisciplinarity in twentieth-century science. Reports from the American National Research Council and the National Academy of Sciences are littered with references to disciplinary "borderlands" and "borderline research." Frank, "'Interdisciplinary,'" 140–41.

87. Gregg had, in fact, partially paid for Penfield's trip to Breslau by way of a small grant. Entin, *Edward Archibald*, 112.

88. WP to Richard Pearce, January 10, 26, 1929, A/N 2/2, box 6, WPP.

89. Pearce gave over $1 million to McGill in 1920 to advance medical instruction, reflecting the RF's older tradition of funding medical education rather than research directly. Marianne P. Fedunkiw, *Rockefeller Foundation Funding and Medical Education in Toronto, Montreal, and Halifax* (McGill-Queen's University Press, 2005), 120; Wilder Penfield, *The Difficult Art of Giving: The Epic of Alan Gregg* (Little, Brown, 1967), 233.

90. Penfield and Gregg eventually became close friends, and Penfield wrote the definitive biography of Gregg, which has served as the main secondary source for nearly all historical writing on his tenure at the RF. See Penfield, *No Man Alone*, 294–302, and *The Difficult Art of Giving*, 230–51. See also the documents in A/N 2/3, box 6, WPP.

91. Schneider, "The Men Who Followed Flexner," 44.

92. Brown, "Alan Gregg and the Rockefeller Foundation's Support of Franz Alexander's Psychosomatic Research"; Bonnie Ellen Blustein, "Percival Bailey and Neurology at the University of Chicago, 1928–1939," *Bulletin of the History of Medicine* 66, no. 1 (1992): 90–113.

93. Rockefeller Foundation, "The Rockefeller Foundation Annual Report—1934" (Rockefeller Foundation, 1934), 78–80.

94. Penfield, *No Man Alone*, 129.

95. Penfield, "The Significance of the Montreal Neurological Institute," 18.

96. Harvey Cushing, "Psychiatrists, Neurologists and the Neurosurgeon," *Yale Journal of Biology and Medicine* 7, no. 3 (1935): 191–207.

97. Robert Kohler notes that the scientific patronage of private foundations acted as a precursor to the postwar federal patronage of science, serving as a bridge from the individualistic science of the nineteenth century to the big science of the twentieth. Although he does not discuss the MNI, it would seem a perfect example. R. E. Kohler, *Partners in Science: Foundations and Natural Scientists, 1900–1945* (University of Chicago Press, 1991), 404–6.

98. "Illocality," *New York Times*, April 21, 1932, 20.

Chapter 2

1. Penfield, *No Man Alone*, 202–3.

2. Penfield, *No Man Alone*, 161–206.

3. Penfield, *No Man Alone*, 208.

4. Lewis, *Something Hidden*, 122–23.

5. See Wilder Penfield and Joseph Evans, "The Frontal Lobe in Man: A Clinical Study of Maximum Removals," *Brain* 58, no. 1 (1935): 115–33. See also Lewis, *Something Hidden*, 118–23; and Wilder Penfield and Joseph Evans, "Functional Defects Produced by Cerebral Lobectomies," *Proceedings: Association for Research in Nervous and Mental Diseases* 13 (1934): 352–77.

6. Rachel Elder, Katja Guenther, and Alison Winter have made similar points about Penfield's operating room and the role of patients as coinvestigators in his neuroscientific enterprise, and I am indebted to their scholarship. See Rachel Elder, "Speaking Secrets: Epilepsy, Neurosurgery, and Patient Testimony in the Age of the Explorable Brain, 1934–1960," *Bulletin of the History of Medicine* 89, no. 4 (2015): 761–89; Guenther, *Localization and Its Discontents*; and Alison Winter, *Memory: Fragments of a Modern History* (University of Chicago Press, 2012), 75–102. Here, I wish to pursue a related line of inquiry. How did Penfield's operating theater become a trading zone for different scientific disciplines?

7. Annmarie Adams has written elegantly about the working relationship between Penfield and the architecture firm Ross & Macdonald. Notably, the firm was also responsible for many Canadian architectural landmarks, including Toronto's Royal York Hotel, Montreal's Mount Royal Hotel, and additions to McGill University. Annmarie Adams, "Designing Penfield: Inside the Montreal Neurological Institute," *Bulletin of the History of Medicine* 93, no. 2 (2019): 207–40.

8. This community had a notably international flavor, including surgeons and scientists from the United States, Poland, Norway, and other European countries. Most notably, the English pathologist Dorothy Russell learned in Montreal the staining techniques that later made her the first woman to hold a pathology chair in Western Europe. Jennian F. Geddes, "A Portrait of 'The Lady': A Life of Dorothy Russell," *Journal of the Royal Society of Medicine* 90, no. 8 (1997): 455–61; Joseph P. Evans, "Exciting Beginnings," *Canadian Medical Association Journal* 116, no. 12 (1977): 1367.

9. The story of Penfield's vasomotor theory of epilepsy is complex but displays his scientific turn of mind. Penfield had suspected that forms of epilepsy not obviously caused by a traumatic brain injury (e.g., from birth injury or meningoencephalitis) might result from less obvious scars that caused blood vessels to contract, starving portions of the brain of oxygen, and subsequently causing epileptic attacks. Versions of this theory extended back at least to the work of Claude Bernard in the nineteenth century, but it was given a new lease on life following Penfield's initial research. Unfortunately, animal experimentation conducted by Penfield's research fellow Theodore Erickson elegantly falsified the so-called vasomotor hypothesis. See the excellent discussion of this clinicoexperimental program in Leblanc, *Radical Treatment*, 54–76. See also the elegant discussion in Delia Gavrus, "Epilepsy and the Laboratory Technician: Technique in Histology and Fiction," in *The History of the Brain and Mind Sciences: Technique, Technology, Therapy*, ed. Stephen T. Casper and Delia Gavrus (University of Rochester Press, 2017), 136–63.

10. Brain mapping, initiated in its modern form by Pierre Paul Broca in 1861, developed into a robust research program in the nineteenth century. Gustav Fritsch and Edward Hitzig's discovery of the motor cortex in 1870 spurred other investigators, such as David Ferrier, to make the electric probe a productive laboratory physiology tool. As a paradigm in neurology,

localization reached its apex in the late nineteenth century and the early twentieth in the elaborate diagrams of aphasiology and in the histological maps of Korbinian Brodmann. Perhaps inevitably, localization enthusiasm was followed by a period of holistic reaction in the early twentieth century, characterized by the clinical holism of Henry Head and Kurt Goldstein and the laboratory studies of Karl Lashley. These historical transformations have inspired a rich literature, including Harrington, *Medicine, Mind, and the Double Brain,* and "Beyond Phrenology"; Guenther, *Localization and Its Discontents*; R. M. Young, *Mind, Brain, and Adaptation in the Nineteenth Century: Cerebral Localization and Its Biological Context from Gall to Ferrier* (Oxford University Press, 1990); L. S. Jacyna, *Lost Words: Narratives of Language and the Brain, 1825–1926* (Princeton University Press, 2009); and Star, *Regions of the Mind.*

11. Wilder Penfield and Edwin Boldrey, "Somatic Motor and Sensory Representation in the Cerebral Cortex of Man as Studied by Electrical Stimulation," *Brain: A Journal of Neurology* 60, no. 4 (1937): 389–443.

12. Penfield, "Sir Charles Sherrington, OM, GBE, FRS," 690. For a detailed analysis of Penfield's brain maps, see Guenther, *Localization and Its Discontents.*

13. Wilder Penfield, "The Cerebral Cortex in Man: 1, The Cerebral Cortex and Consciousness," *Archives of Neurology and Psychiatry* 40, no. 3 (1938): 417.

14. Wilder Penfield, "The Radical Treatment of Traumatic Epilepsy and Its Rationale," *Canadian Medical Association Journal* 23, no. 2 (1930): 194. Penfield also relates the story of Hamilton in his autobiography. See Penfield, *No Man Alone,* 156–60 n. 26.

15. Bailey, *Intracranial Tumors,* 432, 433.

16. Penfield and Evans, "The Frontal Lobe in Man," 118.

17. Penfield and Evans, "The Frontal Lobe in Man," 131. Two additional cases, both young men who underwent similar operations, also included psychometric testing by a local psychiatrist who conducted rudimentary examinations and a version of the Binet-Simon IQ test. The results were inconclusive. See ibid., 120–28.

18. Penfield and Evans, "Functional Defects Produced by Cerebral Lobectomies."

19. John Fulton to WP, December 30, 1932, C/D 16, "Fulton," box 29, WPP.

20. Penfield and Evans, "The Frontal Lobe in Man." More will be said about Penfield's frontal lobe studies in chaps. 4 and 5.

21. To the extent that American psychologists thought about the nervous system, they were mainly reliant on the "conditioned reflexes," attributed to Ivan Pavlov, and employed in the behaviorist psychology of J. B. Watson. As is discussed in chap. 4, by the time Penfield's institute was in full swing, this "switchboard" nervous system would have been considered out-of-date by most neurophysiologists. On behaviorism, see Rebecca Lemov, *World as Laboratory: Experiments with Mice, Mazes, and Men* (Hill & Wang, 2005); John M. O'Donnell, *The Origins of Behaviorism: American Psychology, 1870–1920* (New York University Press, 1985); Daniel P. Todes, "From the Machine to the Ghost Within: Pavlov's Transition from Digestive Physiology to Conditional Reflexes," *American Psychologist* 52, no. 9 (September 1997): 947–55; J. A. Mills, *Control: A History of Behavioral Psychology* (New York University Press, 2000); and Diego Zilio, "Who, What, and When: Skinner's Critiques of Neuroscience and His Main Targets," *Behavior Analyst* 39, no. 2 (October 2016): 197–218. One American psychologist who retained an interest in the brain was Karl Lashley (discussed below). However, Lashley's holistic theory of brain function was diametrically opposed to the localizing tendencies of Penfield and would eventually bring the two men into conflict.

22. Michael M. Sokal, "The Gestalt Psychologists in Behaviorist America," *American Historical Review* 89, no. 5 (1984): 1240.

23. WP to Wolfgang Kohler, March 6, 1936, file C/G 36 K-L, box 53, WPP.

24. WP to Grant Fleming, January 11, 1937, file A/M 2/2-2, box 1, WPP.

25. WP to Grant Fleming, February 11, 1937, file A/M 2/2-2, box 1, WPP.

26. See, e.g., D. Searls, *The Inkblots: Hermann Rorschach, His Iconic Test, and the Power of Seeing* (Crown, 2018), 198–207.

27. Molly Harrower, "Thursday's Child" (n.d.), and "Inkblots and Poems," in *The History of Clinical Psychology in Autobiography*, ed. C. Eugene Walker (Brooks/Cole, 1991), 1:130.

28. Harrower, "Thursday's Child."

29. Harrower, "Inkblots and Poems."

30. Harrower, "Inkblots and Poems," 133, and "Thursday's Child," 56.

31. Sokal, "The Gestalt Psychologists in Behaviorist America"; Mitchell G. Ash, *Gestalt Psychology in German Culture, 1890–1967: Holism and the Quest for Objectivity* (Cambridge University Press, 1995); Harrington, *Reenchanted Science*, 103–39.

32. M. R. Harrower, "Organization in Higher Mental Processes," *Psychologische Forschung* 17 (1932): 56–120. See also Harrower, "Inkblots and Poems," 132.

33. I am greatly indebted to Jonathan Swinton, who shared drafts of his forthcoming biography of Harrower, which tells her remarkable life story in much greater detail.

34. Molly Harrower to Kurt Koffka, December 23, 1933, folder 6, 1933 Dec, M3225, MHP.

35. Molly Harrower, folder 6, Original Proposal 1934, M3201, MHP.

36. WP to Molly Harrower, May 4, 1934, folder 29, Wilder Penfield (Montreal Neurological Institute) (1934–46), M3219, MHP.

37. William B. Barr, "Historical Development of the Neuropsychological Test Battery," in *Textbook of Clinical Neuropsychology*, ed. J. E. Morgan and J. H. Ricker (Taylor & Francis, 2016), 3–4, 5–10.

38. Anne Harrington, "Kurt Goldstein's Neurology of Healing and Wholeness: A Weimar Story," in *Greater Than the Parts: Holism in Biomedicine, 1920–1950*, ed. George Weisz (Oxford University Press, 1998), 25–45; and *Reenchanted Science*, 140–74.

39. Harrower, "Thursday's Child."

40. WP to Molly Harrower, November 12, 1936, folder 29, Wilder Penfield (Montreal Neurological Institute) (1934–46), M3219, MHP.

41. Molly Harrower to Kurt Koffka, December 9, 1936, folder 7, KK-MH Correspondence December 1936, M3228, MHP.

42. Molly Harrower to Kurt Koffka, January 24, 1937, folder 8, KK-MH Correspondence January 1937, M3228, MHP. Elsewhere, Harrower put the matter differently: "It may also happen that the Rockefeller people will insist on sending me to the good boy Cobb, and not the bad boy Penfield, Cobb who never makes a fuss or asks for what he wants, and Penfield who always wants more. However, I tell you, bad boy or not I would a thousand times rather work with him than Cobb who froze the very marrow of my bones!" Harrower to Koffka, December 16, 1936, folder 7, KK-MH Correspondence December 1936, M3228, MHP.

43. Molly Harrower to WP, January 31, 1937, folder 29, Wilder Penfield (Montreal Neurological Institute) (1934–46), M3219, MHP. The phrase "lack of initiative" is a clear reference to the condition Penfield observed in his sister and other frontal lobe patients.

44. James H. Capshew, *Psychologists on the March: Science, Practice, and Professional Identity in America, 1929–1969* (Cambridge University Press, 1999), 21, 30–31; Laurel Furumoto, "On the Margins: Women and the Professionalization of Psychology in the United States, 1890–1940," in *Psychology in Twentieth-Century Thought and Society*, ed. Mitchell G. Ash and William R. Woodward (Cambridge University Press, 1987), 93–113.

45. Harrower, "Inkblots and Poems," 137.

46. Harrower, "Thursday's Child," 142.

47. Harrower's proposal came at the same time that the American psychologist Ward Halstead was building a neuropsychological laboratory connected to Percival Bailey's neurosurgical clinic at the University of Chicago. Notably, Halstead was interested in how brain injury might reveal an underlying "biological intelligence" that could be quantified. By contrast, Harrower's approach centered on the mental and intellectual difficulties of individual patients. Harrower, too, would integrate her practice more closely into the actual surgical operations at Penfield's clinic. While Halstead's efforts resulted in the Halstead-Reitan neuropsychological test battery, the underlying theory of intelligence was largely discredited. Ralph M. Reitan, "Ward Halstead's Contributions to Neuropsychology and the Halstead-Reitan Neuropsychological Test Battery," *Journal of Clinical Psychology* 50, no. 1 (1994): 47–70; W. C. Halstead, *Brain and Intelligence: A Quantitative Study of the Frontal Lobes* (University of Chicago Press, 1949).

48. Molly Harrower to Kurt Koffka, October 4, 1937, folder 9, KK-MH Correspondence October 1937, M3229, MHP.

49. Harrower, "Thursday's Child," 148.

50. Harrower, "Inkblots and Poems," 137.

51. Harrower, "Inkblots and Poems," 140.

52. Molly Harrower to Kurt Koffka, January 26, 1938, folder 5, M3230, MHP. It was later disclosed to Harrower that that operation was "the worst of its kind" and "the most drastic initiation possible." Molly Harrower to Kurt Koffka, January 30, 1938, folder 5, M3230, MHP.

53. Molly Harrower to Kurt Koffka, February 2, 1939, folder 4, MH-KK Correspondence February 1939, M3232, MHP. Harrower's pessimism was premature. Penfield's surgical effectiveness improved notably in the next decade, particularly as Jasper's EEG work was integrated.

54. Molly Harrower to Kurt Koffka, February 16, 1938, folder 7, KK-MH Correspondence February 1938 Cont'd, M3230, MHP.

55. Molly Harrower to Kurt Koffka, January 28, 1938, folder 5, MH-KK Correspondence January 1938 Cont'd, M3230, MHP.

56. Harrower, "Thursday's Child," 153, 154–55.

57. Molly Harrower to Kurt Koffka, February 19, 1940, folder 13, MH-KK Correspondence January and February 1940, M3232, MHP.

58. Harrower, "Thursday's Child," 155.

59. Molly R. Harrower, "Changes in Figure-Ground Perception in Patients with Cortical Lesions," *British Journal of Psychology* 30, no. 1 (1939): 51.

60. Molly Harrower to Kurt Koffka, January 14, 1939, folder 3, MH-KK Correspondence January 1939, M3232, MHP.

61. Harrower, "Thursday's Child," 166.

62. WP to Alan Gregg, February 23, 1939, file C/G 5-3/3 Rockefeller Gregg 1935–1955, box 44, WPP.

63. Molly Harrower to Kurt Koffka, September 23, 1939, folder 11, MH-KK Correspondence September and October 1939, M3232, MHP.

64. Molly Harrower to Kurt Koffka, November 26, 1939, folder 12, MH-KK Correspondence November and December 1939, M3232, MHP.

65. Harrower, "Thursday's Child," 157–58; Molly Harrower-Erickson, "Personality Studies in Cases of Focal Epilepsy," *Bulletin of the Canadian Psychological Association* 5 (1941): 19–21; M. R. Harrower-Erickson and F. R. Miale, "Personality Changes Accompanying Organic Brain Lesions: 3, A Study of Preadolescent Children," *Pedagogical Seminary and Journal of Genetic*

Psychology 58, no. 2 (1941): 391–405; Molly Harrower-Erickson, Wilder Penfield, and Theodore Charles Erickson, "Psychological Studies of Patients with Epileptic Seizures," in *Epilepsy and Cerebral Localization: A Study of the Mechanism, Treatment and Prevention of Epileptic Seizures, by Wilder Penfield and Theodore Charles Erickson* (Charles C. Thomas, 1941), 546–74.

66. To modern readers, the scientific value of Harrower's work may seem questionable. Indeed, many later questioned the reliability of psychological tests to distinguish between neurological and psychiatric conditions. One might also wonder whether Harrower's status was the product of her results suiting Penfield's surgical agenda. While the value of her work is open to question, the claim made here is merely that Harrower established an institutional precedent and amicable climate for psychologists within a surgical institute. Julian B. Rotter, "A Historical and Theoretical Analysis of Some Broad Trends in Clinical Psychology," in *Psychology—a Study of a Science: Study II: Empirical Substructure and Relations with Other Sciences*, vol. 5, *The Process Areas, the Person, and Some Applied Fields: Their Place in Psychology and in Science* (McGraw-Hill, 1962), 780–830; George Frank, "Research on the Clinical Usefulness of the Rorschach: 2, The Assessment of Cerebral Dysfunction," *Perceptual and Motor Skills* 72, no. 1 (1991): 103–11.

67. Molly Harrower, *The Psychologist at Work* (K. Paul, Trench, Trubner, 1937).

68. Harrower, "Thursday's Child," 166.

69. Harrower, "Inkblots and Poems," 143.

70. "War Work in the Montreal Neurological Institute," file A/N 16-2, box 14, WPP; "Address of the Director," file W/U 68, box 156, WPP; Lewis, *Something Hidden*, 164.

71. WP to R. E. Wodehouse, C/G 39 W, box 55, WPP.

72. M. R. Harrower-Erickson and M. E. Steiner, *Large Scale Rorschach Techniques: A Manual for the Group Rorschach and Multiple Choice Test* (Charles C. Thomas, 1945); Searls, *The Inkblots*, 198–201.

73. Harrower's post-MNI career is worthy of a brief digression. Unfortunately, Wisconsin's "nepotism ruling prohibited husband and wife from both being associated with the university," and Harrower-Erickson found herself unemployed and depressed. The marriage did not survive, and the two divorced in 1945. Her copyright on the group Rorschach technique ensured financial stability, allowing her to move to New York, where she was one of only two women regularly involved in the Macy Foundation meetings that served as the origin point for the postwar cybernetics movement (the other was Margaret Mead). Having discovered psychoanalysis in the interim, she retrained as a psychotherapist and pursued a continued interest in poetry. She and Penfield carried on a lengthy correspondence, well into the 1960s. Her career trajectory was in many respects atypical; unlike many female psychologists, she was not barred from productive war work. More typical was the impermanent career path followed by many female psychologists who did not have permanent academic appointments because of the applied nature of their work. Molly Harrower, "Changing Horses in Mid-Stream: An Experimentalist Becomes a Clinician," in *The Psychologists: Autobiographies of Distinguished Living Psychologists*, ed. T. S. Krawiec (Clinical Psychology Publishing Co., 1978), 3:85–104, "Inkblots and Poems," 143, and "Thursday's Child"; Capshew, *Psychologists on the March*, 74.

74. Harrower, "Inkblots and Poems," 140.

75. Wilder Penfield and Theodore C. Erickson, *Epilepsy and Cerebral Localization: A Study of the Mechanism, Treatment and Prevention of Epileptic Seizures* (Charles C. Thomas, 1941), 281.

76. Penfield and Erickson, *Epilepsy and Cerebral Localization*, 301; Wilder Penfield and Harry Steelman, "The Treatment of Focal Epilepsy by Cortical Excision," *Annals of Surgery* 126, no. 5 (1947): 740–62.

77. Wilder Penfield and Herman Flanigin, "Surgical Therapy of Temporal Lobe Seizures," *Archives of Neurology and Psychiatry* 64, no. 4 (1950): 491–500; Herbert Jasper and John

Kershman, "Electroencephalographic Classification of the Epilepsies," *Archives of Neurology and Psychiatry* 45, no. 6 (June 1941): 903–43; Herbert Jasper, Bernard Pertuisset, and Herman Flanigin, "EEG and Cortical Electrograms in Patients with Temporal Lobe Seizures," *Archives of Neurology and Psychiatry* 65, no. 3 (1951): 272–90.

78. Heinrich Klüver and Paul C. Bucy, "An Analysis of Certain Effects of Bilateral Temporal Lobectomy in the Rhesus Monkey, with Special Reference to 'Psychic Blindness,'" *Journal of Psychology* 5, no. 1 (January 1938): 33–54, "Preliminary Analysis of Functions of the Temporal Lobes in Monkeys," *Archives of Neurology and Psychiatry* 42, no. 6 (1939): 979–1000, and "' Psychic Blindness' and Other Symptoms Following Bilateral Temporal Lobectomy in Rhesus Monkeys," *American Journal of Physiology*, 1937.

79. John Hughlings Jackson, "On a Particular Variety of Epilepsy ('Intellectual Aura'), One Case with Symptoms of Organic Brain Disease," *Brain* 11, no. 2 (1888): 179–207; J. Hughlings Jackson and Walter S. Colman, "Case of Epilepsy with Tasting Movements and 'Dreamy State'— Very Small Patch of Softening in the Left Uncinate Gyrus," *Brain* 21, no. 4 (1898): 580–90; Prasad Vannemreddy, James L. Stone, Siddharth Vannemreddy, and Konstantin V. Slavin, "Psychomotor Seizures, Penfield, Gibbs, Bailey and the Development of Anterior Temporal Lobectomy: A Historical Vignette," *Annals of Indian Academy of Neurology* 13, no. 2 (2010): 103–7.

80. Erna L. Gibbs, Frederic A. Gibbs, and Bartolome Fuster, "Psychomotor Epilepsy," *Archives of Neurology and Psychiatry* 60, no. 4 (1948): 331–39.

81. Penfield and Flanigin, "Surgical Therapy of Temporal Lobe Seizures," 491.

82. Jasper, Pertuisset, and Flanigin, "EEG and Cortical Electrograms in Patients with Temporal Lobe Seizures"; Wilder Penfield and Maitland Baldwin, "Temporal Lobe Seizures and the Technic of Subtotal Temporal Lobectomy," *Annals of Surgery* 136, no. 4 (1952): 625–34; William Feindel and Wilder G. Penfield, "Localization of Discharge in Temporal Lobe Automatism," *Archives of Neurology and Psychiatry* 72, no. 5 (November 1, 1954): 605–30; K. M. Earle, M. Baldwin, and W. Penfield, "Temporal Lobe Seizures: The Anatomy and Pathology of the Probable Cause," *Journal of Neuropathology and Experimental Neurology* 12, no. 1 (1953): 98–99; Kenneth Martin Earle, Maitland Baldwin, and Wilder Penfield, "Incisural Sclerosis and Temporal Lobe Seizures Produced by Hippocampal Herniation at Birth," *Archives of Neurology and Psychiatry* 69, no. 1 (1953): 27–42.

83. Penfield and Baldwin, "Temporal Lobe Seizures and the Technic of Subtotal Temporal Lobectomy"; Antonio Nogueira de Almeida, Manoel Jacoben Teixeira, and William Howard Feindel, "From Lateral to Mesial: The Quest for a Surgical Cure for Temporal Lobe Epilepsy," *Epilepsia* 49, no. 1 (January 2008): 98–107.

84. Montreal Neurological Institute, "Montreal Neurological Institute, 14th Annual Report, 1948–1949" (1949), 5, Wilder Penfield Digital Collection, Osler Library of the History of Medicine, McGill University, http://digital.library.mcgill.ca/penfieldfonds/fullrecord.php?ID=10475.

85. WP to John Bassett, February 9, 1949, A/N 16-2, box 14, WPP.

86. Montreal Neurological Institute, "Montreal Neurological Institute, 15th Annual Report, 1949–1950" (1950), 5, Wilder Penfield Digital Collection, Osler Library of the History of Medicine, McGill University, http://digital.library.mcgill.ca/penfieldfonds/fullrecord.php?ID=10476.

87. Wilder Penfield, "Memory Mechanisms," *Archives of Neurology and Psychiatry* 67, no. 2 (1952): 184, 184–86.

88. Lashley quoted in Penfield, "Memory Mechanisms," 195–96. Nadine Weidman has persuasively argued that Lashley's commitment to equipotentiality was the product of a deeper commitment to the heritability of intelligence, which itself played into his racial politics. See Nadine M. Weidman, *Constructing Scientific Psychology: Karl Lashley's Mind-Brain Debates* (Cambridge University Press, 1999). See also Karl S. Lashley, *Brain Mechanisms and Intelligence:*

A Quantitative Study of Injuries to the Brain (University of Chicago Press, 1929), and "In Search of the Engram," in *Physiological Mechanisms in Animal Behaviour*, Society of Experimental Biology Symposium (Cambridge University Press, 1950), 454–82.

89. Brenda Milner, "Brenda Milner," in *The History of Neuroscience in Autobiography*, ed. Larry R. Squire and Society for Neuroscience (Society for Neuroscience, 1998), 2:278–305; *The FCIHR Video History of Medicine in Canada Project: Brenda Milner and Memory* (Friends of Canadian Institutes for Health Research, 2013).

90. Winter, *Memory*, 197–223.

91. Milner, "Brenda Milner," 280. See also Alan F. Collins, "An Intimate Connection: Oliver Zangwill and the Emergence of Neuropsychology in Britain," *History of Psychology* 9, no. 2 (2006): 89–112.

92. Milner, "Brenda Milner."

93. *FCIHR Video History . . . Brenda Milner and Memory.*

94. Milner, "Brenda Milner," 282.

95. *FCIHR Video History . . . Brenda Milner and Memory.* See also Milner, "Brenda Milner."

96. WP to Robert S. Morrison, October 10, 1952, A/N 16-1, box 14, WPP.

97. *FCIHR Video History . . . Brenda Milner and Memory.*

98. *FCIHR Video History . . . Brenda Milner and Memory.*

99. Milner, "Brenda Milner," 285–89.

100. *FCIHR Video History . . . Brenda Milner and Memory.* See also Milner, "Brenda Milner," 285–89.

101. "Meeting on the Temporal Lobe—January 31st, 1953," and "Temporal Lobe Research Project—February 9, 1953," folder A/N 16-1, box 14, WPP. Milner ultimately confirmed this hypothesis a decade later when P.B. died of a pulmonary embolism. Autopsy results revealed a long-standing atrophy of the right hippocampus. Wilder Penfield and Gordon Mathieson, "Memory: Autopsy Findings and Comments on the Role of Hippocampus in Experiential Recall," *Archives of Neurology* 31, no. 3 (September 1, 1974): 145–54.

102. On logical positivism and the postwar unity-of-science movement, see Galison, "The Americanization of Unity."

103. Folder W/P 226, box 144, WPP.

104. Brenda Milner and Wilder G. Penfield, "The Effect of Hippocampal Lesions on Recent Memory," *Transactions of the American Neurological Association* 80 (1955): 42–48; Wilder Penfield and Brenda Milner, "Memory Deficit Produced by Bilateral Lesions in the Hippocampal Zone," *Archives of Neurology and Psychiatry* 79, no. 5 (1958): 475–97.

105. *FCIHR Video History . . . Brenda Milner and Memory.* See also W. B. Scoville and B. Milner, "Loss of Recent Memory after Bilateral Hippocampal Lesions," *Journal of Neurology, Neurosurgery and Psychiatry* 20, no. 1 (1957): 11–21; Brenda Milner, "Memory Mechanisms," *Canadian Medical Association Journal* 116, no. 12 (1977): 1374–76, and "Brenda Milner"; Luke Dittrich, *Patient H.M.: A Story of Memory, Madness, and Family Secrets* (Random House, 2017).

106. Suzanne Corkin, *Permanent Present Tense: The Unforgettable Life of the Amnesic Patient, H.M.* (Basic, 2013); Dittrich, *Patient H.M.*; Larry R. Squire, "The Legacy of Patient H.M. for Neuroscience," *Neuron* 61, no. 1 (January 2009): 6–9.

107. Hospital realities continued to inform Milner's work. The original impetus for her studies of memory—Penfield's temporal lobe operations and the memory deficits they caused—demanded a more permanent solution. In the 1950s and 1960s, Milner adapted a test for language localization introduced by her MNI colleague Juhn Wada (the so-called amobarbital or Wada test) to determine whether an epileptic patient would be susceptible to the kind of

memory loss suffered by earlier patients. Crucially, she worked with H.M. to develop a psychological testing procedure that would separate memory deficits from language deficits. Her adaptation of the Wada test became standard procedure for the surgical centers that would eventually take up the Montreal procedure. Brenda Milner, "Amobarbital Memory Testing: Some Personal Reflections," *Brain and Cognition* 33, no. 1 (1997): 14–17.

Chapter 3

1. Unless otherwise stated, biographical details for Jasper are drawn from Herbert H. Jasper, "Herbert H. Jasper," in *The History of Neuroscience in Autobiography*, ed. Larry R. Squire and Society for Neuroscience (Society for Neuroscience, 1996), 1:320–46; and "Memoirs—Autobiography of Seventy Years Adventures in Neuroscience Research and with Neuroscientists," Jasper's unpublished autobiography, available at the WFF (a copy was generously lent me by Richard LeBlanc). For "restless genes," see "Memoirs," 19. For "saturated" and "philosophical problems," see "Herbert H. Jasper," 322, 321.

2. Jasper, "Herbert H. Jasper," 332–34.

3. The issue of student suicide was a prominent one in the 1920s and led many students to embrace laboratory psychology as a possible solution. See C. P. Loss, *Between Citizens and the State: The Politics of American Higher Education in the 20th Century* (Princeton University Press, 2012), 49–50. Jasper addressed this concern directly in his undergraduate thesis, noting the "inadequacy of intelligence tests in diagnosing student difficulties": "'The evidence from such cases as the Leopold and Loeb case in Chicago and the recent attention being drawn to student suicides . . . is impressing educators of today with the fact . . . that their problem is not solved when they have classified their students on an intellectual basis." See folder 320, 1, box 6, HJF. Jasper's thesis resulted in his earliest scientific publications: Herbert H. Jasper, "Optimism and Pessimism in College Environments," *American Journal of Sociology 34*, no. 5 (1929): 856–73, and "The Measurement of Depression-Elation and Its Relation to a Measure of Extraversion-Introversion," *Journal of Abnormal and Social Psychology* 25, no. 3 (1930): 307–18.

4. According to Jasper, Travis inherited from his mentor, Samuel T. Orton, the idea that "stutterers lacked sufficient hemispheric lateral dominance for effective speech and language development": "It was proposed that stuttering resulted from mixed lateral dominance, often due to the attempt to force a child to be right handed when they were naturally left handed." Growing from post-Broca notions of language lateralization in the left hemisphere of the brain, according to Jasper, Orton and Travis's experimental program aimed to "[develop] electronic amplifiers and oscilloscopes for the recording of action currents from muscle and nerve": "Together we developed techniques for studying the electrical excitability of human muscles with the method of chronaxie determination. . . . This was my first introduction to electrophysiological techniques which fascinated me from the beginning and were to play such a very important role in my future career. Travis was interested in using muscle action potentials to analyze the abnormalities of muscle activity during stuttering, as part of his work in Speech Pathology." Jasper, "Memoirs," 52. As a psychiatric problem that encouraged the growth of electrophysiology in the United States, stuttering may have been nearly as influential as epilepsy. Horace Winchell Magoun and Louise H. Marshall, *American Neuroscience in the Twentieth Century: Confluence of the Neural, Behavioral, and Communicative Streams* (A. A. Balkema, 2003), 398–404.

5. Robert G. Frank, "Instruments, Nerve Action, and the All-or-None Principle," *Osiris* 9 (1994): 208–35.

6. Frank, "Instruments, Nerve Action, and the All-or-None Principle."

7. Jasper, "Herbert H. Jasper," 326–27, and "Memoirs," 55–83.

8. Louise H. Marshall, "Instruments, Techniques, and Social Units in American Neurophysiology, 1870–1950," in *Physiology in the American Context, 1850–1940, ed. Gerald L. Geison* (Springer, 1987), 358–61.

9. Jasper, "Memoirs," 89.

10. Cornelius Borck, "Recording the Brain at Work: The Visible, the Readable, and the Invisible in Electroencephalography," *Journal of the History of the Neurosciences* 17, no. 3 (July 16, 2008): 367–79. The complex and revealing path that Berger took to the EEG and the contextual factors that isolated him from the main currents of neurophysiology are expertly explored by Cornelius Borck. In particular, Borck documents the exacting care and voluminous data that Berger collected to assuage his own doubts about the reality of brain waves and the subtle transformations in his experimental program. Cornelius Borck, *Brainwaves: A Cultural History of Electroencephalography* (Taylor & Francis, 2018), 25–75.

11. Borck, "Recording the Brain at Work."

12. Jasper, "Herbert H. Jasper," 328; Cornelius Borck, "Between Local Cultures and National Styles: Units of Analysis in the History of Electroencephalography," *Comptes rendus biologies* 329, nos. 5–6 (May 2006): 450–59. Jasper and his technician Leonard Carmichael published the first American confirmation of the EEG from work begun before they learned of Adrian's attempts to do the same. Herbert H. Jasper and Leonard Carmichael, "Electrical Potentials from the Intact Human Brain," *Science 81*, no. 2089 (1935): 51–53. See also Jasper, "Memoirs," 90–94.

13. Jasper and Carmichael, "Electrical Potentials from the Intact Human Brain," 53.

14. Frederic A. Gibbs, Erna L. Gibbs, and William G. Lennox, "Epilepsy: A Paroxysmal Cerebral Dysrhythmia," *Brain: A Journal of Neurology* 60, no. 4 (1937): 381–82.

15. Hallowel Davis, Frederic A. Gibbs, and William G. Lennox, "The Electro-Encephalogram in Epilepsy and in Conditions of Impaired Consciousness," *Archives of Neurology and Psychiatry* 34, no. 6 (1935): 1134.

16. H. H. Jasper, "Localized Analyses of the Function of the Human Brain by the Electro-Encephalogram," *Archives of Neurology and Psychiatry* 36 (1936): 1131.

17. Wilder Penfield, "Herbert Jasper," in *Recent Contributions to Neurophysiology: International Symposium in Neurosciences in Honor of Herbert H. Jasper/Colloque international en sciences neurologiques en hommage à Herbert H. Jasper*, ed. Jean-Pierre Cordeau and Pierre Gloor (Elsevier, 1972), 11. See also Jasper, "Memoirs," 121–28.

18. Jasper, "Memoirs," 118–27.

19. "McGill Will Open Brain Laboratory," *Montreal Gazette*, February 24, 1939.

20. Jasper averaged ten to twelve EEG examinations per day, along with two to three intraoperative EEG examinations during brain operations. Jasper, "Memoirs," 114.

21. Jasper and Kershman, "Electroencephalographic Classification of the Epilepsies"; Penfield and Erickson, *Epilepsy and Cerebral Localization*.

22. Jasper, "Memoir," 158.

23. WP to Lyman Duff, February 27, 1950, folder A/M 3 2/3, box 2, WPP.

24. K. A. C. Elliott, "An Unorthodox Career," *Bulletin of the Canadian Biochemical Society* 17 (1980): 12–25, and "Neurochemistry," *Canadian Medical Association Journal* 116, no. 12 (1977): 1372–73.

25. On the discovery of neurotransmitters, see E. S. Valenstein, *The War of the Soups and the Sparks: The Discovery of Neurotransmitters and the Dispute over How Nerves Communicate* (Columbia University Press, 2013). As Valenstein notes, while neurotransmitters between nerves

and muscles were established by the 1940s and chemical transmission within the spinal cord was established in the 1950s, it was not until the 1960s and 1970s that neurotransmitters within the brain itself became an accepted scientific fact. In this respect, Elliott and Jasper's research was impressively forward-thinking.

26. Elliott, "Neurochemistry."

27. Herbert H. Jasper, "The Saga of K. A. C. Elliott and GABA," *Neurochemical Research* 9, no. 3 (1984): 450.

28. Jasper, "The Saga of K. A. C. Elliott and GABA"; Kenneth Allan Campbell Elliott, Irvine H. Page, and J. H. Quastel, eds., *Neurochemistry: The Chemical Dynamics of Brain and Nerve* (Charles C. Thomas, 1955).

29. Herbert H. Jasper, "Electrical Signs of Epileptic Discharge," *Electroencephalography and Clinical Neurophysiology* 1, no. 1 (1949): 11–18; K. A. C. Elliott, "Biochemical Approaches in the Study of Epilepsy," *Electroencephalography and Clinical Neurophysiology* 1, no. 1 (1949): 29–31; Giuseppe Moruzzi and Horace W. Magoun, "Brain Stem Reticular Formation and Activation of the EEG," *Electroencephalography and Clinical Neurophysiology* 1, no. 1 (1949): 455–73.

30. Jerome Engle to Herbert Jasper, August 27, 1993, folder 145, box 3, HJF.

31. Committee on Neurobiology, *Survey of Neurobiology* (National Research Council, 1952), 2.

32. Paul Maclean to WP, November 25, 1951, C/G 51 M, box 61, WPP.

33. J. Droogleever-Fortuyn, "Reticular Formation and Choice of Behavior," in *The Collected Works of Warren S. McCulloch*, ed. Rook McCulloch (Intersystems, 1989), 4:1302.

34. Penfield, "Memory Mechanisms," 184. Penfield's central integrating system ideas appeared as early as 1938 but expanded rapidly in his collaborations with Jasper and acquired crucial new supporting evidence from EEG studies. Penfield, "The Cerebral Cortex in Man"; Wilder Penfield and Herbert H. Jasper, *Epilepsy and the Functional Anatomy of the Human Brain* (Little, Brown, 1954), and "Highest Level Seizures," in *Epilepsy: Proceedings of the Association, Held Jointly with the International League against Epilepsy, December 13 and 14, 1946, New York* (Research Publications—Association for Research in Nervous and Mental Disease), vol. 26, ed. William G. Lennox (Williams & Wilkins, 1947), 252–69; Herbert Jasper and Wilder Penfield, "Electrocorticograms in Man: Effect of Voluntary Movement upon the Electrical Activity of the Precentral Gyrus," *Archiv für Psychiatrie und Nervenkrankheiten* 183, nos. 1–2 (1949): 163–74.

35. Margaret B. Rheinberger and Herbert H. Jasper, "Electrical Activity of the Cerebral Cortex in the Unanesthetized Cat," *American Journal of Physiology—Legacy Content* 119, no. 1 (1937): 186–96; Moruzzi and Magoun, "Brain Stem Reticular Formation and Activation of the EEG."

36. Magoun quoted in Herbert H. Jasper and Frederic Bremer, eds., *Brain Mechanisms and Consciousness* (Charles C. Thomas, 1954), 307.

37. Jasper and Bremer, eds., *Brain Mechanisms and Consciousness*.

38. Boleslav L. Lichterman, "The Moscow Colloquium on Electroencephalography of Higher Nervous Activity and Its Impact on International Brain Research," *Journal of the History of the Neurosciences* 19, no. 4 (October 6, 2010): 313–32.

39. G. D. Smirnov, "Soviet Report on Montreal International Physiological Congress," *Bulletin of the Academy of Sciences of the USSR* 24, no. 1 (January 1954): 82.

40. Richard Leblanc, "The White Paper: Wilder Penfield, the Stream of Consciousness, and the Physiology of Mind," *Journal of the History of the Neurosciences* 28, no. 4 (October 2, 2019): 416–36.

41. Loren Graham notes that the reticular formation was objectionable to the Soviets for other reasons. For instance, in addition to implying "spontaneous activities (output without

corresponding input)," its filtering and shaping influence on sensory processes made it difficult to maintain a "'copy-theory' of knowledge, as most dialectical materialists [had] done." Moreover, atheistic Marxists thought it might function to reinsert religious discourse in physiology, acting as a Cartesian "seat of the soul." Loren R. Graham, *Science and Philosophy in the Soviet Union* (Knopf, 1972), 404–7. See also Luciano Mecacci, *Brain and History: The Relationship between Neurophysiology and Psychology in Soviet Research* (Brunner/Mazel, 1979).

42. Charles Shagass and Herbert H. Jasper, "Conditioning of the Occipital Alpha Rhythm in Man," *Journal of Experimental Psychology* 28, no. 5 (1941): 373–88; Choh-Luh Li, Hugh McLennan, and Herbert Jasper, "Brain Waves and Unit Discharge in Cerebral Cortex," *Science* 116, no. 3024 (1952): 656–57; Herbert Jasper, Peter Gloor, and Brenda Milner, "Higher Functions of the Nervous System," *Annual Review of Physiology* 18, no. 1 (1956): 359–86; Herbert H. Jasper and Benjamin Doane, "Neurophysiological Mechanisms in Learning," *Progress in Physiological Psychology* 2 (1968): 79–117; H. Jasper, G. Ricci, and B. Doane, "Microelectrode Analysis of Cortical Cell Discharge during Avoidance Conditioning in the Monkey," in *The Moscow Colloquium on Electroencephalography of Higher Nervous Activity: Sponsored by the Academy of Sciences of the USSR with the Cooperation of the International Federation of Societies for Electroencephalography and Clinical Neurophysiology, Moscow, October 6–11, 1958*, ed. H. Jasper and G. D. Smirnov (EEG Journal, 1960), 137–56, and "Patterns of Cortical Neuronal Discharge during Conditioned Responses in Monkeys," in *Ciba Foundation Symposium on the Neurological Basis of Behaviour*, ed. G. E. W. Wolstenholme and Cecilia M. O'Connor (Little, Brown, 1958), 277–94; David H. Hubel, "David H. Hubel," in Larry R. Squire and Society for Neuroscience, eds., *The History of Neuroscience in Autobiography*, 1:298–317; Pierre Gloor, "H. H. Jasper, Neuroscientist of Our Century," in *Neurotransmitters and Cortical Function*, ed. Massimo Avoli, Tomas A. Reader, Robert W. Dykes, and Pierre Gloor (Springer, 1988), 1–13.

43. *Herbert H. Jasper, MD Interviewed by Andre Olivier, MD*, 1993, https://youtu.be/q_ytTKzJ_UI.

44. Lichterman, "The Moscow Colloquium on Electroencephalography of Higher Nervous Activity and Its Impact on International Brain Research"; Céline Cherici, "Scientific Circulations between East and West in the Cold War Period: The International Moscow Colloquium (1958)," in *History of the Neurosciences in France and Russia: From Charcot and Sechenov to IBRO*, ed. Jean-Gaël Barbara, Jean-Claude Dupont, and Irina Sirotkina (Hermann, 2011), 243–58; Pierre Buser, Jean-Gaël Barbara, Boleslav Lichterman, and François Clarac, "The International Brain Research Organization from Its Conception to Adulthood," in ibid., 299–314; Louise H. Marshall, W. A. Rosenblith, P. Gloor, G. Krauthamer, C. Blakemore, and S. Cozzens, "Early History of IBRO: The Birth of Organized Neuroscience," *Neuroscience* 72, no. 1 (1996): 283–306.

45. Jasper kept an extensive diary during his trip to Russia. See Herbert Jasper, diary, 4, file 372, box 8, HJF.

46. Jean-Gaël Barbara, "French Neurophysiology between East and West: Polemics on Pavlovian Heritage and the Reception of Cybernetics in the USSR," in Barbara, Dupont, and Sirotkina, eds., *History of the Neurosciences in France and Russia*, 225–41.

47. Jasper, diary, 15, file 372, box 8, HJF.

48. Jasper, Ricci, and Doane, "Microelectrode Analysis of Cortical Cell Discharge during Avoidance Conditioning in the Monkey." See also Cherici, "Scientific Circulations between East and West in the Cold War Period."

49. Jasper, "Memoirs," 207.

50. Buser, Barbara, Lichterman, and Clarac, "The International Brain Research Organization from Its Conception to Adulthood"; Jasper, "Memoirs," 208.

51. Herbert Jasper to Rudolfo Llinas, October 7, 1993, folder 163, box 3, HJF.

52. The first effort, the International Brain Commission, was launched in London in 1903 by an assorted group of European neurologists, physiologists, anatomists, and psychiatrists as part of the International Association of Academies (IAA). The IAA was a transatlantic organization that aimed to promote international cooperation in sciences that could particularly benefit from such coordinated global work, such as geophysics and astronomy. In that spirit, the founders of the International Brain Commission hoped to coordinate different centers around the world to contribute to an international brain atlas that would require coordinated anatomical work. Ambitious, and populated by key scientific figures, including Santiago Ramón y Cajal, Wilhelm His, Constantin von Monakow, and Paul Fleschig, the organization could not survive the international disaster of World War I. Attempts to revive it in the 1920s were unsuccessful. Jochen Richter, "The Brain Commission of the International Association of Academies: The First International Society of Neurosciences," *Brain Research Bulletin* 52, no. 6 (2000): 445–57. Notably, after World War II, the Frenchman Alfred Fessard tried unsuccessfully to convince the United Nations to establish a brain research institute.

53. See also Marshall et al., "Early History of IBRO."

54. Biographical material is drawn from Francis O. Schmitt, "Adventures in Molecular Biology," *Annual Review of Biophysics and Biophysical Chemistry* 14, no. 1 (1985): 1–23, and *The Never-Ceasing Search* (American Philosophical Society, 1990).

55. Schmitt, *The Never-Ceasing Search*, 5–19.

56. Schmitt, "Adventures in Molecular Biology," 3.

57. Schmitt, "Adventures in Molecular Biology," 6–8.

58. Warren Weaver, "Molecular Biology: Origin of the Term," *Science* 170, no. 3958 (1970): 581–82. See also Kohler, *Partners in Science*, 316–18; and Lily E. Kay, *The Molecular Vision of Life: Caltech, the Rockefeller Foundation, and the Rise of the New Biology* (Oxford University Press, 1993).

59. Kohler, *Partners in Science*, 317.

60. The F. O. Schmitt Oral History Collection at MIT is replete with reminiscences of the Schmitty Vereins and other recollections of his time at Washington University. A representative example is an interview conducted with Viktor Hamburger, a renowned embryologist who was skeptical of Schmitt's reductionist program. See "Viktor Hamburger," box 1, SOHC; and Schmitt, *The Never-Ceasing Search*, 118–22.

61. Schmitt, "Adventures in Molecular Biology," 12.

62. Schmitt, *The Never-Ceasing Search*, 131.

63. "David Robertson," box 2, SOHC.

64. Schmitt, *The Never-Ceasing Search*, 145–76; Nicolas Rasmussen, *Picture Control: The Electron Microscope and the Transformation of Biology in America, 1940–1960* (Stanford University Press, 1999), 153–96. An elegant discussion of Schmitt and his appetite for squid can be found in Kathryn Maxson Jones, "Francis O. Schmitt: At the Intersection of Neuroscience and Squid," in *Why Study Biology by the Sea?*, ed. Karl S. Matlin, Jane Maienschein, and Rachel A. Ankeny (University of Chicago Press, 2020), 187–210.

65. J. L. Oncley, R. C. Williams, F. O. Schmitt, and R. H. Bolt, eds., *Biophysical Science: A Study Program* (Wiley, 1959). See also Schmitt, *The Never-Ceasing Search*, 188–99.

66. F. O. Schmitt, "Acknowledgements," in Oncley, Williams, Schmitt, and Bolt, eds., *Biophysical Science*, v. See also Schmitt, *The Never-Ceasing Search*, 198; and Nicolas Rasmussen, "The Mid-Century Biophysics Bubble: Hiroshima and the Biological Revolution in America, Revisited," *History of Science* 35, no. 3 (September 1997): 245–93.

67. "Theodore Melnechuk," 12–13, box 1, SOHC. Schmitt's wets and dries mapped onto a larger battle being waged, sometimes referred to as the *war of the soups and the sparks*. This decades-long debate over the existence of neurotransmitters pitted neurophysiologists (sparks) against pharmacologists (soups). The dries roughly map onto the "sparks," but Schmitt also held most "soups" in contempt as well as their attention was directed to chemical interactions *between* nerve cells, a theory not in line with his own developing ideas. See Valenstein, *The War of the Soups and the Sparks*.

68. F. O. Schmitt, "Molecular Organization of the Nerve Fiber," in Oncley, Williams, Schmitt, and Bolt, eds., *Biophysical Science*, 455.

69. The notion had first been proposed in Alexander Forbes, "The Interpretation of Spinal Reflexes in Terms of Present Knowledge of Nerve Conduction," *Physiological Reviews* 2, no. 3 (1922): 361–414. See also Rafael Lorente de Nó, "Analysis of the Activity of the Chains of Internuncial Neurons," *Journal of Neurophysiology* 1, no. 3 (May 1938): 207–44.

70. Indeed, even the existence of neurotransmitters in the brain itself was not considered until the late 1960s. See Valenstein, *The War of the Soups and the Sparks*, 157–80.

71. Swazey, "Forging a Neuroscience Community," 529.

72. Manfred Eigen and Leo de Maeyer, "Self-Dissociation and Protonic Charge Transport in Water and Ice," *Proceedings of the Royal Society of London: Series A, Mathematical and Physical Sciences* 247, no. 1251 (October 21, 1958): 505–33.

73. Schmitt quoted in Swazey, "Forging a Neuroscience Community," 530.

74. S. P. R. Rose, "Holger Hyden and the Biochemistry of Memory," *Brain Research Bulletin* 50, nos. 5–6 (November 1999): 443; T. J. [Trevor J.] Pinch and H. M. [Harry M.] Collins, "Edible Knowledge: The Chemical Transfer of Memory," in *The Golem: What You Should Know about Science*, 2nd ed. (Cambridge University Press, 1998), 5–26. For more on the relationship between information theory and the genetic code, see Lily E. Kay, *Who Wrote the Book of Life? A History of the Genetic Code* (Stanford University Press, 2000).

75. Francis O. Schmitt, ed., *Macromolecular Specificity and Biological Memory* (MIT Press, 1962), 1.

76. Notably, a review of *Macromolecular Specificity and Biological Memory* by the English neurophysiologist J. Z. Young began: "There can be few serious scientific publications that contain such orgies of speculation as this one." J. Z. Young, "The Mechanism of Memory," *Endeavour* 72, no. 86 (May 1963): 101.

77. Schmitt, "A Proposal from the Massachusetts Institute of Technology for Support from the National Aeronautics and Space Administration for the Neurosciences Research Program," 6, folder 11, box 64, FOSP.

78. Francis O. Schmitt, "The Biomedical Communication Problem," *Technology Review* 66, no. 5 (March 1964): 40.

79. Francis O. Schmitt, "Life, Science and Inner Commitment," *Technology Review* 63, no. 9 (July 1961): 43–45, 70–72.

80. The full text of Schmitt's private note is contained in "Nerve Ideas, 1959–1963," folder 30, box 2, FOSP. Schmitt was also involved in a number of religious initiatives, including the New England Lutheran Faculty Conference and the Science-Theology Discussion Group at the Union Theological Seminary in New York. For more on his religious history, see Schmitt, *The Never-Ceasing Search*, 335–56.

81. A proposal for an international brain institute had been submitted to UNESCO as early as 1947, but E. D. Adrian, the Nobel Prize–winning British physiologist, had "pour[ed] cold water" on it in the early 1950s. See E. D. Adrian to Herbert H. Jasper, April 13, 1964, folder 132, box

3, HJF; UN Department of Social Affairs, *The Question of Establishing United Nations Research Laboratories* (United Nations, 1948), 234–66.

82. Herbert Jasper to Antoine Rémond, December 2, 1958, folder 187, box 4, HJF.

83. IBRO, "IBRO Programme Prospectus, 1961–1964," 11–12, folder 240, box 5, HJF.

84. Colin Blakemore, the "early participant," is quoted in Marshall et al., "Early History of IBRO," 299.

85. Marshall et al., "Early History of IBRO," 293–98.

86. Schmitt quoted in Swazey, "Forging a Neuroscience Community," 533.

87. "Theodore Melnechuk," 1–10, box 1, SOHC.

88. "Theodore Melnechuk," 12, box 1, SOHC.

89. "Herbert Jasper," 5, box 1, SOHC.

90. Walle Nauta, "Some Brain Structures and Functions Related to Memory," *Neuroscience Research Program Bulletin* 2, no. 5 (October 1964): 1–35.

91. According to Frederic Worden, an NRP participant and late president of the organization, Milner displayed a personal dislike for Schmitt and the entire agenda of the NRP: "I remember for example, Brenda Milner, who was also a leading neuropsychologist, who had an intense dislike of Frank. She thought he was arrogant. She had an intense dislike of N.R.P." Worden also noted that many attacked the NRP because its product "wasn't laboratory research" and "was intangible, conceptual progress, and this kind of stuff": "Then of course, another strong theme was in the stated meetings, where it's all very well for those few scientists who are in it, but what about the vast world of neuroscientists who aren't in it? We'd, of course, say, 'Well, they get to read the books.'" "Frederic Worden September 1985," 18, box 2, SOHC.

92. "Theodore Melnechuk," 15, box 1, SOHC.

93. Bernard Davis, "Metabolic Regulation and Information Storage in Bacteria," in *The Neurosciences: A Study Program*, ed. Gardner C. Quarton, Francis O. Schmitt, and Theodore Melnechuk (Rockefeller University Press, 1967), 121–22.

94. "Theodore Melnechuk," 15–16, 14, 21, box 1, SOHC.

95. Robert J. Frank, Louise H. Marshall, and Horace W. Magoun, "The Neurosciences," in *Advances in American Medicine: Essays at the Bicentennial*, ed. John Z. Bowers and Elizabeth F. Purcell (Josiah Macy Jr. Foundation, 1976), 2:604; International Brain Research Organization, *IBRO Survey of Research Facilities and Manpower in Brain Sciences in the United States* (National Academy of Sciences, 1968).

96. "Herbert Jasper," p. 6, box 1, SOHC. See also Robert W. Doty, "Neuroscience," in *History of the American Physiological Society: The First Century, 1887–1987*, ed. Toby A. Appel, John R Brobeck, and Orr E. Reynolds (The Society, 1987), 429; Frank, Marshall, and Magoun, "The Neurosciences," 603–4; and Society for Neuroscience, "The Creation of Neuroscience: The Society for Neuroscience and the Quest for Disciplinary Unity, 1969–1995" (Society for Neuroscience, 2019), 7–10, https://www.sfn.org/About/History-of-SfN/1969-1995/~/media/SfN/Images/History%20of%20SfN/pdf/HistoryofSfN.ashx.

97. Society for Neuroscience, "The Creation of Neuroscience," 18–79.

98. IBRO, "IBRO at Crossroads: Retrospect and Prospects; Tenth Anniversary Report" (UNESCO, 1971), 4, https://unesdoc.unesco.org/ark:/48223/pf0000001113.locale=en.

Chapter 4

1. D. O. Hebb, *The Organization of Behavior: A Neuropsychological Theory* (Wiley, 1949), 62.

2. Paul Adams, "Hebb and Darwin," *Journal of Theoretical Biology* 195, no. 4 (1998): 419–38.

3. Steven J. Haggbloom et al., "The 100 Most Eminent Psychologists of the 20th Century," *Review of General Psychology* 6, no. 2 (2002): 139–52.

4. John C. Fentress, "D. O. Hebb and the Developmental Organization of Behavior," *Developmental Psychobiology* 20, no. 2 (March 1987): 105.

5. D. O. Hebb, "D. O. Hebb," in *A History of Psychology in Autobiography*, ed. G. Lindzey (W. H. Freeman, 1980), 7:273–79.

6. On Meynert, see Guenther, *Localization and Its Discontents*, 13–38; William James, *The Principles of Psychology*, 2 vols. (1890; Macmillan, 1910), 1:12–80; Todes, "From the Machine to the Ghost Within"; K. W. Buckley, *Mechanical Man: John Broadus Watson and the Beginnings of Behaviorism* (Guilford, 1989); O'Donnell, *The Origins of Behaviorism*; and Mills, *Control*.

7. Donald O. Hebb, "Elementary School Methods," *Teacher's Magazine* 12, no. 51 (1930): 23–26, and "D. O. Hebb," 273, 80–82.

8. Hebb, "D. O. Hebb," 283.

9. D. O. Hebb, "Conditioned and Unconditioned Reflexes and Inhibition" (1930), DHF.

10. Hebb, "D. O. Hebb," 282–83.

11. Donald A. Dewsbury, "The Chicago Five: A Family Group of Integrative Psychobiologists," *History of Psychology* 5, no. 1 (2002): 16–37; Louis Leon Thurstone, *The Nature of Intelligence* (Routledge, 1924), and *The Vectors of Mind: Multiple-Factor Analysis for the Isolation of Primary Traits* (University of Chicago Press, 1935).

12. The definitive work on Lashley remains Weidman, *Constructing Scientific Psychology*.

13. D. O. Hebb, "The Interpretation of Experimental Data on Neural Action" (1934), accession no. 0000-2364.01.188, DHF.

14. D. O. Hebb, "The Innate Organization of Visual Activity: 1, Perception of Figures by Rats Reared in Total Darkness," *Pedagogical Seminary and Journal of Genetic Psychology* 51, no. 1 (1937): 101–2.

15. Hebb, "D. O. Hebb," 287.

16. Hebb, "D. O. Hebb," 290.

17. Molly Harrower to Kurt Koffka, January 23, 1938, folder 5, KK-MH Correspondence January 1938 Cont'd, M3230, MHP.

18. D. O. Hebb to Molly Harrower, May 16, 1979, folder 15, Penfield Obituary, M3201, MHP.

19. G. Jefferson, "Removal of Right or Left Frontal Lobes in Man," *British Medical Journal* 2, no. 3995 (July 31, 1937): 3095.

20. R. M. Brickner, *The Intellectual Functions of the Frontal Lobes: A Study Based upon Observation of a Man after Partial Bilateral Frontal Lobectomy* (Macmillan, 1936).

21. D. O. Hebb and W. Penfield, "Human Behavior after Extensive Bilateral Removal from the Frontal Lobes," *Archives of Neurology and Psychiatry* 44, no. 2 (1940): 421–22.

22. Hebb and Penfield, "Human Behavior after Extensive Bilateral Removal from the Frontal Lobes," 421.

23. D. O. Hebb to Ira Nichols, December 23, 1940, accession no. 0000-2364.01.6, DHF.

24. Hebb and Penfield, "Human Behavior after Extensive Bilateral Removal from the Frontal Lobes," 426.

25. Harrower, "Thursday's Child," 160.

26. Hebb and Penfield, "Human Behavior after Extensive Bilateral Removal from the Frontal Lobes," 421.

27. Hebb and Penfield, "Human Behavior after Extensive Bilateral Removal from the Frontal Lobes," 433–34.

28. D. O. Hebb to WP, October 24, 1940, C/G 40 H-I-J, box 55, WPP.

29. D. O. Hebb, "The Effect of Early and Late Brain Injury upon Test Scores, and the Nature of Normal Adult Intelligence," *Proceedings of the American Philosophical Society* 85, no. 3 (February 25, 1942): 286, 290. Richard Brown has argued persuasively that Raymond Cattell's more well-known theory of "fluid" and "crystallized" intelligences was, in fact, borrowed from Hebb. See Richard E. Brown, "Hebb and Cattell: The Genesis of the Theory of Fluid and Crystallized Intelligence," *Frontiers in Human Neuroscience* 10 (December 15, 2016): 1–10.

30. Hebb, "D. O. Hebb," 292, 293.

31. D. O. Hebb to Frank Beach, August 10, 1966, accession no. 1991-0040.01, DHF.

32. Hebb, "D. O. Hebb," 293–94.

33. Lemov, *World as Laboratory*, 74.

34. Sokal, "The Gestalt Psychologists in Behaviorist America"; David W. Carroll, *Purpose and Cognition: Edward Tolman and the Transformation of American Psychology* (Cambridge University Press, 2017); Edward C. Tolman, "Cognitive Maps in Rats and Men," *Psychological Review* 55, no. 4 (1948): 189–208, and *Purposive Behavior in Animals and Man* (Century, 1932).

35. Sokal, "The Gestalt Psychologists in Behaviorist America." On Gestalt, see Harrington, *Reenchanted Science*, 103–39; and Ash, *Gestalt Psychology in German Culture*.

36. Donald Olding Hebb, "A Neuropsychological Theory," in *Psychology: A Study of a Science*, vol. 1, *Sensory, Perceptual, and Physiological Formulations*, ed. S. Koch (McGraw-Hill, 1959), 626–27.

37. Hebb, "D. O. Hebb," 295.

38. D. O. Hebb, "Intelligence, Brain Function and the Theory of Mind," *Brain* 82, no. 2 (June 1, 1959): 266.

39. Hebb, *The Organization of Behavior*, 62.

40. Hebb, *The Organization of Behavior*, 31.

41. Hebb, *The Organization of Behavior*, 74.

42. The most detailed intellectual history of Hebb's book is Richard E. Brown, "Donald O. Hebb and the Organization of Behavior: 17 Years in the Writing," *Molecular Brain* 13, no. 1 (December 2020): 1–28. Brown documents the breadth and depth of Hebb's use of other neurological and psychological data, such as Henry Head's notion of a neural "schema" and Frank Beach's ideas about the neural basis of sexual excitement.

43. Henry Nissen to D. O. Hebb, May 11, 1940, accession no. 0000-2364.01.6, Hebb Papers.

44. D. O. Hebb to Henry Nissen, May 19, 1948, accession no. 0000-2364.01.6, Hebb Papers.

45. Much of Lashley's criticism was of Hebb's "synaptic theory of learning" and his notion of a cell "lattice," which later became the cell assembly. For Lashley's notes, see accession no. 0000-2364.01.197.1, DHF.

46. D. O. Hebb to Frank Beach, August 10, 1966, DHF.

47. Lashley, "In Search of the Engram," 473.

48. Karl S. Lashley, "The Problem of Serial Order in Behavior," in *Cerebral Mechanisms in Behavior* (Hixon Symposium), ed. Lloyd A. [Lloyd Alexander] Jeffress (Wiley, 1951), 113.

49. A note on the original manuscript of *Organization of Behavior* made by Lashley may be relevant here. In the manuscript, Hebb claimed that the cell assembly "relates the individual nerve cell to psychological phenomena" and that "a bridge has been thrown across the great gap between the details of neurophysiology and the molar conceptions of psychology." Lashley scribbled in the margins that Hebb's bridge "lacks the central span." Hebb noted that Lashley later retracted this criticism, suggesting the he eventually conceded the plausibility of Hebb's theory. Accession no. 0000-2364.01.197.1, DHF.

50. Lashley, "The Problem of Serial Order in Behavior," 133.

51. See, e.g., Karl S. Lashley, "The Behavioristic Interpretation of Consciousness: 1," *Psychological Review* 30, no. 4 (1923): 237–72, and "The Behavioristic Interpretation of Consciousness: 2," *Psychological Review* 30, no. 5 (1923): 329–53.

52. A small cottage industry of scholarship has grown examining the relationship between Hebb and Lashley, with former Hebb and Lashley students arguing about the provenance of the cell-assembly idea. The most extensive examination of the issue is J. Orbach, *The Neuropsychological Theories of Lashley and Hebb: Contemporary Perspectives Fifty Years After Hebb's The Organization of Behavior: Vanuxem Lectures and Selected Theoretical Papers of Lashley* (University Press of America, 1998), 71. Orbach, a student of Lashley's, makes a compelling case that many of Hebb's ideas were derived from concerns and work done by Lashley, a notion that is, perhaps, unsurprising given that Lashley was Hebb's mentor. Nevertheless, Orbach concedes that the most important aspect of Hebb's theory—the "empirically assembled nerve net"—was exclusively Hebb's.

53. Hebb, "D. O. Hebb," 301.

54. E. G. Boring to D. O. Hebb, January 4, 1950, DHF.

55. R. Leeper, review of *The Organization of Behavior: A Neuropsychological Theory*, by D. O. Hebb, *Journal of Abnormal and Social Psychology* 45 (1950): 768.

56. H. E. Gardner, *The Mind's New Science: A History of the Cognitive Revolution* (Basic, 1984), 271.

57. Milner, "Brenda Milner," 282.

58. Hebb, "A Neuropsychological Theory," 632.

59. Mortimer Mishkin and Donald G. Forgays, "Word Recognition as a Function of Retinal Locus," *Journal of Experimental Psychology* 43, no. 1 (1952): 43; Jacob Orbach, "Retinal Locus as a Factor in the Recognition of Visually Perceived Words," *American Journal of Psychology* 65, no. 4 (October 1952): 555–62; Woodburn Heron, "Perception as a Function of Retinal Locus and Attention," *American Journal of Psychology* 70, no. 1 (March 1957): 38; Vera D. Hunton, "The Recognition of Inverted Pictures by Children," *Journal of Genetic Psychology* 86, no. 2 (1955): 281–88; Bernard Hymovitch, "The Effects of Experimental Variations on Problem Solving in the Rat," *Journal of Comparative and Physiological Psychology* 45, no. 4 (1952): 313–21; R. S. Clarke, W. Heron, M. L. Fetherstonhaugh, D. G. Forgays, and D. O. Hebb, "Individual Differences in Dogs: Preliminary Report on the Effects of Early Experience," *Canadian Journal of Psychology/Revue canadienne de psychologie* 5, no. 4 (1951): 150–56; William R. Thompson and Woodburn Heron, "The Effects of Restricting Early Experience on the Problem-Solving Capacity of Dogs," *Canadian Journal of Psychology/Revue canadienne de psychologie* 8, no. 1 (1954): 17–31; Ronald Melzack, "The Genesis of Emotional Behavior: An Experimental Study of the Dog," *Journal of Comparative and Physiological Psychology* 47, no. 2 (1954): 166–68; Helen Mahut, "Breed Differences in the Dog's Emotional Behaviour," *Canadian Journal of Psychology/Revue canadienne de psychologie* 12, no. 1 (1958): 35–44.

60. Lemov, *World as Laboratory*; J. D. Marks, *The Search for the "Manchurian Candidate": The CIA and Mind Control* (Times Books, 1979); Alfred W. McCoy, *A Question of Torture: CIA Interrogation, from the Cold War to the War on Terror* (Henry Holt, 2006).

61. Hebb quoted in George Cooper, ed., *Appendices to the Opinion of George Cooper, Q.C., regarding Canadian Government Funding of the Allan Memorial Institute in the 1950's and 1960's,* 3 vols. (Department of Justice, 1986), vol. 1, app. 21, p. 2.

62. Alfred W. McCoy, "Science in Dachau's Shadow: Hebb, Beecher, and the Development of CIA Psychological Torture and Modern Medical Ethics," *Journal of the History of the Behavioral*

Sciences 43, no. 4 (2007): 401–17, and *A Question of Torture*; Richard E. Brown, "Alfred McCoy, Hebb, the CIA and Torture," *Journal of the History of the Behavioral Sciences* 43, no. 2 (2007): 205–13.

63. Hebb (and his daughters) had even conducted a preliminary experiment in 1947, raising rats in sensory-rich environments, and then testing for improved performance in maze-running tasks. Donald O. Hebb, "The Effects of Early Experience on Problem Solving at Maturity," *American Psychologist* 2 (1947): 306–7, and *Organization of Behavior*, 235–74.

64. D. O. Hebb to Robert Morrison, July 17, 1951, accession no. 0000-2364.01.36, DHF.

65. Woodburn Heron, "The Pathology of Boredom," *Scientific American* 196, no. 1 (January 1957): 52–56.

66. William H. Bexton, Woodburn Heron, and Thomas H. Scott, "Effects of Decreased Variation in the Sensory Environment," *Canadian Journal of Psychology/Revue canadienne de psychologie* 8, no. 2 (1954): 71.

67. Heron, "The Pathology of Boredom," 53.

68. Heron, "The Pathology of Boredom," 54.

69. Brown, "Alfred McCoy, Hebb, the CIA and Torture."

70. Bexton, Heron, and Scott, "Effects of Decreased Variation in the Sensory Environment," 70.

71. Donald Olding Hebb, "Drives and the CNS (Conceptual Nervous System)," *Psychological Review* 62, no. 4 (1955): 243–54.

72. McCoy, *A Question of Torture*; Marks, *The Search for the "Manchurian Candidate."*

73. D. O. Hebb, "Sensory Deprivation: Facts in Search of a Theory," *Journal of Nervous and Mental Disease* 132 (1961): 40–43.

74. Peter M. Milner, "Peter M. Milner," in *The History of Neuroscience in Autobiography*, ed. Larry R. Squire and Society for Neuroscience (Society for Neuroscience, 2012), 8:290–323.

75. Richard F. Thompson, "James Olds—1922–1976," *Biographical Memoirs, National Academy of Sciences* 77 (1999): 151–52; T. Parsons, E. Shils, and E. C. Tolman, *Toward a General Theory of Action* (Harvard University Press, 1951); J. Olds, *The Growth and Structure of Motives: Psychological Studies in the Theory of Action* (Free Press, 1956).

76. Hebb, "A Neuropsychological Theory," 625.

77. Milner, "Peter M. Milner," 309–14.

78. James Olds, "Pleasure Centers in the Brain," *Scientific American* 195, no. 4 (October 1956): 108–17, 108.

79. Quoted in Richard F. Thompson, "James Olds: 1922–1976," *American Journal of Psychology* 92, no. 1 (1979): 152.

80. Olds, "Pleasure Centers in the Brain," 110.

81. James Olds and Peter Milner, "Positive Reinforcement Produced by Electrical Stimulation of Septal Area and Other Regions of Rat Brain," *Journal of Comparative and Physiological Psychology* 47, no. 6 (1954): 419–27.

82. Olds, "Pleasure Centers in the Brain."

83. James Olds, "Self-Stimulation of the Brain: Its Use to Study Local Effects of Hunger, Sex, and Drugs," *Science* 127, no. 3294 (February 14, 1958): 315–24.

84. Otniel E. Dror, "Cold War 'Super-Pleasure': Insatiability, Self-Stimulation, and the Postwar Brain," *Osiris* 31, no. 1 (July 2016): 227–49.

85. Elliot S Valenstein, *Brain Control: A Critical Examination of Brain Stimulation and Psychosurgery* (Wiley, 1974), 64–146.

86. Hebb, "Annual Report to the National Research Council of Canada," January 4, 1953, accession no. 0000-2364.01.205, DHF.

87. Martin S. Pernick, *A Calculus of Suffering: Pain, Professionalism, and Anesthesia in Nineteenth-Century America* (Columbia University Press, 1985); R. Rey, *The History of Pain*, trans. L. E. Wallace, J. A. Cadden, and S. W. Cadden (Harvard University Press, 1998), 147–280.

88. Ronald Melzack, "Pain: Past, Present, and Future," in *Pain: New Perspectives in Therapy and Research*, ed. Matisyohu Weisenberg and Bernard Tursky (Plenum, 1976), 136.

89. Rey, *The History of Pain*, 147–346; I. Baszanger, *Inventing Pain Medicine: From the Laboratory to the Clinic* (Rutgers University Press, 1998), 19–62; Keith Wailoo, *Pain: A Political History* (Johns Hopkins University Press, 2014), 57–97; Matthew Soleiman, "Mechanisms of Experience: Cognitivism, Cybernetics, and the Postwar Science of Pain," *Isis* 116, no. 1 (2025): 23–42.

90. Hebb, *Organization of Behavior*, 182–90.

91. Harold Merskey, ed., *Thoughts and Findings on Pain: The Hebb-Bishop Correspondence and a Selection of Papers* (Canadian Pain Society, 1996), 9–52.

92. Mary-Ellen Jeans, Joel Katz, and Paul Taenzer, "Remembering Ronald Melzack (1929–2019)," *Canadian Journal of Pain* 4, no. 1 (January 1, 2020): 122–24; Joel Katz, "Ronald Melzack (1929–2019)," *American Psychologist* 75, no. 7 (October 2020): 1022–23.

93. John C. Liebeskind, "Oral History Interview with Ronald Melzack" (October 16, 1993), 8, John C. Liebeskind History of Pain Collection, History and Special Collections Division, Louise M. Darling Biomedical Library, University of California, Los Angeles.

94. Ronald Melzack and William R. Thompson, "Early Environment," *Scientific American* 194, no. 1 (1956): 41.

95. Ronald Melzack, "The McGill Pain Questionnaire: From Description to Measurement," *Anesthesiology* 103, no. 1 (July 1, 2005): 199–202.

96. Ronald Melzack to D. O. Hebb, May 26, 1959, accession no. 0000-2364.01.110.1, DHF.

97. Ronald Melzack and Patrick D. Wall, "Pain Mechanisms: A New Theory," *Science* 150, no. 3699 (November 19, 1965): 971–79.

98. L. A. Reynolds, *Innovation in Pain Management* (Wellcome Trust Centre for the History of Medicine, University College London, 2004), 79.

99. Melzack and Wall, "Pain Mechanisms," 978.

100. Wailoo, *Pain*, 57–97.

101. Melzack, "The McGill Pain Questionnaire."

102. E. Scarry, *The Body in Pain: The Making and Unmaking of the World* (Oxford University Press, 1987), v, 7–9, 17.

103. Nathanial Rochester to D. O. Hebb, January 13, 1954, accession no. 0000-2364.01.43, DHF.

104. Warren S. McCulloch and Walter Pitts, "A Logical Calculus of the Ideas Immanent in Nervous Activity," *Bulletin of Mathematical Biophysics* 5, no. 4 (1943): 115.

105. Tara H. Abraham, "(Physio) Logical Circuits: The Intellectual Origins of the McCulloch-Pitts Neural Networks," *Journal of the History of the Behavioral Sciences* 38, no. 1 (2002): 3–25, and *Rebel Genius: Warren S. McCulloch's Transdisciplinary Life in Science* (MIT Press, 2016).

106. For a discussion of the reaction of biological scientists to cybernetics, see Tara H. Abraham, "Transcending Disciplines: Scientific Styles in Studies of the Brain in Mid-Twentieth Century America," *Studies in History and Philosophy of Science Part C: Studies in History and Philosophy of Biological and Biomedical Sciences* 43, no. 2 (2012): 552–68.

107. "Oral-History: Nathaniel Rochester," Engineering and Technology History Wiki, 1991, https://ethw.org/Oral-History:Nathaniel_Rochester; Nathaniel Rochester, "The 701 Project as Seen by Its Chief Architect," *IEEE Annals of the History of Computing* 5, no. 2 (April 1983): 115–17.

108. John McCarthy, Marvin Minsky, Claude Shannon, and Nathaniel Rochester, "A Proposal for the Dartmouth Summer Research Project on Artificial Intelligence," August 31, 1955, 13–15.

109. Milner, "Peter M. Milner," 314–16.

110. Milner noted the similarities between a seizure and the uncontrolled fission reactions that he worked to prevent at the Chalk River reactor. Milner, "Peter M. Milner," 314–17.

111. N. Rochester, J. Holland, L. Haibt, and W. Duda, "Tests on a Cell Assembly Theory of the Action of the Brain, Using a Large Digital Computer," *IEEE Transactions on Information Theory* 2, no. 3 (September 1956): 88.

112. McCarthy, Minsky, Shannon, and Rochester, "A Proposal for the Dartmouth Summer Research Project on Artificial Intelligence"; Ronald Kline, "Cybernetics, Automata Studies, and the Dartmouth Conference on Artificial Intelligence," *IEEE Annals of the History of Computing* 33, no. 4 (April 2011): 5–16.

113. P. N. Edwards, *The Closed World: Computers and the Politics of Discourse in Cold War America* (MIT Press, 1997), 239–73.

114. Mikel Olazaran, "A Sociological History of the Neural Network Controversy," in *Advances in Computers*, vol. 37, ed. M. C. Yovits (Elsevier, 1993), 335–425.

115. Paul Werbos to D. O. Hebb, [before October 29, 1964], accession no. 0000-2364.01.48, DHF.

116. D. O. Hebb to Paul Werbos, October 29, 1964, accession no. 0000-2364.01.48, DHF.

117. Paul Werbos to D. O. Hebb, October 4, 1972, accession no. 0000-2364.01.48, DHF.

118. Werbos recounts meeting Minsky and noting that "neurons work this way," likely mentioning the updated Hebb model. Evidently, Minsky dismissed this, saying: "Every neural modeller in the business knows that it follows McCulloch and Pitts." See Olazaran, "A Sociological History of the Neural Network Controversy," 393.

119. The key paper here is David E. Rumelhart, Geoffrey E. Hinton, and Ronald J. Williams, "Learning Representations by Back-Propagating Errors," *Nature* 323, no. 6088 (1986): 533–36.

120. As Rumelhart and his colleagues put it: "In the beginning . . . it was very hard to see how anything resembling a neural network could learn at all. . . . After McCulloch and Pitts (1943) showed how neural-like networks could compute, the main problem then facing workers in this area was to understand how such networks could learn. The first set of ideas that really got the enterprise going were contained in Donald Hebb's *Organization of Behavior* (1949). . . . The essence of Hebb's ideas still persists today in many learning paradigms." David E. Rumelhart, James L. McClelland, and PDP Research Group, *Parallel Distributed Processing: Explorations in the Microstructure of Cognition*, vol. 1, *Foundations* (MIT Press, 1986), 152–53. See also Daniel Crevier, *AI: The Tumultuous History of the Search for Artificial Intelligence* (Basic, 1993), 214–16.

121. Hebb, *The Organization of Behavior*, 275.

122. Hebb, *The Organization of Behavior*, 300, 303.

123. Lashley quoted in Weidman, *Constructing Scientific Psychology*, 172.

124. Lashley quoted in Weidman, *Constructing Scientific Psychology*, 164.

125. The complex uptake of Hebb's sensory deprivation studies into education policy is expertly explored in Mical Raz, *What's Wrong with the Poor? Psychiatry, Race, and the War on Poverty*, Studies in Social Medicine (University of North Carolina Press, 2013), 1–111.

126. J. Shurkin, *Broken Genius: The Rise and Fall of William Shockley, Creator of the Electronic Age* (Palgrave Macmillan, 2006).

127. William Shockley to D. O. Hebb, May 19, 1968, and Hebb to Shockley, June 6, 1968, accession no. 0000-2364.01.44, DHF.

128. Arthur R. Jensen, "How Much Can We Boost IQ and Scholastic Achievement?," *Harvard Educational Review* 39, no. 1 (1969): 1–123.

129. D. O. Hebb, "A Return to Jensen and His Social Science Critics," *American Psychologist* 25, no. 6 (1970): 568.

130. D. O. Hebb, "Open Letter: To a Friend Who Thinks the IQ Is a Social Evil," *American Psychologist* 33, no. 12 (1978): 1144.

131. D. O. Hebb to Ulric Neisser, February 6, 1974, accession no. 0000-2364.01.41, DHF.

132. Ulric Neisser to D. O. Hebb, March 21, 1974, accession no. 0000-2364.01.41, DHF.

133. Historians have typically argued that the cognitive revolution was linked to the growing ubiquity of computers. This argument is made most forcefully by P. N. Edwards. However, the story of Hebb and K.M. suggests that neurology and patient experience were as crucial to the so-called cognitive revolution of the 1950s as any computer. P. N. Edwards, *The Closed World*, 175–302.

134. D. O. Hebb, *Essay on Mind* (Erlbaum, 1980), 79.

135. Malcolm Macmillan, *An Odd Kind of Fame: Stories of Phineas Gage* (MIT Press, 2000).

136. See Kurt Goldstein, "The Mental Changes due to Frontal Lobe Damage," *Journal of Psychology* 17, no. 2 (April 1944): 187–208; and Donald O. Hebb, "Man's Frontal Lobes: A Critical Review," *Archives of Neurology and Psychiatry* 54, no. 1 (1945): 10–24. Hebb and Goldstein ultimately disagreed on whether injury to the frontal lobes produced characteristic symptoms, as Goldstein claimed, or whether they produced no discernible symptoms that could be shown with psychological tests, as Hebb (and, to a lesser degree, Penfield) claimed. As was often the case in scientific disputes, the issue resolved into arguments about methods and materials. Goldstein claimed that Hebb's cases (particularly K.M.) were not numerous enough to counter his own (and others') extensive experience with frontal lobe cases. Hebb countered that the frontal lobe cases of Goldstein and others lacked proper controls. More importantly, those cases almost always had complicating factors, such as extensive damage due to compression from cerebral tumor. By contrast, the case of K.M. was a clean excision of scar tissue that produced no subsequent lesion or cerebral disorder (as measured by electrocorticogram). Hebb therefore contended that the case of K.M. showed that, when complicating factors were removed, the function of the frontal lobes could not be said to be "abstract attitude," as Goldstein had claimed. For his part, Ward Halstead agreed, writing Hebb privately: "It has never been clear to me what [Goldstein] means by abstract attitude. . . . Personally, I deplore the use of such global concepts and terminology unless they can be demonstrated to be absolutely necessary." Ward Halstead to D. O. Hebb, February 21, 1944, accession no. 0000-2364.01.7, DHF. Although their disagreement was both wide and deep, Hebb and Goldstein carried on an amiable correspondence. See, e.g., Hebb to Kurt Goldstein, November 19, 1948, and Goldstein to Hebb, December 11, 1948, accession no. 0000-2364.01.7, DHF.

137. D. O. Hebb to WP, May 26, 1944, accession no. 0000-2364.01.6, DHF.

138. "Penfield and the Frontal Lobes," October 29, 1976, accession no. 0000-2364.01.65, DHF.

139. *FCIHR Video History . . . Brenda Milner and Memory*.

Chapter 5

1. Herbert Jasper, "Neurology and Psychiatry: Two Solitudes?," address delivered in November 1963 at the Allan Memorial Institute of Psychiatry, McGill University, folder 11, box 1, HJF.

2. Linda Leith, *Introducing Hugh MacLennan's Two Solitudes: A Reader's Guide* (ECW, 1990).

3. On Meyer's influence on American psychiatry, see Lamb, *Pathologist of the Mind*; and Pressman, *Last Resort*.

4. Penfield, *No Man Alone*, 75.

5. WP to Adolf Meyer, June 1, 1924, C/G 25 M, box 51, WPP.

6. Penfield, *No Man Alone*, 296.

7. Penfield, *No Man Alone*, 75.

8. WP to Eldridge H. Campbell, September 3, 1937, folder C/G 37 C, box 54, WPP.

9. "Memorandum regarding Proposals to Create an Institute of Psychiatry," folder A/M 10-1 McGill Misc 1941–44 Psychiatry, box 4, WPP.

10. WP to J. C. Meakins, June 12, 1940, folder A/M 10-1 McGill Misc 1941–44 Psychiatry, box 4, WPP.

11. WP to Adolf Meyer, January 2, 1942, folder C/D 2-1 Meyer, box 24, WPP (emphasis added).

12. Adolf Meyer to WP, January 8, 1942, folder C/D 2-1 Meyer, box 24, WPP.

13. William Sargant, "Ewen Cameron, M.D., F.R.C.P.(C.) D.P.M.," *British Medical Journal* 3, no. 5568 (1967): 803–4.

14. Ewen Cameron, *Objective and Experimental Psychiatry* (Macmillan, 1935), v–vi.

15. WP to D. Ewen Cameron, March 20, 1944, folder A/M 10-1 McGill Misc 1941–44 Psychiatry, box 4, WPP.

16. WP to J. R. Fraser, March 30, 1944, folder A/M 2-2/4, box 1, WPP. A quarter century later, Penfield's interaction with Cameron still generated sadness. The letter that Penfield sent Cameron was eventually returned to him by a successor at the Allan, generating the following response from Penfield: "I remembered our interview and the disappointment that came to me at that time. I had been quite instrumental in his being selected as Professor and I looked forward with joy to the possibility, at last, of having a constructive inter-relationship with Psychiatry so that his cold and calculating interview, which took place in the Neurological Institute in my office here on the sixth floor, filled me with dismay." WP to Theodore Rasmussen, November 2, 1971, folder A/M 10 1/2, box 4, WPP.

17. Lamb, *Pathologist of the Mind*; Pressman, *Last Resort*, 18–46. The phrase *found itself* appears frequently in Adolf Meyer, "The Contributions of Psychiatry to the Understanding of Life Problems," in *The Collected Papers of Adolf Meyer*, ed. Alexander H. Leighton, vol. 4, *Mental Hygiene* (Johns Hopkins University Press, 1952), 1–16.

18. "General Administrative Policies," June 25, 1964, accession no. 387, Ref 38-220-1-9, MG 1098, DECF.

19. Carlyle F. Jacobsen, J. B. Wolfe, and T. A. Jackson. "An Experimental Analysis of the Functions of the Frontal Association Areas in Primates." *Journal of Nervous and Mental Disease* 82, no. 1 (1935): 10.

20. Elliot S. Valenstein, *Great and Desperate Cures: The Rise and Decline of Psychosurgery and Other Radical Treatments for Mental Illness* (Basic, 1986), 62–121. Pressman complicates the traditional origin story of lobotomy by pointing to the broader context of frontal lobe research, in which Penfield's studies played a prominent role. See Jack D. Pressman, "Sufficient Promise: John F. Fulton and the Origins of Psychosurgery," *Bulletin of the History of Medicine* 62, no. 1 (1988): 1–22, and *Last Resort*, 47–101.

21. John F. Fulton to WP, December 30, 1932, C/D 16, box 29, WPP.

22. Walter Freeman to WP, December 7, 1936, C/G 36 E-F, box 53, WPP.

23. Penfield quoted in Alan Gregg, diary, December 8, 1938, C/G 5-3/5, "Gregg Diaries Extracts—1931–50," box 44, WPP.

24. Walter Penfield and Stanley Cobb, "Experimentation in Clinical Medicine," C/D 14-6, box 29, WPP. The other procedure that Penfield discussed was insulin coma treatment for schizophrenia.

25. Penfield and Cobb, "Experimentation in Clinical Medicine."

26. WP to F. Golla, May 6, 1940, C/G 40 G, box 55, WPP.

27. Hebb and Penfield, "Human Behavior after Extensive Bilateral Removal from the Frontal Lobes."

28. Walter Freeman, *Psychosurgery: Intelligence, Emotion and Social Behavior Following Prefrontal Lobotomy for Mental Disorders* (Charles C. Thomas, 1942).

29. D. O. Hebb to WP, May 14, 1940, WP to Hebb, May 16, 1940, and Hebb to WP, May 18, 1940, W/P 109, box 141, WPP.

30. Stanley Cobb, "Review of Neuropsychiatry for 1940," *Archives of Internal Medicine* 66, no. 6 (1940): 1352.

31. WP to Stanley Cobb, October 3, 1940, C/D 14-6, box 29, WPP.

32. WP to J. R. Fraser, November 5, 1942, folder A/M 10-1, box 4, WPP.

33. WP to George E. Reed, October 4, 1944, C/G 44 R, box 58, WPP.

34. W. Penfield, "Bilateral Frontal Gyrectomy and Postoperative Intelligence," *Research Publications—Association for Research in Nervous and Mental Disease* 27 (1948): 519. This H.M. is not to be confused with the famous amnesiac patient H.M. later studied closely by the MNI neuropsychologist Brenda Milner.

35. Penfield, "Bilateral Frontal Gyrectomy and Postoperative Intelligence," 522.

36. Penfield, "Bilateral Frontal Gyrectomy and Postoperative Intelligence."

37. Pressman, *Last Resort*, 194–235.

38. WP to D. O. Hebb, May 26–October 16, 1944, C/G 44 H-I-J, box 57, WPP.

39. WP to D. O. Hebb, January 28, 1946, C/G 46 H-I-J, box 58, WPP.

40. Robert Malmo to Herbert Jasper, May 17, 1993, folder 116, box 2, HJF.

41. Herbert H. Jasper, "A Historical Perspective: The Rise and Fall of Prefrontal Lobotomy," *Advances in Neurology* 66 (1995): 97–114.

42. Brown, "Alan Gregg and the Rockefeller Foundation's Support of Franz Alexander's Psychosomatic Research"; Pressman, "Human Understanding."

43. WP to W. S. Ross, May 15, 1945, folder C/G 45 R, box 58, WPP.

44. Pool and Elsberg, *The Neurological Institute of New York*, 68; J. Lawrence Pool, "Topectomy: The Treatment of Mental Illness by Frontal Gyrectomy or Bilateral Subtotal Ablation of Frontal Cortex," *The Lancet* 254, no. 6583 (October 29, 1949): 776–81.

45. WP to Ewen Cameron, October 2, 1947, C/G 47 M, box 59, WPP.

46. Pressman, *Last Resort*, 318–62.

47. Penfield, "Bilateral Frontal Gyrectomy and Postoperative Intelligence," 534.

48. R. B. Malmo, "Psychological Aspects of Frontal Gyrectomy and Frontal Lobotomy in Mental Patients," *Research Publications—Association for Research in Nervous and Mental Disease* 27 (1948): 537–64.

49. D. E. Cameron and M. D. Prados, "Bilateral Frontal Gyrectomy: Psychiatric Results," *Research Publications—Association for Research in Nervous and Mental Disease* 27 (1948): 537.

50. Ewen Cameron to WP, April 7, 1948, folder C/G 48 C, box 59, WPP.

51. See also Walter Freeman and James Watts to WP, July 7, 1947, WP to Freeman, July 14, 1947, and WP to Henry A. Riley, July 14, 1947, Freeman to WP, July 21, 1947, and Riley to WP, August 1, 1947, C/G 47 E-F, box 59, WPP.

52. WP to C. Sidney Burwell, May 7, 1949, C/G 49 B, box 60, WPP (emphasis added).

53. WP to Ewen Cameron, April 23, 1952, C/G 52 C, box 61, WPP.

54. Ewen Cameron to WP, May 2, 1952, C/G 52 C, box 61, WPP.

55. WP to Carl-Eugen Erdmann, December 20, 1954, C/G E-F, box 62, WPP.

56. Eric Hutton, "Penfield," *Maclean's*, February 18, 1956, 68.

57. Molly Harrower, "The Birth of Neuropsychiatry: The Influence of Kurt Goldstein and Wilder Penfield," folder 17, Harrower on Goldstein 1937–1938, M3201, MHP.

58. Anne Collins, *In the Sleep Room: The Story of the CIA Brainwashing Experiments in Canada* (Key Porter, 1997), 11.

59. Robert A. Cleghorn, "The McGill Experience of Robert A. Cleghorn, MD: Recollections of D. Ewen Cameron," *Canadian Bulletin of Medical History* 7, no. 1 (April 1990): 70, 73.

60. D. Ewen Cameron, "The Day Hospital: An Experimental Form of Hospitalization for Psychiatric Patients," *Modern Hospital* 69, no. 3 (1947): 60–63. See also, e.g., papers by Cameron on day hospitalization in MG 1098, accession no. 387 Ref 38-220-1-2 and 1-3, DECF.

61. D. Ewen Cameron, Leonard Levy, Thomas Ban, and Leonard Rubenstein, "Automation of Psychotherapy," *Comprehensive Psychiatry* 5, no. 1 (1964): 1–14; Rebecca Lemov, "Brainwashing's Avatar: The Curious Career of Dr. Ewen Cameron," *Grey Room* 45 (2011): 60–87.

62. D. Ewen Cameron, "Psychic Driving," *American Journal of Psychiatry* 112, no. 7 (1956): 503.

63. D. Ewen Cameron, John G. Lohrenz, and K. A. Handcock, "The Depatterning Treatment of Schizophrenia," *Comprehensive Psychiatry* 3, no. 2 (1962): 65.

64. Cameron, "Psychic Driving," 504.

65. Although the details of Cameron's treatments were available in his published writings, the full extent of their horror became clear only in the 1970s, following the revelations by Marks of the CIA's brainwashing experiments. See Marks, *The Search for the "Manchurian Candidate*," 133–56; Collins, *In the Sleep Room*; and Lemov, "Brainwashing's Avatar," and *World as Laboratory*, 211–18.

66. Marks, *The Search for the "Manchurian Candidate."*

67. On the changing image of Cameron, see Lemov, "Brainwashing's Avatar."

68. Cameron, "Psychic Driving," 504.

69. Marks, *The Search for the "Manchurian Candidate*," 118.

70. D. O. Hebb to Cyril James, November 29, 1955, folder 0000-2039.01.3, DHF.

71. D. Ewen Cameron, "The Processes of Remembering," *British Journal of Psychiatry* 109, no. 460 (1963): 325–40.

72. On McConnell and the saga of planarian learning, see Pinch and Collins, "Edible Knowledge."

73. Cameron, "The Processes of Remembering," 329.

74. Cameron, "The Processes of Remembering," 330.

75. Collins, *In the Sleep Room*, 104.

76. Jasper, "Neurology & Psychiatry: Two Solitudes?"

77. Cleghorn, "The McGill Experience of Robert A. Cleghorn," 73.

78. Lemov, *World as Laboratory*, 217–18; Cleghorn, "The McGill Experience of Robert A. Cleghorn."

79. Robert A. Cleghorn, "Pitfalls in Thinking Big—Megalomania," *Psychiatric Quarterly* 38, no. 1–4 (1964): 607.

80. David Healy, *The Psychopharmacologists: Interviews by David Healey* (Taylor & Francis, 1998), 159–86; E. Valenstein, *Blaming the Brain: The Truth about Drugs and Mental Health* (Free Press, 1998), 9–57.

81. Anne Harrington, *Mind Fixers: Psychiatry's Troubled Search for the Biology of Mental Illness* (Norton, 2019), 74–106.

Chapter 6

1. Hubel, "David H. Hubel," 298.

2. Hubel, "David H. Hubel."

3. Hubel, "David H. Hubel," 297, 300.

4. Hubel, "David H. Hubel," 300.

5. Charles G. Gross, "From Imhotep to Hubel and Wiesel," in *Cerebral Cortex*, vol. 12, *Extrastriate Cortex in Primates*, ed. Kathleen S. Rockland, Jon H. Kaas, and Alan Peters (Springer Science + Business Media, 1997), 1–58. See also N. Pastore, *Selective History of Theories of Visual Perception, 1650–1950* (Oxford University Press, 1971), 169–319.

6. H. Keffer Hartline, Henry G. Wagner, and Edward Ford MacNichol, "The Peripheral Origin of Nervous Activity in the Visual System," in *Cold Spring Harbor Symposia on Quantitative Biology*, vol. 17, *The Neuron* (Cold Spring Harbor Laboratory Press, 1952), 125–41.

7. S. W. Kuffler, "Neurons in the Retina: Organization, Inhibition and Excitation Problems," in *Cold Spring Harbor Symposia on Quantitative Biology*, vol. 17, *The Neuron*, 281–92.

8. Hartline, Wagner, and MacNichol, "The Peripheral Origin of Nervous Activity in the Visual System"; Kuffler, "Neurons in the Retina: Organization, Inhibition and Excitation Problems."

9. Hubel, "David H. Hubel."

10. Robert Cohen, Edwin Weinstein, and David Marlow, "David McKenzie Rioch, MD, 1900–1985: Three Eulogies," *Psychiatry* 49, no. 2 (May 1986): 180–84. Notably, while he participated in the Macy conferences on cybernetics, Rioch remained aloof to the value of cybernetics for neurophysiology. According to one colleague, Rioch "was one of those who never fell into the trap of looking to the theory of information in its mathematical form as a short-cut to the development of imposing theoretical structures in fields other than those for which it was developed." See Donald M. Mackay, "David Mckenzie Rioch—a Chiel O' Mony Pairts," in *Principles, Practices, and Positions in Neuropsychiatric Research, ed. Joseph V. Brady and Walle J. H. Nauta* (Elsevier, 1972), x.

11. Hubel quoted in Nikolas S. Rose and Joelle M. Abi-Rached, *Neuro: The New Brain Sciences and the Management of the Mind* (Princeton University Press, 2013), 26.

12. In the 1920s, Rioch had done voluminous anatomical research on brain stem structures, particularly in dogs, and was a key figure in the burgeoning area of comparative neuroanatomy. Because of this early research on the brain stem and its relationship to emotion, aggression, and sexuality, he also took a keen interest in the discoveries of Peter Milner and James Olds (see chap. 4) related to the phenomenon of self-stimulation of the brain; in point of fact, it was his group at Walter Reed that dubbed the phenomenon *self-stimulation* in 1958. Magoun and Marshall, *American Neuroscience in the Twentieth Century*, 68–69, 237–38.

13. Gordon M. Shepherd, *Creating Modern Neuroscience: The Revolutionary 1950s* (Oxford University Press, 2009), 69–83.

14. Allan H. Bretag, "The Glass Micropipette Electrode: A History of Its Inventors and Users to 1950," *Journal of General Physiology* 149, no. 4 (April 3, 2017): 417–30.

15. Li, McLennan, and Jasper, "Brain Waves and Unit Discharge in Cerebral Cortex," 7; Choh-Luh Li and Herbert Jasper, "Microelectrode Studies of the Electrical Activity of the

Cerebral Cortex in the Cat," *Journal of Physiology* 121, no. 1 (1953): 117; Gloor, "H. H. Jasper, Neuroscientist of Our Century," 7; David H. Hubel and Torsten N. Wiesel, *Brain and Visual Perception: The Story of a 25-Year Collaboration* (Oxford University Press, 2005), 38.

16. Hubel, "David H. Hubel," 303–4. See also Hubel and Wiesel, *Brain and Visual Perception*, 18–19.

17. Hubel, "David H. Hubel," 304; H. Jasper, G. Ricci, and B. Doane, "Microelectrode Studies of Conditioning: Technique and Preliminary Results," file 5, box 1, HJF.

18. Hubel, "David H. Hubel," 304; Hubel and Wiesel, *Brain and Visual Perception*, 19.

19. Hubel's research report on his electrodes, made as part of WRAIR Project 6-60-10-017, Stress, Substack 5, "Electrophysiological Studies of the Nervous System under Conditions of Stress, Fatigue and Disease," noted: "The electrode to be described was developed to fill the need for an easily-made, sturdy device capable of resolving single neuron action potentials at least as well as the commonly used micropipette. It was developed for use not only in acute animal experiments in the central nervous system, but also in cases where pipettes may be especially prone to break, such as in chronic unrestrained preparations, in muscle, and in the human being during neurosurgical procedures. . . . Since steel wire becomes too fragile near the tip when thus sharpened, and also requires too thick a shaft, tungsten was selected for its stiffness and availability." "Tungsten Electrode 1957," box 1, DHP.

20. Hubel and Wiesel, *Brain and Visual Perception*, 20.

21. Hubel and Wiesel, *Brain and Visual Perception*, 20–21.

22. David H. Hubel, "Tungsten Microelectrode for Recording from Single Units," *Science* 125, no. 3247 (1957): 549–50.

23. Hubel and Wiesel, *Brain and Visual Perception*, 30.

24. Hubel and Wiesel, *Brain and Visual Perception*, 22.

25. Hubel, "David H. Hubel," 308.

26. Gross, "From Imhotep to Hubel and Wiesel"; Peter H. Schiller, "Past and Present Ideas about How the Visual Scene Is Analyzed by the Brain," in *Cerebral Cortex*, vol. 12, *Extrastriate Cortex in Primates*, ed. Kathleen S. Rockland, Jon H. Kaas, and Alan Peters (Springer, 1997), 59–90.

27. Pastore, *Selective History of Theories of Visual Perception*, 320–49; Abraham, *Rebel Genius*, 163–64; Luciano Mecacci, "Pathways of Perception," in *The Enchanted Loom: Chapters in the History of Neuroscience* (History of Neuroscience 4), ed. Pietro Corsi (Oxford University Press, 1991), 272–88.

28. Wall quoted in Reynolds, *Innovation in Pain Management*, 75.

29. Abraham, *Rebel Genius*, 161–64.

30. J. Lettvin, H. Maturana, W. McCulloch, and W. Pitts, "What the Frog's Eye Tells the Frog's Brain," *Proceedings of the IRE* 47, no. 11 (November 1959): 1950; David H. Hubel, "The Visual Cortex of the Brain," *Scientific American* 209, no. 5 (November 1963): 63. Hubel noted a similar phenomenon in describing his work in 1962: "In all of this work we have been particularly encouraged to find that the areas we study can be understood in terms of comparatively simple concepts such as the nerve impulse, convergence of many nerves on a single cell, excitation and inhibition." He added, in what could be read as a slight against cybernetics: "[M]oreover . . . one can conclude that at least some parts of the brain can be followed relatively easily, without necessarily requiring higher mathematics, computers or a knowledge of network theories."

31. Hubel and Wiesel, *Brain and Visual Perception*, 79. For Hubel and Wiesel's commentary on Lettvin and Maturana, see David H. Hubel and Torsten N. Wiesel, "Receptive Fields, Binocular Interaction and Functional Architecture in the Cat's Visual Cortex," *Journal of Physiology* 160, no. 1 (1962): 149–50.

32. Eric Kandel quoted in U. J. McMahan and B. Katz, *Steve: Remembrances of Stephen W. Kuffler* (Sinauer Associates, 1990), 46.

33. Hubel and Wiesel, *Brain and Visual Perception*, 369. Notably, von Senden's 1932 *Raum- und Gestaltauffassung bei operierten Blindgeborenen vor und nach der Operation* (Space and shape perception in blind babies before and after the operation) examined sixty-six cases of children who underwent cataract operations. The perceptual problems that these children experienced after having their sight restored formed a major source for Hebb's theory of cell assemblies. The impact of this work and its origins have received surprisingly little attention from historians.

34. Hubel and Wiesel, *Brain and Visual Perception*, 369, 370.

35. Hubel and Wiesel, *Brain and Visual Perception*, 370–72.

36. This experimental work is reviewed, along with the relevant papers, in Hubel and Wiesel, *Brain and Visual Perception*, 369–594.

37. Hubel and Wiesel, *Brain and Visual Perception*, 406.

38. Ulric Neisser, *Cognitive Psychology* (Meredith, 1967), 77–78.

39. Noam Chomsky, *Reflections on Language* (Pantheon, 1975), 8–9.

40. Kunihiko Fukushima, "Neocognitron: A Self-Organizing Neural Network Model for a Mechanism of Pattern Recognition Unaffected by Shift in Position," *Biological Cybernetics* 36, no. 4 (April 1980): 193–202; Yann LeCun, Yoshua Bengio, and Geoffrey Hinton, "Deep Learning," *Nature* 521, no. 7553 (May 28, 2015): 439.

41. Hubel and Wiesel, *Brain and Visual Perception*, 31.

42. Hubel and Wiesel, *Brain and Visual Perception*, 708.

Conclusion

1. Herbert Jasper to David Hubel, October 13, 1981, folder 154, box 3, HJF.

2. David Hubel to Herbert Jasper, November 2, 1981, folder 154, box 3, HJF.

3. David H. Hubel, *Eye, Brain, and Vision* (Scientific American Library, 1988), 8.

4. Milton Silverman, "Now They're Exploring the Brain," *Saturday Evening Post*, October 8, 1949, 26–27, 80–86. On the MNI's influence on popular culture, see Winter, *Memory*, 75–102.

5. On Penfield's second career as a social commentator and expert on language, see Lewis, *Something Hidden*, 242–304.

6. The details surrounding Herbert Jasper's departure are discussed in Feindel and Leblanc, *The Wounded Brain Healed*, 360–62.

7. *Herbert H. Jasper, MD Interviewed by Andre Olivier, MD.*

8. Michel Trepanier, "Science and Technology: The Coming of Age of a City of Knowledge," in *Montreal: The History of a North American City*, ed. Dany Fougères and Roderick Macleod, (McGill-Queen's University Press, 2018), 2:246–312.

9. On the continued scientific achievements of the MNI, see Feindel and Leblanc, *The Wounded Brain Healed*, 321–440.

10. Dittrich, *Patient H.M.*, 267–92; Corkin, *Permanent Present Tense*.

11. Valenstein, *The War of the Soups and the Sparks*, 157–80, and *Blaming the Brain*, 9–94. On the psychopharmacological revolution and its effect on theories of brain function, see David Healy, *The Creation of Psychopharmacology* (Harvard University Press, 2002).

12. Perhaps the most famous patient discussed in this book, H.M., provides an important counterpoint. In 2016, the grandson of William Beecher Scoville, a journalist name Luke
Dittrich, produced a startling exposé of the life of patient H.M. and the questionable behavior
of Corkin and MIT in relation to issues of informed consent, proper storage of research records,
and possession of H.M.'s brain after his death in 2008. Dittrich, *Patient H.M.*

13. Lorraine Daston, *Rivals: How Scientists Learned to Cooperate* (Columbia Global Reports,
2023).

14. George Adelman to Herbert Jasper, October 25, 1991, folder 151, box 3, HJF.

Bibliography

Abi-Rached, Joelle M. "From Brain to Neuro: The Brain Research Association and the Making of British Neuroscience, 1965–1996." *Journal of the History of the Neurosciences* 21, no. 2 (April 2012): 189–213.

Abraham, Tara H. "(Physio) Logical Circuits: The Intellectual Origins of the McCulloch-Pitts Neural Networks." *Journal of the History of the Behavioral Sciences* 38, no. 1 (2002): 3–25.

———. "Transcending Disciplines: Scientific Styles in Studies of the Brain in Mid-Twentieth Century America." *Studies in History and Philosophy of Science Part C: Studies in History and Philosophy of Biological and Biomedical Sciences* 43, no. 2 (2012): 552–68.

———. *Rebel Genius: Warren S. McCulloch's Transdisciplinary Life in Science.* MIT Press, 2016.

Adams, Annmarie. "Designing Penfield: Inside the Montreal Neurological Institute." *Bulletin of the History of Medicine* 93, no. 2 (2019): 207–40.

Adams, Paul. "Hebb and Darwin." *Journal of Theoretical Biology* 195, no. 4 (1998): 419–38.

Adelman, George. "The Neurosciences Research Program at MIT and the Beginning of the Modern Field of Neuroscience." *Journal of the History of the Neurosciences* 19, no. 1 (January 15, 2010): 15–23.

Almeida, Antonio Nogueira de, Manoel Jacoben Teixeira, and William Howard Feindel. "From Lateral to Mesial: The Quest for a Surgical Cure for Temporal Lobe Epilepsy." *Epilepsia* 49, no. 1 (January 2008): 98–107.

Ash, Mitchell G. *Gestalt Psychology in German Culture, 1890–1967: Holism and the Quest for Objectivity.* Cambridge University Press, 1995.

Bailey, Percival. *Intracranial Tumors.* Charles C. Thomas, 1933.

Bailey, Percival, and Paul C. Bucy. "Oligodendrogliomas of the Brain." *Journal of Pathology and Bacteriology* 32, no. 4 (1929): 735–51.

Barbara, Jean-Gaël. "French Neurophysiology between East and West: Polemics on Pavlovian Heritage and the Reception of Cybernetics in the USSR." In *History of the Neurosciences in France and Russia: From Charcot and Sechenov to IBRO*, ed. Jean-Gaël Barbara, Jean-Claude Dupont, and Irina Sirotkina, 225–41. Hermann, 2011.

Barr, William B. "Historical Development of the Neuropsychological Test Battery." In *Textbook of Clinical Neuropsychology*, ed. J. E. Morgan and J. H. Ricker, 3–17. Taylor & Francis, 2016.

Baszanger, I. *Inventing Pain Medicine: From the Laboratory to the Clinic*. Rutgers University Press, 1998.

Bazett, H. Cuthbert, and Wilder Penfield. "A Study of the Sherrington Decerebrate Animal in the Chronic As Well As the Acute Condition." *Brain* 45, no. 2 (1922): 185–265.

Bexton, William H., Woodburn Heron, and Thomas H. Scott. "Effects of Decreased Variation in the Sensory Environment." *Canadian Journal of Psychology/Revue canadienne de psychologie* 8, no. 2 (1954): 70.

Bliss, Michael. *William Osler: A Life in Medicine*. University of Toronto Press, 2002.

———. *Harvey Cushing: A Life in Surgery*. Oxford University Press, 2007.

Blustein, Bonnie Ellen. "Percival Bailey and Neurology at the University of Chicago, 1928–1939." *Bulletin of the History of Medicine* 66, no. 1 (1992): 90–113.

Borck, Cornelius. "Between Local Cultures and National Styles: Units of Analysis in the History of Electroencephalography." *Comptes rendus biologies* 329, nos. 5–6 (May 2006): 450–59.

———. "Recording the Brain at Work: The Visible, the Readable, and the Invisible in Electroencephalography." *Journal of the History of the Neurosciences* 17, no. 3 (July 16, 2008): 367–79.

———. *Brainwaves: A Cultural History of Electroencephalography*. Taylor & Francis, 2018.

Bracegirdle, Brian. "The Microscopical Tradition." In *Companion Encyclopedia of the History of Medicine*, ed. Roy Porter and W. F. [William F.] Bynum, 1:102–19. Routledge, 1993.

Bretag, Allan H. "The Glass Micropipette Electrode: A History of Its Inventors and Users to 1950." *Journal of General Physiology* 149, no. 4 (April 3, 2017): 417–30.

Brickner, R. M. *The Intellectual Functions of the Frontal Lobes: A Study Based upon Observation of a Man After Partial Bilateral Frontal Lobectomy*. Macmillan, 1936.

Brieger, Gert H. "From Conservative to Radical Surgery in Late Nineteenth-Century America." In *Medical Theory, Surgical Practice: Studies in the History of Surgery*, ed. Christopher Lawrence, 226–27. Routledge, 1992.

Brown, Richard E. "Alfred McCoy, Hebb, the CIA and Torture." *Journal of the History of the Behavioral Sciences* 43, no. 2 (2007): 205–13.

———. "Hebb and Cattell: The Genesis of the Theory of Fluid and Crystallized Intelligence." *Frontiers in Human Neuroscience* 10 (December 15, 2016): 1–10.

———. "Donald O. Hebb and the Organization of Behavior: 17 Years in the Writing." *Molecular Brain* 13, no. 1 (December 2020): 1–28.

Brown, Theodore M. "Alan Gregg and the Rockefeller Foundation's Support of Franz Alexander's Psychosomatic Research." *Bulletin of the History of Medicine* 61, no. 2 (1987): 155–82.

Buckley, K. W. *Mechanical Man: John Broadus Watson and the Beginnings of Behaviorism*. Guilford, 1989.

Buser, Pierre, Jean-Gaël Barbara, Boleslav Lichterman, and François Clarac. "The International Brain Research Organization from Its Conception to Adulthood." In *History of the Neurosciences in France and Russia: From Charcot and Sechenov to IBRO*, ed. Jean-Gaël Barbara, Jean-Claude Dupont, and Irina Sirotkina, 299–314. Hermann, 2011.

Cajal, Santiago Ramón y. *Recollections of My Life*. MIT Press, 1989.

Cameron, D. Ewen. *Objective and Experimental Psychiatry*. Macmillan, 1935.

———. "The Day Hospital: An Experimental Form of Hospitalization for Psychiatric Patients." *Modern Hospital* 69, no. 3 (1947): 60–63.

———. "Psychic Driving." *American Journal of Psychiatry* 112, no. 7 (1956): 502–9.

———. "The Processes of Remembering." *British Journal of Psychiatry* 109, no. 460 (1963): 325–40.

Cameron, D. Ewen, Leonard Levy, Thomas Ban, and Leonard Rubenstein. "Automation of Psychotherapy." *Comprehensive Psychiatry* 5, no. 1 (1964): 1–14.

Cameron, D. Ewen, John G. Lohrenz, and K. A. Handcock. "The Depatterning Treatment of Schizophrenia." *Comprehensive Psychiatry* 3, no. 2 (1962): 65–76.

Cameron, D. E., and M. D. Prados. "Bilateral Frontal Gyrectomy: Psychiatric Results." *Research Publications—Association for Research in Nervous and Mental Disease* 27 (1948): 534–37.

Capshew, James H. *Psychologists on the March: Science, Practice, and Professional Identity in America, 1929–1969.* Cambridge University Press, 1999.

Carroll, David W. *Purpose and Cognition: Edward Tolman and the Transformation of American Psychology.* Cambridge University Press, 2017.

Cherici, Céline. "Scientific Circulations between East and West in the Cold War Period: The International Moscow Colloquium (1958)." In *History of the Neurosciences in France and Russia: From Charcot and Sechenov to IBRO,* ed. Jean-Gaël Barbara, Jean-Claude Dupont, and Irina Sirotkina, 243–58. Hermann, 2011.

Chomsky, Noam. *Reflections on Language.* Pantheon, 1975.

Clarke, E., and L. S. Jacyna. *Nineteenth-Century Origins of Neuroscientific Concepts.* University of California Press, 1992.

Clarke, R. S., W. Heron, M. L. Fetherstonhaugh, D. G. Forgays, and D. O. Hebb. "Individual Differences in Dogs: Preliminary Report on the Effects of Early Experience." *Canadian Journal of Psychology/Revue canadienne de psychologie* 5, no. 4 (1951): 150–56.

Cleghorn, Robert A. "Pitfalls in Thinking Big—Megalomania." *Psychiatric Quarterly* 38, nos. 1–4 (1964): 607–18.

———. "The McGill Experience of Robert A. Cleghorn, MD: Recollections of D. Ewen Cameron." *Canadian Bulletin of Medical History* 7, no. 1 (April 1990): 53–76.

Cobb, Stanley. "Review of Neuropsychiatry for 1940." *Archives of Internal Medicine* 66, no. 6 (1940): 1341–54.

———. *Borderlands of Psychiatry.* Harvard University Press, 1943.

Cohen, Robert, Edwin Weinstein, and David Marlow. "David McKenzie Rioch, MD, 1900–1985: Three Eulogies." *Psychiatry* 49, no. 2 (May 1986): 180–84.

Collins, Alan F. "An Intimate Connection: Oliver Zangwill and the Emergence of Neuropsychology in Britain." *History of Psychology* 9, no. 2 (2006): 89–112.

Collins, Anne. *In the Sleep Room: The Story of the CIA Brainwashing Experiments in Canada.* Key Porter, 1997.

Committee on Neurobiology. *Survey of Neurobiology.* National Research Council, 1952.

Cooper, George, ed. *Appendices to the Opinion of George Cooper, Q.C., regarding Canadian Government Funding of the Allan Memorial Institute in the 1950's and 1960's.* 3 vols. Department of Justice, 1986.

Corkin, Suzanne. *Permanent Present Tense: The Unforgettable Life of the Amnesic Patient, H.M.* Basic, 2013.

Crawford, Elisabeth A. "The Universe of International Science, 1880–1939." In *Solomon's House Revisited: The Organization and Institutionalization of Science,* ed. Tore Frängsmyr, 251–69. Science History Publications, 1990.

Crevier, Daniel. *AI: The Tumultuous History of the Search for Artificial Intelligence.* Basic, 1993.

Cunningham, A., and P. Williams. *The Laboratory Revolution in Medicine.* Cambridge University Press, 2002.

Cushing, Harvey. "The Special Field of Neurological Surgery." *Bulletin of the Johns Hopkins Hospital* 16, no. 168 (March 1905): 77–87.

———. "Psychiatrists, Neurologists and the Neurosurgeon." *Yale Journal of Biology and Medicine* 7, no. 3 (1935): 191–207.

Daston, Lorraine. *Rivals: How Scientists Learned to Cooperate*. Columbia Global Reports, 2023.

Daston, Lorraine, and Peter Galison. *Objectivity*. Zone, 2007.

Davis, Bernard. "Metabolic Regulation and Information Storage in Bacteria." In *The Neurosciences: A Study Program*, ed. Gardner C. Quarton, Francis O. Schmitt, and Theodore Melnechuk, 113–22. Rockefeller University Press, 1967.

Davis, Hallowel, Frederic A. Gibbs, and William G. Lennox. "The Electro-Encephalogram in Epilepsy and in Conditions of Impaired Consciousness." *Archives of Neurology and Psychiatry* 34, no. 6 (1935): 1133–48.

De Sio, Fabio. "One, No-One and a Hundred Thousand Brains: J. C. Eccles, J. Z. Young and the Establishment of the Neurosciences (1930s–1960s)." In *Progress in Brain Research*, vol. 243, *Imagining the Brain: Episodes in the History of Brain Research*, ed. Chaira Ambrosio and William MacLehose, 257–98. Elsevier, 2018.

Dewsbury, Donald A. "The Chicago Five: A Family Group of Integrative Psychobiologists." *History of Psychology* 5, no. 1 (2002): 16–37.

Dittrich, Luke. *Patient H.M.: A Story of Memory, Madness, and Family Secrets*. Random House, 2017.

Doty, Robert W. "Neuroscience." In *History of the American Physiological Society: The First Century, 1887–1987*, ed. Toby A. Appel, John R. Brobeck, and Orr E. Reynolds, 427–34. The Society, 1987.

Droogleever-Fortuyn, J. "Reticular Formation and Choice of Behavior." In *The Collected Works of Warren S. McCulloch*, ed. Rook McCulloch, 4:1302–5. Intersystems, 1989.

Dror, Otniel E. "Cold War 'Super-Pleasure': Insatiability, Self-Stimulation, and the Postwar Brain." *Osiris* 31, no. 1 (July 2016): 227–49.

Dwyer, Ellen. "Stories of Epilepsy, 1880–1930." In *Framing Disease: Studies in Cultural History*, ed. C. E. Rosenberg and J. L. Golden, 248–72. Rutgers University Press, 1992.

Earle, Kenneth Martin, Maitland Baldwin, and Wilder Penfield. "Incisural Sclerosis and Temporal Lobe Seizures Produced by Hippocampal Herniation at Birth." *Archives of Neurology and Psychiatry* 69, no. 1 (1953): 27–42.

———. "Temporal Lobe Seizures: The Anatomy and Pathology of the Probable Cause." *Journal of Neuropathology and Experimental Neurology* 12, no. 1 (1953): 98–99.

Edwards, P. N. *The Closed World: Computers and the Politics of Discourse in Cold War America*. MIT Press, 1997.

Eigen, Manfred, and Leo de Maeyer. "Self-Dissociation and Protonic Charge Transport in Water and Ice." *Proceedings of the Royal Society of London: Series A, Mathematical and Physical Sciences* 247, no. 1251 (October 21, 1958): 505–33.

Elder, Rachel. "Speaking Secrets: Epilepsy, Neurosurgery, and Patient Testimony in the Age of the Explorable Brain, 1934–1960." *Bulletin of the History of Medicine* 89, no. 4 (2015): 761–89.

Elliott, K. A. C. "Biochemical Approaches in the Study of Epilepsy." *Electroencephalography and Clinical Neurophysiology* 1, no. 1 (1949): 29–31.

———. "Neurochemistry." *Canadian Medical Association Journal* 116, no. 12 (1977): 1372–73.

———. "An Unorthodox Career." *Bulletin of the Canadian Biochemical Society* 17 (1980): 12–25.

Elliott, Kenneth Allan Campbell, Irvine H. Page, and J. H. Quastel, eds. *Neurochemistry: The Chemical Dynamics of Brain and Nerve*. Charles C. Thomas, 1955.

Entin, M. A. *Edward Archibald: Surgeon of the Royal Vic*. McGill-Queen's University Press, 2004.

Evans, Joseph P. "Exciting Beginnings." *Canadian Medical Association Journal* 116, no. 12 (1977): 1367.

The FCIHR Video History of Medicine in Canada Project: Brenda Milner and Memory. Friends of Canadian Institutes for Health Research, 2013.

Fedunkiw, Marianne P. *Rockefeller Foundation Funding and Medical Education in Toronto, Montreal, and Halifax.* McGill-Queen's University Press, 2005.

Feindel, William. "Osler and the 'Medico-Chirurgical Neurologists': Horsley, Cushing, and Penfield." *Journal of Neurosurgery* 99, no. 1 (2003): 188–99.

———. "The Physiologist and the Neurosurgeon: The Enduring Influence of Charles Sherrington on the Career of Wilder Penfield." *Brain* 130, no. 11 (April 5, 2007): 2758–65.

Feindel, William, and Richard Leblanc. *The Wounded Brain Healed: The Golden Age of the Montreal Neurological Institute, 1934–1993.* McGill-Queen's University Press, 2016.

Feindel, William, Richard Leblanc, and Jean-Guy Villemure. "History of the Surgical Treatment of Epilepsy." In *A History of Neurosurgery: In Its Scientific and Professional Contexts*, ed. Samuel H. Greenblatt, T. Forcht Dagi, and Mel H. Epstein, 465–88. American Association of Neurological Surgeons, 1997.

Feindel, William, and Wilder G. Penfield. "Localization of Discharge in Temporal Lobe Automatism." *Archives of Neurology and Psychiatry* 72, no. 5 (November 1, 1954): 605–30.

Fentress, John C. "D. O. Hebb and the Developmental Organization of Behavior." *Developmental Psychobiology* 20, no. 2 (March 1987): 103–9.

Ferguson, Sherise, and Maciej S. Lesniak. "Percival Bailey and the Classification of Brain Tumors." *Neurosurgical Focus* 18, no. 4 (April 2005): 1–6.

Foerster, Otfrid, and Wilder Penfield. "The Structural Basis of Traumatic Epilepsy and Results of Radical Operation." *Brain* 53 (July 1930): 99–119.

Forbes, Alexander. "The Interpretation of Spinal Reflexes in Terms of Present Knowledge of Nerve Conduction." *Physiological Reviews* 2, no. 3 (1922): 361–414.

Forbes, Alexander, and Alan Gregg. "Electrical Studies in Mammalian Reflexes: 1, The Flexion Reflex." *American Journal of Physiology* 37, no. 1 (1915): 118–76.

Frank, George. "Research on the Clinical Usefulness of the Rorschach: 2, The Assessment of Cerebral Dysfunction." *Perceptual and Motor Skills* 72, no. 1 (1991): 103–11.

Frank, Robert G. "Instruments, Nerve Action, and the All-or-None Principle." *Osiris* 9 (1994): 208–35.

Frank, Robert J., Louise H. Marshall, and Horace W. Magoun. "The Neurosciences." In *Advances in American Medicine: Essays at the Bicentennial*, ed. John Z. Bowers and Elizabeth F. Purcell, 2:552–613. Josiah Macy Jr. Foundation, 1976.

Frank, Roberta. "'Interdisciplinary': The First Half Century." *Issues in Interdisciplinary Studies* 6 (1988): 139–51.

Freeman, Walter. *Psychosurgery: Intelligence, Emotion and Social Behavior Following Prefrontal Lobotomy for Mental Disorders.* Charles C. Thomas, 1942.

Fukushima, Kunihiko. "Neocognitron: A Self-Organizing Neural Network Model for a Mechanism of Pattern Recognition Unaffected by Shift in Position." *Biological Cybernetics* 36, no. 4 (April 1980): 193–202.

Fulton, John F. "Sir Charles Scott Sherrington, O.M. (1857–1952)." *Journal of Neurophysiology* 15, no. 3 (May 1, 1952): 167–90.

———. "The Historical Contribution of Physiology to Neurology." In *Science, Medicine, and History: Essays on the Evolution of Scientific Thought and Medical Practice Written in Honour of Charles Singer*, ed. E. Ashworth Underwood, 537–44. Oxford University Press, 1953.

Furumoto, Laurel. "On the Margins: Women and the Professionalization of Psychology in the United States, 1890–1940." In *Psychology in Twentieth-Century Thought and Society*, ed. Mitchell G. Ash and William R. Woodward, 93–113. Cambridge University Press, 1987.

Galison, Peter. *Image and Logic: A Material Culture of Microphysics*. University of Chicago Press, 1997.

———. "The Americanization of Unity." *Daedalus* 127, no. 1 (1998): 45–71.

———. "Trading with the Enemy." In *Trading Zones and International Expertise: Creating New Kinds of Collaboration*, ed. Michael E. Gorman, 25–52. MIT Press, 2010.

———. "Meanings of Scientific Unity: The Law, the Orchestra, the Pyramid, the Quilt and the Ring." In *Pursuing the Unity of Science*, 12–29. Routledge, 2016.

Gardner, H. E. *The Mind's New Science: A History of the Cognitive Revolution*. Basic, 1984.

Gardner, Martha, and Allan M. Brandt. "The Golden Age of Medicine?" In *Companion to Medicine in the Twentieth Century*, ed. Roger Cooter and John V. Pickstone, 21–37. Routledge, 2000.

Gavrus, Delia. "Epilepsy and the Laboratory Technician: Technique in Histology and Fiction." In *The History of the Brain and Mind Sciences: Technique, Technology, Therapy*, ed. Stephen T. Casper and Delia Gavrus, 136–63. University of Rochester Press, 2017.

———. "Opening the Skull: Neurosurgery as a Case Study of Surgical Specialisation." In *The Palgrave Handbook of the History of Surgery*, ed. Thomas Schlich, 435–55. Palgrave Macmillan, 2018.

Geddes, Jennian F. "A Portrait of 'The Lady': A Life of Dorothy Russell." *Journal of the Royal Society of Medicine* 90, no. 8 (1997): 455–61.

Geison, Gerald L. *Michael Foster and the Cambridge School of Physiology: The Scientific Enterprise in Late Victorian Society*. Princeton University Press, 1978.

Gibbs, Erna L., Frederic A. Gibbs, and Bartolome Fuster. "Psychomotor Epilepsy." *Archives of Neurology and Psychiatry* 60, no. 4 (1948): 331–39.

Gibbs, Frederic A., Erna L. Gibbs, and William G. Lennox. "Epilepsy: A Paroxysmal Cerebral Dysrhythmia." *Brain: A Journal of Neurology* 60, no. 4 (1937): 377–88.

Gill, Amandip S., and Devin K. Binder. "Wilder Penfield, Pío del Río-Hortega, and the Discovery of Oligodendroglia." *Neurosurgery* 60, no. 5 (May 2007): 940–48.

Gloor, Pierre. "H. H. Jasper, Neuroscientist of Our Century." In *Neurotransmitters and Cortical Function*, ed. Massimo Avoli, Tomas A. Reader, Robert W. Dykes, and Pierre Gloor, 1–13. Springer, 1988.

Goldstein, Kurt. "The Mental Changes due to Frontal Lobe Damage." *Journal of Psychology* 17, no. 2 (April 1944): 187–208.

Gordin, Michael D. *Scientific Babel: How Science Was Done Before and After Global English*. University of Chicago Press, 2015.

Graff, Harvey J. *Undisciplining Knowledge: Interdisciplinarity in the Twentieth Century*. Johns Hopkins University Press, 2015.

Graham, Loren R. *Science and Philosophy in the Soviet Union*. Knopf, 1972.

Granovetter, Mark S. "The Strength of Weak Ties." *American Journal of Sociology* 78, no. 6 (1973): 1360–80.

Greenblatt, Samuel H. "The Emergence of Cushing's Leadership: 1901 to 1920." In *A History of Neurosurgery: In Its Scientific and Professional Contexts*, ed. Samuel H. Greenblatt, T. Forcht Dagi, and Mel H. Epstein, 167–90. American Association of Neurological Surgeons, 1997.

Greenblatt, Samuel H., T. Forcht Dagi, and Mel H. Epstein, eds. *A History of Neurosurgery: In Its Scientific and Professional Contexts*. American Association of Neurological Surgeons, 1997.

Greene, Mott T. "Writing Scientific Biography." *Journal of the History of Biology* 40, no. 4 (December 3, 2007): 727–59.

Gross, Charles G. "From Imhotep to Hubel and Wiesel." In *Cerebral Cortex*, vol. 12, *Extrastriate Cortex in Primates*, ed. Kathleen S. Rockland, Jon H. Kaas, and Alan Peters, 1–58. Springer Science + Business Media, 1997.

Guenther, Katja. *Localization and Its Discontents: A Genealogy of Psychoanalysis and the Neuro Disciplines*. University of Chicago Press, 2015.

Haggbloom, Steven J., Renee Warnick, Jason E. Warnick, Vinessa K. Jones, Gary L. Yarbrough, Tenea M. Russell, Chris M. Borecky, Reagan McGahhey, John L. Powell, and Jamie Beavers. "The 100 Most Eminent Psychologists of the 20th Century." *Review of General Psychology* 6, no. 2 (2002): 139–52.

Hagner, Michael. "The Electrical Excitability of the Brain: Toward the Emergence of an Experiment." *Journal of the History of the Neurosciences* 21, no. 3 (July 2012): 237–49.

Halstead, W. C. *Brain and Intelligence: A Quantitative Study of the Frontal Lobes*. University of Chicago Press, 1949.

Hanaway, J., and R. L. Cruess. *McGill Medicine: The Second Half Century, 1885–1936*. McGill-Queen's University Press, 2006.

Harrington, Anne. "Beyond Phrenology: Localization Theory in the Modern Era." In *The Enchanted Loom: Chapters in the History of Neuroscience* (History of Neuroscience 4), ed. Pietro Corsi, 207–34. Oxford University Press, 1991.

———. *Medicine, Mind, and the Double Brain: A Study in Nineteenth-Century Thought*. Princeton University Press, 1989.

———. *Reenchanted Science: Holism in German Culture from Wilhelm II to Hitler*. Princeton University Press, 1996.

———. "Kurt Goldstein's Neurology of Healing and Wholeness: A Weimar Story." In *Greater Than the Parts: Holism in Biomedicine, 1920–1950*, ed. George Weisz, 25–45. Oxford University Press, 1998.

———. *Mind Fixers: Psychiatry's Troubled Search for the Biology of Mental Illness*. Norton, 2019.

Harrower, M. R. "Organization in Higher Mental Processes." *Psychologische Forschung* 17 (1932): 56–120.

———. *The Psychologist at Work*. K. Paul, Trench, Trubner, 1937.

———. "Changes in Figure-Ground Perception in Patients with Cortical Lesions." *British Journal of Psychology* 30, no. 1 (1939): 47–51.

———. "Personality Studies in Cases of Focal Epilepsy." *Bulletin of the Canadian Psychological Association* 2 (1941): 19–21.

———. "Changing Horses in Mid-Stream: An Experimentalist Becomes a Clinician." In *The Psychologists: Autobiographies of Distinguished Living Psychologists*, ed. T. S. Krawiec, 3:85–104. Clinical Psychology Publishing Co., 1978.

———. "Inkblots and Poems." In *The History of Clinical Psychology in Autobiography*, ed. C. Eugene Walker, 1:125–70. Brooks/Cole, 1991.

———. "Thursday's Child." n.d. Molly Harrower Papers (M842), Cummins Center for the History of Psychology, University of Akron.

Harrower-Erickson, M. R., and F. R. Miale. "Personality Changes Accompanying Organic Brain Lesions: 3, A Study of Preadolescent Children." *Pedagogical Seminary and Journal of Genetic Psychology* 58, no. 2 (1941): 391–405.

Harrower-Erickson, Molly, Wilder Penfield, and Theodore Charles Erickson. "Psychological Studies of Patients with Epileptic Seizures." In *Epilepsy and Cerebral Localization: A Study of the Mechanism, Treatment and Prevention of Epileptic Seizures*, by Wilder Penfield and Theodore Charles Erickson, 546–74. Charles C. Thomas, 1941.

Harrower-Erickson, M. R., and M. E. Steiner. *Large Scale Rorschach Techniques: A Manual for the Group Rorschach and Multiple Choice Test*. Charles C. Thomas, 1945.

Hartline, H. Keffer, Henry G. Wagner, and Edward Ford MacNichol. "The Peripheral Origin of Nervous Activity in the Visual System." In *Cold Spring Harbor Symposia on Quantitative Biology*, vol. 17, *The Neuron*, 125–41. Cold Spring Harbor Laboratory Press, 1952.

Healy, David. *The Psychopharmacologists: Interviews by David Healey*. Taylor & Francis, 1998.

———. *The Creation of Psychopharmacology*. Harvard University Press, 2002.

Hebb, Donald O. "Elementary School Methods." *Teacher's Magazine* 12, no. 51 (1930): 23–26.

———. "The Innate Organization of Visual Activity: 1, Perception of Figures by Rats Reared in Total Darkness." *Pedagogical Seminary and Journal of Genetic Psychology* 51, no. 1 (1937): 101–26.

———. "The Effect of Early and Late Brain Injury upon Test Scores, and the Nature of Normal Adult Intelligence." *Proceedings of the American Philosophical Society* 85, no. 3 (February 25, 1942): 275–92.

———. "Man's Frontal Lobes: A Critical Review." *Archives of Neurology and Psychiatry* 54, no. 1 (1945): 10–24.

———. "The Effects of Early Experience on Problem Solving at Maturity." *American Psychologist* 2 (1947): 306–7.

———. *The Organization of Behavior: A Neuropsychological Theory*. Wiley, 1949.

———. "Drives and the CNS (Conceptual Nervous System)." *Psychological Review* 62, no. 4 (1955): 243–54.

———. "A Neuropsychological Theory." In *Psychology: A Study of a Science*, vol. 1, *Sensory, Perceptual, and Physiological Formulations*, ed. S. Koch, 622–43. McGraw-Hill, 1959.

———. "Intelligence, Brain Function and the Theory of Mind." *Brain* 82, no. 2 (June 1, 1959): 260–75.

———. "Sensory Deprivation: Facts in Search of a Theory." *Journal of Nervous and Mental Disease* 132 (1961): 40–43.

———. "A Return to Jensen and His Social Science Critics." *American Psychologist* 25, no. 6 (1970): 568.

———. "Open Letter: To a Friend Who Thinks the IQ Is a Social Evil." *American Psychologist* 33, no. 12 (1978): 1143–44.

———. "D. O. Hebb." In *A History of Psychology in Autobiography*, ed. G. Lindzey, 7:273–309. W. H. Freeman, 1980.

———. *Essay on Mind*. Erlbaum, 1980.

Hebb, D. O., and W. Penfield. "Human Behavior after Extensive Bilateral Removal from the Frontal Lobes." *Archives of Neurology and Psychiatry* 44, no. 2 (1940): 421–38.

Herbert H. Jasper, MD Interviewed by Andre Olivier, MD. American Association of Neurological Surgeons, 1993. https://youtu.be/q_ytTKzJ_UI.

Heritage Minutes: Wilder Penfield. Historica Canada, 2016. https://www.youtube.com/watch?v=pUOG2g4hj8s.

Heron, Woodburn. "The Pathology of Boredom." *Scientific American* 196, no. 1 (January 1957): 52–56.

———. "Perception as a Function of Retinal Locus and Attention." *American Journal of Psychology* 70, no. 1 (March 1957): 38.

Hubel, David H. "Tungsten Microelectrode for Recording from Single Units." *Science* 125, no. 3247 (1957): 549–50.

———. "The Visual Cortex of the Brain." *Scientific American* 209, no. 5 (November 1963): 54–63.

———. *Eye, Brain, and Vision.* Scientific American Library, 1988.

———. "David H. Hubel." In *The History of Neuroscience in Autobiography*, ed. Larry R. Squire and Society for Neuroscience, 1:298–317. Society for Neuroscience, 1996.

Hubel, David H., and Torsten N. "Receptive Fields, Binocular Interaction and Functional Architecture in the Cat's Visual Cortex." *Journal of Physiology* 160, no. 1 (1962): 106–54.

———. Wiesel. *Brain and Visual Perception: The Story of a 25-Year Collaboration.* Oxford University Press, 2005.

Hunton, Vera D. "The Recognition of Inverted Pictures by Children." *Journal of Genetic Psychology* 86, no. 2 (1955): 281–88.

Hutton, Eric. "Penfield." *Maclean's*, February 18, 1956, 11–15, 68–74.

Hymovitch, Bernard. "The Effects of Experimental Variations on Problem Solving in the Rat." *Journal of Comparative and Physiological Psychology* 45, no. 4 (1952): 313–21.

"Illocality." *New York Times*, April 21, 1932.

International Brain Research Organization. *IBRO Survey of Research Facilities and Manpower in Brain Sciences in the United States.* National Academy of Sciences, 1968.

———. "IBRO at Crossroads: Retrospect and Prospects: Tenth Anniversary Report." UNESCO, 1971. https://unesdoc.unesco.org/ark:/48223/pf0000001113.locale=en.

Jackson, John Hughlings. "On the Anatomical, Physiological and Pathological Investigation of Epilepsies." *West Riding Lunatic Asylum Medical Reports* 3 (1873): 315–39.

Jackson, John Hughlings. "On a Particular Variety of Epilepsy ('Intellectual Aura'), One Case with Symptoms of Organic Brain Disease." *Brain* 11, no. 2 (1888): 179–207.

Jackson, J. Hughlings, and Walter S. Colman. "Case of Epilepsy with Tasting Movements and 'Dreamy State'—Very Small Patch of Softening in the Left Uncinate Gyrus." *Brain* 21, no. 4 (1898): 580–90.

Jacobsen, Carlyle F., J. B. Wolfe, and T. A. Jackson. "An Experimental Analysis of the Functions of the Frontal Association Areas in Primates." *Journal of Nervous and Mental Disease* 82, no. 1 (1935): 1–14.

Jacyna, L. S. *Lost Words: Narratives of Language and the Brain, 1825–1926.* Princeton University Press, 2009.

James, William. *The Principles of Psychology.* 1890. 2 vols. Macmillan, 1910.

Jasper, H. H. "Optimism and Pessimism in College Environments." *American Journal of Sociology* 34, no. 5 (1929): 856–73.

———. "The Measurement of Depression-Elation and Its Relation to a Measure of Extraversion-Introversion." *Journal of Abnormal and Social Psychology* 25, no. 3 (1930): 307–18.

———. "Localized Analyses of the Function of the Human Brain by the Electro-Encephalogram." *Archives of Neurology and Psychiatry* 36 (1936): 1131–34.

———. "Electrical Signs of Epileptic Discharge." *Electroencephalography and Clinical Neurophysiology* 1, no. 1 (1949): 11–18.

———. "The Saga of K. A. C. Elliott and GABA." *Neurochemical Research* 9, no. 3 (1984): 449–60.

———. "A Historical Perspective: The Rise and Fall of Prefrontal Lobotomy." *Advances in Neurology* 66 (1995): 97–114.

———. "Herbert H. Jasper." In *The History of Neuroscience in Autobiography*, ed. Larry R. Squire and Society for Neuroscience, 1:320–46. Society for Neuroscience, 1996.

———. "Memoirs—Autobiography of Seventy Years Adventures in Neuroscience Research and with Neuroscientists." n.d. WFF.

Jasper, Herbert H., and Frederic Bremer, eds. *Brain Mechanisms and Consciousness.* Charles C. Thomas, 1954.

Jasper, Herbert H., and Leonard Carmichael. "Electrical Potentials from the Intact Human Brain." *Science 81,* no. 2089 (1935): 51–3.

Jasper, Herbert H., and Benjamin Doane. "Neurophysiological Mechanisms in Learning." *Progress in Physiological Psychology* 2 (1968): 79–117.

Jasper, Herbert, Peter Gloor, and Brenda Milner. "Higher Functions of the Nervous System." *Annual Review of Physiology* 18, no. 1 (1956): 359–86.

Jasper, Herbert, and John Kershman. "Electroencephalographic Classification of the Epilepsies." *Archives of Neurology and Psychiatry* 45, no. 6 (June 1941): 903–43.

Jasper, Herbert, and Wilder Penfield. "Electrocorticograms in Man: Effect of Voluntary Movement upon the Electrical Activity of the Precentral Gyrus." *Archiv für Psychiatrie und Nervenkrankheiten* 183, nos. 1–2 (1949): 163–74.

Jasper, Herbert, Bernard Pertuisset, and Herman Flanigin. "EEG and Cortical Electrograms in Patients with Temporal Lobe Seizures." *Archives of Neurology and Psychiatry* 65, no. 3 (1951): 272–90.

Jasper, H., G. F. Ricci, and B. Doane. "Patterns of Cortical Neuronal Discharge during Conditioned Responses in Monkeys." In *Ciba Foundation Symposium on the Neurological Basis of Behaviour,* ed. G. E. W. Wolstenholme and Cecilia M. O'Connor, 277–94. Little, Brown, 1958.

———. "Microelectrode Analysis of Cortical Cell Discharge during Avoidance Conditioning in the Monkey." In *The Moscow Colloquium on Electroencephalography of Higher Nervous Activity: Sponsored by the Academy of Sciences of the USSR with the Cooperation of the International Federation of Societies for Electroencephalography and Clinical Neurophysiology, Moscow, October 6–11, 1958,* ed. H. Jasper and G. D. Smirnov, 137–56. EEG Journal, 1960.

Jasper, H. H., and G. D. Smirnov, eds. *The Moscow Colloquium on Electroencephalography of Higher Nervous Activity: Sponsored by the Academy of Sciences of the USSR with the Cooperation of the International Federation of Societies for Electroencephalography and Clinical Neurophysiology, Moscow, October 6–11, 1958.* EEG Journal, 1960.

Jeans, Mary-Ellen, Joel Katz, and Paul Taenzer. "Remembering Ronald Melzack (1929–2019)." *Canadian Journal of Pain* 4, no. 1 (January 1, 2020): 122–24.

Jefferson, G. "Removal of Right or Left Frontal Lobes in Man." *British Medical Journal* 2, no. 3995 (July 31, 1937): 199–206.

Jensen, Arthur R. "How Much Can We Boost IQ and Scholastic Achievement?" *Harvard Educational Review* 39, no. 1 (1969): 1–123.

Katz, Joel. "Ronald Melzack (1929–2019)." *American Psychologist* 75, no. 7 (October 2020): 1022–23.

Kay, Lily E. *The Molecular Vision of Life: Caltech, the Rockefeller Foundation, and the Rise of the New Biology.* Oxford University Press, 1993.

———. *Who Wrote the Book of Life? A History of the Genetic Code.* Stanford University Press, 2000.

Kernahan, Peter. "Franklin Martin and the Standardization of American Surgery, 1890–1940." PhD diss., University of Minnesota, 2010.

Klein, Julie Thompson. *Interdisciplinarity: History, Theory, and Practice.* Wayne State University Press, 1990.

Kline, Ronald. "Cybernetics, Automata Studies, and the Dartmouth Conference on Artificial Intelligence." *IEEE Annals of the History of Computing* 33, no. 4 (April 2011): 5–16.

Klüver, Heinrich, and Paul C. Bucy. "'Psychic Blindness' and Other Symptoms Following Bilateral Temporal Lobectomy in Rhesus Monkeys." *American Journal of Physiology* 119, no. 2 (1937): 352–53.

———. "An Analysis of Certain Effects of Bilateral Temporal Lobectomy in the Rhesus Monkey, with Special Reference to 'Psychic Blindness.'" *Journal of Psychology* 5, no. 1 (January 1938): 33–54.

———. "Preliminary Analysis of Functions of the Temporal Lobes in Monkeys." *Archives of Neurology and Psychiatry* 42, no. 6 (1939): 979–1000.

Kohler, R. E. *Partners in Science: Foundations and Natural Scientists, 1900–1945.* University of Chicago Press, 1991.

———. *Landscapes and Labscapes: Exploring the Lab-Field Border in Biology.* University of Chicago Press, 2010.

Kubie, Lawrence S. "Some Implications for Psychoanalysis of Modern Concepts of the Organization of the Brain." *Psychoanalytic Quarterly* 22, no. 1 (1953): 21–68.

Kuffler, S. W. "Neurons in the Retina: Organization, Inhibition and Excitation Problems." In *Cold Spring Harbor Symposia on Quantitative Biology*, vol. 17, *The Neuron*, 281–92. Cold Spring Harbor Laboratory Press, 1952.

Lamb, S. D. *Pathologist of the Mind: Adolf Meyer and the Origins of American Psychiatry.* Johns Hopkins University Press, 2014.

Landström, Catharina. "Internationalism between Two Wars." In *Internationalism and Science*, ed. Aant Elzinga and Catharina Landström, 46–77. Taylor Graham, 1996.

Lashley, Karl S. "The Behavioristic Interpretation of Consciousness: 1." *Psychological Review* 30, no. 4 (1923): 237–72.

———. "The Behavioristic Interpretation of Consciousness: 2." *Psychological Review* 30, no. 5 (1923): 329–53.

———. *Brain Mechanisms and Intelligence: A Quantitative Study of Injuries to the Brain.* University of Chicago Press, 1929.

———. "In Search of the Engram." In *Physiological Mechanisms in Animal Behaviour* (Society of Experimental Biology Symposium), 454–82. Cambridge University Press, 1950.

———. "The Problem of Serial Order in Behavior." In *Cerebral Mechanisms in Behavior* (Hixon Symposium), ed. Lloyd A. [Lloyd Alexander] Jeffress, 112–46. Wiley, 1951.

Lattes, Raffaele. "Surgical Pathology at the College of Physicians and Surgeons of Columbia University." In *Guiding the Surgeon's Hand: The History of American Surgical Pathology*, ed. Juan Rosai, 41–60. American Registry of Pathology, 1997.

Leblanc, Richard. "The White Paper: Wilder Penfield, the Stream of Consciousness, and the Physiology of Mind." *Journal of the History of the Neurosciences* 28, no. 4 (October 2, 2019): 416–36.

———. *Radical Treatment: Wilder Penfield's Life in Neuroscience.* McGill-Queen's University Press, 2020.

LeCun, Yann, Yoshua Bengio, and Geoffrey Hinton. "Deep Learning." *Nature* 521, no. 7553 (May 28, 2015): 436–44.

Leeper, R. Review of *The Organization of Behavior: A Neuropsychological Theory*, by D. O. Hebb. *Journal of Abnormal and Social Psychology* 45 (1950): 768–75.

Leith, Linda. *Introducing Hugh MacLennan's Two Solitudes: A Reader's Guide.* ECW, 1990.

Lemov, Rebecca. *World as Laboratory: Experiments with Mice, Mazes, and Men.* Hill & Wang, 2005.

————. "Brainwashing's Avatar: The Curious Career of Dr. Ewen Cameron." *Grey Room* 45 (2011): 60–87.

Lettvin, J., H. Maturana, W. McCulloch, and W. Pitts. "What the Frog's Eye Tells the Frog's Brain." *Proceedings of the IRE* 47, no. 11 (November 1959): 1940–51.

Lewis, J. *Something Hidden: A Biography of Wilder Penfield.* Formac, 1983.

Li, Choh-Luh, and Herbert Jasper. "Microelectrode Studies of the Electrical Activity of the Cerebral Cortex in the Cat." *Journal of Physiology* 121, no. 1 (1953): 117.

Li, Choh-Luh, Hugh McLennan, and Herbert Jasper. "Brain Waves and Unit Discharge in Cerebral Cortex." *Science* 116, no. 3024 (1952): 656–57.

Lichterman, Boleslav L. "The Moscow Colloquium on Electroencephalography of Higher Nervous Activity and Its Impact on International Brain Research." *Journal of the History of the Neurosciences* 19, no. 4 (October 6, 2010): 313–32.

Liebeskind, John C. "Oral History Interview with Ronald Melzack." October 16, 1993. John C. Liebeskind History of Pain Collection, History and Special Collections Division, Louise M. Darling Biomedical Library, University of California, Los Angeles.

Lorente de Nó, Rafael. "Analysis of the Activity of the Chains of Internuncial Neurons." *Journal of Neurophysiology* 1, no. 3 (May 1938): 207–44.

Loss, C. P. *Between Citizens and the State: The Politics of American Higher Education in the 20th Century.* Princeton University Press, 2012.

Löwel, Siegrid, and Wolf Singer. "Selection of Intrinsic Horizontal Connections in the Visual Cortex by Correlated Neuronal Activity." *Science* 255, no. 5041 (1992): 209–12.

Mackay, Donald M. "David Mckenzie Rioch—a Chiel O' Mony Pairts." In *Principles, Practices, and Positions in Neuropsychiatric Research,* ed. Joseph V. Brady and Walle J. H. Nauta, ix–xi. Elsevier, 1972.

Macmillan, Malcolm. *An Odd Kind of Fame: Stories of Phineas Gage.* MIT Press, 2000.

Magoun, Horace Winchell, and Louise H. Marshall. *American Neuroscience in the Twentieth Century: Confluence of the Neural, Behavioral, and Communicative Streams.* A. A. Balkema, 2003.

Mahut, Helen. "Breed Differences in the Dog's Emotional Behaviour." *Canadian Journal of Psychology/Revue canadienne de psychologie* 12, no. 1 (1958): 35–44.

Malmo, R. B. "Psychological Aspects of Frontal Gyrectomy and Frontal Lobotomy in Mental Patients." *Research Publications—Association for Research in Nervous and Mental Disease* 27 (1948): 537–64.

Marks, J. D. *The Search for the "Manchurian Candidate": The CIA and Mind Control.* Times Books, 1979.

Marshall, Louise H. "Instruments, Techniques, and Social Units in American Neurophysiology, 1870–1950." In *Physiology in the American Context, 1850–1940,* ed. Gerald L. Geison, 351–69. Springer, 1987.

Marshall, Louise H., W. A. Rosenblith, P. Gloor, G. Krauthamer, C. Blakemore, and S. Cozzens. "Early History of IBRO: The Birth of Organized Neuroscience." *Neuroscience* 72, no. 1 (1996): 283–306.

Maxson Jones, Kathryn. "Francis O. Schmitt: At the Intersection of Neuroscience and Squid." In *Why Study Biology by the Sea?,* ed. Karl S. Matlin, Jane Maienschein, and Rachel A. Ankeny, 187–210. University of Chicago Press, 2020.

McCarthy, John, Marvin Minsky, Claude Shannon, and Nathaniel Rochester. "A Proposal for the Dartmouth Summer Research Project on Artificial Intelligence." August 31, 1955. https://dimes.rockarch.org/objects/ZbZohZDEsi9kKVuERRAC9A/view.

McCoy, Alfred W. *A Question of Torture: CIA Interrogation, from the Cold War to the War on Terror*. Henry Holt, 2006.

———. "Science in Dachau's Shadow: Hebb, Beecher, and the Development of CIA Psychological Torture and Modern Medical Ethics." *Journal of the History of the Behavioral Sciences* 43, no. 4 (2007): 401–17.

McCulloch, Warren S., and Walter Pitts. "A Logical Calculus of the Ideas Immanent in Nervous Activity." *Bulletin of Mathematical Biophysics* 5, no. 4 (1943): 115–33.

"McGill Will Open Brain Laboratory." *Montreal Gazette*, February 24, 1939.

McMahan, U. J., and B. Katz. *Steve: Remembrances of Stephen W. Kuffler*. Sinauer Associates, 1990.

Mecacci, Luciano. *Brain and History: The Relationship between Neurophysiology and Psychology in Soviet Research*. Brunner/Mazel, 1979.

———. "Pathways of Perception." In *The Enchanted Loom: Chapters in the History of Neuroscience (History of Neuroscience, no. 4)*, ed. Pietro Corsi, 272–88. Oxford University Press, 1991.

Melzack, Ronald. "The Genesis of Emotional Behavior: An Experimental Study of the Dog." *Journal of Comparative and Physiological Psychology* 47, no. 2 (1954): 166–68.

———. "Pain: Past, Present, and Future." In *Pain: New Perspectives in Therapy and Research*, ed. Matisyohu Weisenberg and Bernard Tursky, 135–46. Plenum, 1976.

———. "The McGill Pain Questionnaire: From Description to Measurement." *Anesthesiology* 103, no. 1 (July 1, 2005): 199–202.

Melzack, Ronald, and William R. Thompson. "Early Environment." *Scientific American* 194, no. 1 (1956): 38–43.

Melzack, Ronald, and Patrick D. Wall. "Pain Mechanisms: A New Theory." *Science* 150, no. 3699 (November 19, 1965): 971–79.

Merskey, Harold, ed. *Thoughts and Findings on Pain: The Hebb-Bishop Correspondence and a Selection of Papers*. Canadian Pain Society, 1996.

Meyer, Adolf. "The Contributions of Psychiatry to the Understanding of Life Problems." In *The Collected Papers of Adolf Meyer*, vol. 4, *Mental Hygiene*, ed. Alexander H. Leighton, 1–16. Johns Hopkins University Press, 1952.

Mills, J. A. *Control: A History of Behavioral Psychology*. New York University Press, 2000.

Milner, Brenda. "Memory Mechanisms." *Canadian Medical Association Journal* 116, no. 12 (1977): 1374–76.

———. "Amobarbital Memory Testing: Some Personal Reflections." *Brain and Cognition* 33, no. 1 (1997): 14–17.

———. "Brenda Milner." In *The History of Neuroscience in Autobiography*, ed. Larry R. Squire and Society for Neuroscience, 2:276–305. Society for Neuroscience, 1998.

Milner, Brenda, and Wilder G. Penfield. "The Effect of Hippocampal Lesions on Recent Memory." *Transactions of the American Neurological Association* 80 (1955): 42–48.

———. "The Cell Assembly: Mark II." *Psychological Review* 64, no. 4 (1957): 242–52.

Milner, Peter M. "Peter M. Milner." In *The History of Neuroscience in Autobiography*, ed. Larry R. Squire and Society for Neuroscience, 8:290–323. Society for Neuroscience, 2012.

Mishkin, Mortimer, and Donald G. Forgays. "Word Recognition as a Function of Retinal Locus." *Journal of Experimental Psychology* 43, no. 1 (1952): 43–48.

Montreal Neurological Institute. "Montreal Neurological Institute, 14th Annual Report, 1948–1949." 1949. Wilder Penfield Digital Collection, Osler Library of the History of Medicine, McGill University. http://digital.library.mcgill.ca/penfieldfonds/fullrecord.php?ID=10475.

———. "Montreal Neurological Institute, 15th Annual Report, 1949–1950." 1950. Wilder Penfield Digital Collection, Osler Library of the History of Medicine, McGill University. http://digital.library.mcgill.ca/penfieldfonds/fullrecord.php?ID=10476.

Moruzzi, Giuseppe, and Horace W. Magoun. "Brain Stem Reticular Formation and Activation of the EEG." *Electroencephalography and Clinical Neurophysiology* 1, no. 1 (1949): 455–73.

Lyons, Arthur E. "The Crucible Years, 1880 to 1900: Macewen to Cushing." In *A History of Neurosurgery: In Its Scientific and Professional Contexts*, ed. S. H. Greenblatt, T. F. Dagi, and M. H. Epstein, 153–66. American Association of Neurological Surgeons, 1997.

Nauta, Walle. "Some Brain Structures and Functions Related to Memory." *Neuroscience Research Program Bulletin* 2, no. 5 (October 1964): 1–35.

Neisser, Ulric. *Cognitive Psychology*. Meredith, 1967.

Nieto-Galan, Agusti. "The History of Science in Spain: A Critical Overview." *Nuncius* 23, no. 2 (2008): 211–36.

Nye, Mary Jo. "Scientific Biography: History of Science by Another Means?" *Isis* 97, no. 2 (June 1, 2006): 322–29.

O'Donnell, John M. *The Origins of Behaviorism: American Psychology, 1870–1920*. New York University Press, 1985.

Olazaran, Mikel. "A Sociological History of the Neural Network Controversy." In *Advances in Computers*, vol. 37, ed. M. C. Yovits, 335–425. Elsevier, 1993.

Olds, James. "Pleasure Centers in the Brain." *Scientific American* 195, no. 4 (October 1956): 108–17.

———. *The Growth and Structure of Motives: Psychological Studies in the Theory of Action*. Free Press, 1956.

———. "Self-Stimulation of the Brain: Its Use to Study Local Effects of Hunger, Sex, and Drugs." *Science* 127, no. 3294 (February 14, 1958): 315–24.

Olds, James, and Peter Milner. "Positive Reinforcement Produced by Electrical Stimulation of Septal Area and Other Regions of Rat Brain." *Journal of Comparative and Physiological Psychology* 47, no. 6 (1954): 419–27.

Oncley, J. L., R. C. Williams, F. O. Schmitt, and R. H. Bolt, eds. *Biophysical Science: A Study Program*. Wiley, 1959.

"Oral-History: Nathaniel Rochester." Engineering and Technology History Wiki, 1991. https://ethw.org/Oral-History:Nathaniel_Rochester.

Orbach, Jacob. "Retinal Locus as a Factor in the Recognition of Visually Perceived Words." *American Journal of Psychology* 65, no. 4 (October 1952): 555–62.

Orbach, Jack. *The Neuropsychological Theories of Lashley and Hebb: Contemporary Perspectives Fifty Years After Hebb's The Organization of Behavior: Vanuxem Lectures and Selected Theoretical Papers of Lashley*. University Press of America, 1998.

Osborne, A. E. "Responsibility of Epileptics." *Medico-Legal Journal* 11, no. 2 (1893): 210–28.

Parsons, T., E. Shils, and E. C. Tolman. *Toward a General Theory of Action*. Harvard University Press, 1951.

Pastore, N. *Selective History of Theories of Visual Perception, 1650–1950*. Oxford University Press, 1971.

Penfield, Wilder. "Alterations of the Golgi Apparatus in Nerve Cells." *Brain* 43, no. 3 (November 1, 1920): 290–305.

———. "Oligodendroglia and Its Relation to Classical Neuroglia." *Brain* 47, no. 4 (1924): 430–52.

———. "The Career of Ramón y Cajal." *Archives of Neurology and Psychiatry* 16, no. 2 (1926): 213–20.

———. "The Mechanism of Cicatricial Contraction in the Brain." *Brain* 50, nos. 3–4 (1927): 499–517.

———. "The Radical Treatment of Traumatic Epilepsy and Its Rationale." *Canadian Medical Association Journal* 23, no. 2 (1930): 189–97.

———, ed. *Cytology and Cellular Pathology of the Nervous System.* 3 vols. Paul B. Hoeber, 1932.

———. "The Significance of the Montreal Neurological Institute." In *Montreal Neurological Institute Annual Reports, 1934–1960, 1–18.* Montreal Neurological Institute and Montreal Neurological Hospital, 1934.

———. "The Cerebral Cortex in Man: 1, The Cerebral Cortex and Consciousness." *Archives of Neurology and Psychiatry* 40, no. 3 (1938): 417.

———. "The Epilepsies: With a Note on Radical Therapy." *New England Journal of Medicine* 221, no. 6 (1939): 209–18.

———. "Pío del Río-Hortega, MD, 1882–1945." *Archives of Neurology and Psychiatry* 54, no. 5 (1945): 413–16.

———. "Bilateral Frontal Gyrectomy and Postoperative Intelligence." *Research Publications—Association for Research in Nervous and Mental Disease* 27 (1948): 519–34.

———. "Memory Mechanisms." *Archives of Neurology and Psychiatry* 67, no. 2 (1952): 178–98.

———. "Sir Charles Sherrington, OM, GBE, FRS." *Nature* 169, no. 4304 (April 26, 1952): 690.

———. "Edward Archibald, 1872–1945." *Canadian Journal of Surgery* 1, no. 2 (1958): 167–74.

———. *The Difficult Art of Giving: The Epic of Alan Gregg.* Little, Brown, 1967.

———. "Herbert Jasper." In *Recent Contributions to Neurophysiology: International Symposium in Neurosciences in Honor of Herbert H. Jasper/Colloque international en sciences neurologiques en hommage à Herbert H. Jasper,* ed. Jean-Pierre Cordeau and Pierre Gloor, 9–12. Elsevier, 1972.

———. *No Man Alone: A Neurosurgeon's Life.* Little, Brown, 1977.

Penfield, Wilder, and Maitland Baldwin. "Temporal Lobe Seizures and the Technic of Subtotal Temporal Lobectomy." *Annals of Surgery* 136, no. 4 (1952): 625–34.

Penfield, Wilder, and Edwin Boldrey. "Somatic Motor and Sensory Representation in the Cerebral Cortex of Man as Studied by Electrical Stimulation." *Brain: A Journal of Neurology* 60, no. 4 (1937): 389–443.

Penfield, Wilder, and Richard C. Buckley. "Punctures of the Brain: The Factors Concerned in Gliosis and in Cicatricial Contraction." *Archives of Neurology and Psychiatry* 20, no. 1 (1928): 1–13.

Penfield, Wilder, and Theodore C. Erickson. *Epilepsy and Cerebral Localization: A Study of the Mechanism, Treatment and Prevention of Epileptic Seizures.* Charles C. Thomas, 1941.

Penfield, Wilder, and Joseph Evans. "Functional Defects Produced by Cerebral Lobectomies." *Proceedings: Association for Research in Nervous and Mental Diseases* 13 (1934): 352–77.

———. "The Frontal Lobe in Man: A Clinical Study of Maximum Removals." *Brain* 58, no. 1 (1935): 115–33.

Penfield, Wilder, and Herman Flanigin. "Surgical Therapy of Temporal Lobe Seizures." *Archives of Neurology and Psychiatry* 64, no. 4 (1950): 491–500.

Penfield, Wilder G., and Herbert H. Jasper. "Highest Level Seizures." In *Epilepsy: Proceedings of the Association, Held Jointly with the International League against Epilepsy, December 13 and 14, 1946, New York* (Research Publications—Association for Research in Nervous and Mental Disease), vol. 26, ed. William G. Lennox, 252–69. Williams & Wilkins, 1947.

———. *Epilepsy and the Functional Anatomy of the Human Brain.* Little, Brown, 1954.

Penfield, Wilder, and Gordon Mathieson. "Memory: Autopsy Findings and Comments on the Role of Hippocampus in Experiential Recall." *Archives of Neurology* 31, no. 3 (September 1, 1974): 145–54.

Penfield, Wilder, and Brenda Milner. "Memory Deficit Produced by Bilateral Lesions in the Hippocampal Zone." *Archives of Neurology and Psychiatry* 79, no. 5 (1958): 475–97.

Penfield, Wilder, and Harry Steelman. "The Treatment of Focal Epilepsy by Cortical Excision." *Annals of Surgery* 126, no. 5 (1947): 740–62.

Pernick, Martin S. *A Calculus of Suffering: Pain, Professionalism, and Anesthesia in Nineteenth-Century America.* Columbia University Press, 1985.

Pickstone, J. V. *Medical Innovations in Historical Perspective.* Palgrave Macmillan, 1992.

Pinch, T. J. [Trevor J.], and H. M. [Harry M.] Collins. "Edible Knowledge: The Chemical Transfer of Memory." In *The Golem: What You Should Know about Science*, 2nd ed., 5–26. Cambridge University Press, 1998.

Pitts, Walter, and Warren S. McCulloch. "How We Know Universals: The Perception of Auditory and Visual Forms." *Bulletin of Mathematical Biophysics* 9, no. 3 (September 1947): 127–47.

Pool, J. Lawrence. "Topectomy: The Treatment of Mental Illness by Frontal Gyrectomy or Bilateral Subtotal Ablation of Frontal Cortex." *The Lancet* 254, no. 6583 (October 29, 1949): 776–81.

Pool, James Lawrence, and Charles Albert Elsberg. *The Neurological Institute of New York, 1909–1974: With Personal Anecdotes.* Pocket Knife, 1975.

Porter, Theodore M. "Is the Life of the Scientist a Scientific Unit?" *Isis* 97, no. 2 (June 1, 2006): 314–21.

Prados, Miguel, and William C. Gibson. "Pío del Río-Hortega, 1882–1945." *Journal of Neurosurgery* 3, no. 4 (1946): 275–84.

Pressman, Jack D. "Sufficient Promise: John F. Fulton and the Origins of Psychosurgery." *Bulletin of the History of Medicine* 62, no. 1 (1988): 1–22.

———. "Human Understanding: Psychosomatic Medicine and the Mission of the Rockefeller Foundation." In *Greater Than the Parts: Holism in Biomedicine, 1920–1950*, ed. Christopher Lawrence and G. Weisz, 189–208. Oxford University Press, 1998.

———. *Last Resort: Psychosurgery and the Limits of Medicine.* Cambridge History of Medicine. Cambridge University Press, 1998.

"Public-Health Statesman." *Time*, November 26, 1956.

Rasmussen, Nicolas. "The Mid-Century Biophysics Bubble: Hiroshima and the Biological Revolution in America, Revisited." *History of Science* 35, no. 3 (September 1997): 245–93.

———. *Picture Control: The Electron Microscope and the Transformation of Biology in America, 1940–1960.* Stanford University Press, 1999.

Raz, Mical. *What's Wrong with the Poor? Psychiatry, Race, and the War on Poverty.* Studies in Social Medicine. University of North Carolina Press, 2013.

Reitan, Ralph M. "Ward Halstead's Contributions to Neuropsychology and the Halstead-Reitan Neuropsychological Test Battery." *Journal of Clinical Psychology* 50, no. 1 (1994): 47–70.

Rey, R. *The History of Pain.* Translated by L. E. Wallace, J. A. Cadden, and S. W. Cadden. Harvard University Press, 1998.

Reynolds, L. A. *Innovation in Pain Management.* Wellcome Trust Centre for the History of Medicine, University College London, 2004.

Rheinberger, Margaret B., and Herbert H. Jasper. "Electrical Activity of the Cerebral Cortex in the Unanesthetized Cat." *American Journal of Physiology—Legacy Content* 119, no. 1 (1937): 186–96.

Richards, Joan L. "Introduction: Fragmented Lives." *Isis* 97, no. 2 (June 1, 2006): 302–5.

Richter, Jochen. "The Brain Commission of the International Association of Academies: The First International Society of Neurosciences." *Brain Research Bulletin* 52, no. 6 (2000): 445–57.

Río-Hortega, Pío del, and Wilder Penfield. "Cerebral Cicatrix: The Reaction of Neuroglia and Microglia to Brain Wounds." *Bulletin of the Johns Hopkins Hospital* 41, no. 5 (November 1927): 278–303.

Rochester, Nathaniel. "The 701 Project as Seen by Its Chief Architect." *IEEE Annals of the History of Computing* 5, no. 2 (April 1983): 115–17.

Rochester, N., J. Holland, L. Haibt, and W. Duda. "Tests on a Cell Assembly Theory of the Action of the Brain, Using a Large Digital Computer." *IEEE Transactions on Information Theory* 2, no. 3 (September 1956): 80–93.

Rockefeller Foundation. "The Rockefeller Foundation Annual Report—1934." Rockefeller Foundation, 1934.

Rose, Nikolas S., and Joelle M. Abi-Rached. *Neuro: The New Brain Sciences and the Management of the Mind*. Princeton University Press, 2013.

Rose, S. P. R. "Holger Hyden and the Biochemistry of Memory." *Brain Research Bulletin* 50, nos. 5–6 (November 1999): 443.

Rosenberg, Charles E. *The Care of Strangers: The Rise of America's Hospital System*. Basic, 1987.

Rotter, Julian B. "A Historical and Theoretical Analysis of Some Broad Trends in Clinical Psychology." In *Psychology—a Study of a Science: Study II, Empirical Substructure and Relations with Other Sciences*, vol. 5, *The Process Areas, the Person, and Some Applied Fields: Their Place in Psychology and in Science*, 780–830. McGraw-Hill, 1962.

Rumelhart, David E., Geoffrey E. Hinton, and Ronald J. Williams. "Learning Representations by Back-Propagating Errors." *Nature* 323, no. 6088 (1986): 533–36.

Rumelhart, David E., James L. McClelland, and PDP Research Group. *Parallel Distributed Processing: Explorations in the Microstructure of Cognition*. Vol. 1, *Foundations*. MIT Press, 1986.

Sargant, William. "Ewen Cameron, M.D., F.R.C.P.(C.) D.P.M." *British Medical Journal* 3, no. 5568 (1967): 803–4.

Scarry, E. *The Body in Pain: The Making and Unmaking of the World*. Oxford University Press, 1987.

Schiller, Peter H. "Past and Present Ideas about How the Visual Scene Is Analyzed by the Brain." In *Cerebral Cortex*, vol. 12, *Extrastriate Cortex in Primates*, ed. Kathleen S. Rockland, Jon H. Kaas, and Alan Peters, 59–90. Springer, 1997.

Schlich, Thomas. *The Origins of Organ Transplantation: Surgery and Laboratory Science, 1880–1930*. University of Rochester Press, 2010.

———. "Asepsis and Bacteriology: A Realignment of Surgery and Laboratory Science." *Medical History* 56, no. 3 (July 2012): 308–34.

———. "Negotiating Technologies in Surgery: The Controversy about Surgical Gloves in the 1890s." *Bulletin of the History of Medicine* 87, no. 2 (2013): 170–97.

———. "'One and the Same the World Over': The International Culture of Surgical Exchange in an Age of Globalization, 1870–1914." *Journal of the History of Medicine and Allied Sciences* 71, no. 3 (July 2016): 247–70.

Schmitt, F. O. "Acknowledgements." In *Biophysical Science: A Study Program*, ed. J. L. Oncley, R. C. Williams, F. O. Schmitt, and R. H. Bolt, v. Wiley, 1959.

———. "Molecular Organization of the Nerve Fiber." In *Biophysical Science: A Study Program*, ed. J. L. Oncley, R. C. Williams, F. O. Schmitt, and R. H. Bolt, 455–65. Wiley, 1959.

———. "Life, Science and Inner Commitment." *Technology Review* 63, no. 9 (July 1961): 43–45, 70–72.

———, ed. *Macromolecular Specificity and Biological Memory*. MIT Press, 1962.

———. "The Biomedical Communication Problem." *Technology Review* 66, no. 5 (March 1964): 13–15, 40.

———. "Adventures in Molecular Biology." *Annual Review of Biophysics and Biophysical Chemistry* 14, no. 1 (1985): 1–23.

———. *The Never-Ceasing Search*. American Philosophical Society, 1990.

Schneider, William H. "The Men Who Followed Flexner: Richard Pearce, Alan Gregg, and the Rockefeller Foundation Medical Divisions, 1919–1951." In *Rockefeller Philanthropy and Modern Biomedicine: International Initiatives from World War I to the Cold War*, ed. William H. Schneider, 7–60. Indiana University Press, 2002.

———. "The Model American Foundation Officer: Alan Gregg and the Rockefeller Foundation Medical Divisions." *Minerva* 41, no. 2 (2003): 155–66.

Scoville, W. B., and B. Milner. "Loss of Recent Memory after Bilateral Hippocampal Lesions." *Journal of Neurology, Neurosurgery and Psychiatry* 20, no. 1 (1957): 11–21.

Scull, A. *Madhouse: A Tragic Tale of Megalomania and Modern Medicine*. Yale University Press, 2007.

Searls, D. *The Inkblots: Hermann Rorschach, His Iconic Test, and the Power of Seeing*. Crown, 2018.

Shagass, Charles, and Herbert H. Jasper. "Conditioning of the Occipital Alpha Rhythm in Man." *Journal of Experimental Psychology* 28, no. 5 (1941): 373–88.

Shatz, Carla J. "The Developing Brain." *Scientific American* 267, no. 3 (September 1992): 60–67.

Shepherd, Gordon M. *Foundations of the Neuron Doctrine*. Oxford University Press, 1991.

———. *Creating Modern Neuroscience: The Revolutionary 1950s*. Oxford University Press, 2009.

Sherrington, C. S. *The Integrative Action of the Nervous System*. Scribner, 1906.

———. *Mammalian Physiology: A Course of Practical Exercises*. Clarendon, 1919.

Shurkin, J. *Broken Genius: The Rise and Fall of William Shockley, Creator of the Electronic Age*. Palgrave Macmillan, 2006.

Silverman, Milton. "Now They're Exploring the Brain." *Saturday Evening Post*, October 8, 1949, 26–27, 80–86.

Smirnov, G. D. "Soviet Report on Montreal International Physiological Congress." *Bulletin of the Academy of Sciences of the USSR* 24, no. 1 (January 1954): 80–88.

Smith, R. *Inhibition: History and Meaning in the Sciences of Mind and Brain*. University of California Press, 1992.

Society for Neuroscience. "The Creation of Neuroscience: The Society for Neuroscience and the Quest for Disciplinary Unity, 1969–1995." Society for Neuroscience, 2019. https://www.sfn.org/About/History-of-SfN/1969-1995/~/media/SfN/Images/History%20of%20SfN/pdf/HistoryofSfN.ashx.

Sokal, Michael M. "The Gestalt Psychologists in Behaviorist America." *American Historical Review* 89, no. 5 (1984): 1240.

Soleiman, Matthew. "Mechanisms of Experience: Cognitivism, Cybernetics, and the Postwar Science of Pain." *Isis* 116, no. 1 (2025): 23–42.

Squire, Larry R. "The Legacy of Patient H.M. for Neuroscience." *Neuron* 61, no. 1 (January 2009): 6–9.

Stahnisch, F. W. *A New Field in Mind: A History of Interdisciplinarity in the Early Brain Sciences*. McGill-Queen's University Press, 2020.

Star, Susan Leigh. *Regions of the Mind: Brain Research and the Quest for Scientific Certainty.* Stanford University Press, 1989.

Steinberg, David A., and George K. York. *An Introduction to the Life and Work of John Hughlings Jackson with a Catalogue Raisonné of His Writings.* Wellcome Trust Centre for the History of Medicine, University College London, 2006.

Swazey, Judith P. "Sherrington's Concept of Integrative Action." *Journal of the History of Biology* 1, no. 1 (1968): 57–89.

———. "Forging a Neuroscience Community: A Brief History of the Neurosciences Research Program." In *The Neurosciences: Paths of Discovery*, ed. Frederic G. Worden, Judith P. Swazey, and George Adelman, 529–46. MIT Press, 1975.

Temkin, Owsei. *The Falling Sickness: A History of Epilepsy from the Greeks to the Beginnings of Modern Neurology.* Johns Hopkins University Press, 1994.

Terrall, Mary. "Biography as Cultural History of Science." *Isis* 97, no. 2 (June 1, 2006): 306–13.

Thompson, Richard F. "James Olds: 1922–1976." *American Journal of Psychology* 92, no. 1 (1979): 151–52.

———. "James Olds—1922–1976." *Biographical Memoirs, National Academy of Sciences* 77 (1999): 1–19.

Thompson, William R., and Woodburn Heron. "The Effects of Restricting Early Experience on the Problem-Solving Capacity of Dogs." *Canadian Journal of Psychology/Revue canadienne de psychologie* 8, no. 1 (1954): 17–31.

Thurstone, Louis Leon. *The Nature of Intelligence.* Routledge, 1924.

———. *The Vectors of Mind: Multiple-Factor Analysis for the Isolation of Primary Traits.* University of Chicago Press, 1935.

Todes, Daniel P. "From the Machine to the Ghost Within: Pavlov's Transition from Digestive Physiology to Conditional Reflexes." *American Psychologist* 52, no. 9 (September 1997): 947–55.

Tolman, Edward Chace. *Purposive Behavior in Animals and Man.* Century, 1932.

———. "Cognitive Maps in Rats and Men." *Psychological Review* 55, no. 4 (1948): 189–208.

Trepanier, Michel. "Science and Technology: The Coming of Age of a City of Knowledge." In *Montreal: The History of a North American City*, ed. Dany Fougères and Roderick Macleod, 2:246–312. McGill-Queen's University Press, 2018.

UN Department of Social Affairs. *The Question of Establishing United Nations Research Laboratories.* United Nations, 1948.

Valenstein, Elliot S. *Brain Control: A Critical Examination of Brain Stimulation and Psychosurgery.* Wiley, 1974.

———. *Great and Desperate Cures: The Rise and Decline of Psychosurgery and Other Radical Treatments for Mental Illness.* Basic, 1986.

———. *Blaming the Brain: The Truth about Drugs and Mental Health.* Free Press, 1998.

———. *The War of the Soups and the Sparks: The Discovery of Neurotransmitters and the Dispute over How Nerves Communicate.* Columbia University Press, 2013.

Vannemreddy, Prasad, James L. Stone, Siddharth Vannemreddy, and Konstantin V. Slavin. "Psychomotor Seizures, Penfield, Gibbs, Bailey and the Development of Anterior Temporal Lobectomy: A Historical Vignette." *Annals of Indian Academy of Neurology* 13, no. 2 (2010): 103–7.

Wailoo, Keith. *Pain: A Political History.* Johns Hopkins University Press, 2014.

Weaver, Warren. "Molecular Biology: Origin of the Term." *Science* 170, no. 3958 (1970): 581–82.

Weidman, Nadine M. *Constructing Scientific Psychology: Karl Lashley's Mind-Brain Debates.* Cambridge University Press, 1999.

Wheatley, S. C. *The Politics of Philanthropy: Abraham Flexner and Medical Education.* University of Wisconsin Press, 1988.

Winter, Alison. *Memory: Fragments of a Modern History.* University of Chicago Press, 2012.

Young, J. Z. "The Mechanism of Memory." *Endeavour* 72, no. 86 (May 1963): 101–2.

Young, R. M. *Mind, Brain, and Adaptation in the Nineteenth Century: Cerebral Localization and Its Biological Context from Gall to Ferrier.* Oxford University Press, 1990.

Zilio, Diego. "Who, What, and When: Skinner's Critiques of Neuroscience and His Main Targets." *Behavior Analyst* 39, no. 2 (October 2016): 197–218.

Index

www.ingramcontent.com/pod-product-compliance
Lightning Source LLC
Chambersburg PA
CBHW030456210326
41597CB00013B/694